Computing Supplementum 9

R. Albrecht, G. Alefeld, H. J. Stetter (eds.)

Validation Numerics

Theory and Applications

Springer-Verlag Wien New York

Prof. Dr. R. Albrecht
Institut für Informatik
Universität Innsbruck
Austria

Prof. Dr. G. Alefeld
Institut für Angewandte Mathematik
Universität Karlsruhe
Federal Republic of Germany

Prof. Dr. H. J. Stetter
Institut für Angewandte und Numerische Mathematik
Technische Universität Wien
Austria

© 1993 Springer-Verlag/Wien

Typesetting: Asco Trade Typesetting Limited, Hong Kong

Printed in Austria by Eugen Ketterl GesmbH, A-1180 Wien

Printed on acid-free paper

With 23 Figures

ISSN 0344-8029

ISBN 3-211-82451-0 Springer-Verlag Wien New York
ISBN 0-387-82451-0 Springer-Verlag New York Wien

Preface

This volume is dedicated to Professor Dr. Ulrich Kulisch, Director of the Institute for Applied Mathematics of Karlsruhe University, on the occassion of his 60th birthday.

The contributions give a comprehensive overview of methods which can be effectively used for the validation of the solutions of a variety of problems. These include simultaneous systems of linear and nonlinear equations, differential and integral equations, and certain applications from the technical sciences. Furthermore there are some papers which present improvements of the tools of Validation Numerics.

All papers have been refereed. We wish to thank the contributors for their manuscripts and the referees for their prompt and careful reviews. We are also grateful to the Publisher, Springer-Verlag, Wien for an efficient and quick production of this volume.

Last but not least we want to thank Professor Kulisch to whom we owe so much in the field of Validation Numerics. Without his pioneering work, most of the papers in this volume would not have been written.

Innsbruck/Karlsruhe/Vienna,
February 1993

R. Albrecht
G. Alefeld
H. J. Stetter

Contents

Listed in Current Contents

Computing, Suppl. 9, 1–10 (1993)

On a Unified Concept of Mathematics*

N. Apostolatos, Athens

Dedicated to Professor U. Kulisch on the occasion of his 60th birthday

Abstract — Zusammenfassung

On a Unified Concept of Mathematics. Mathematics is not simply a branch of science, but the unique and conclusive means of expressing and solving any problem in any field whatever. In this paper I wish to submit some thoughts justifing my belief in the unification of Mathematics. Besides that, I make a proposal for replacement of the structureblock for questions (with two or more answers) in structuregrams, which not only makes the structuregrams more representative and comprehensive but also corrects certain shortcomings and makes them more adapted to the structure of programming languages.

AMS Subject Classification: 68Q20, 65G10, 65D99

Key words: Algorithms, finite numerical mathematics, finite approximation theory, iterative methods, interval mathematics, structuregrams.

Über eine einheitliche Auffassung der Mathematik. Mathematik ist nicht einfach ein Zweig der Wissenschaften, sondern das einzige und entscheidende Mittel zum Beschreiben und Lösen von Problemen zu allen Bereichen. In dieser Arbeit möchte ich einigen Gedanken vorlegen, welche meinen Glauben an die einheitliche Mathematik rechtfertigen. Nebenher unterbreite ich einen Vorschlag wie der Strukturblock für Fragen (mit Zwei- bzw. Mehrfacherverzweigungsblöcken) durch Struktogramme ersetzt werden kann, um nicht nur Struktogramme repräsentativer und umfassender zu machen, sondern auch manche Mängel zu beseitigen und diese mehr der Struktur von Programmiersprachen anzupassen.

1. Mathematics

The language, in the broadest sense of the word, is the predominant factor in our life. The need for a strict definition led to its mathematization. Mathematics became the basic instrument for this purpose. In reality Mathematics is a super-language, since it is not simply a language, but constitutes the only language for the strict definition of other languages. In order to define a language we need an alphabet. So far, no one has succeeded in recording the entire alphabet of the language of Mathematics and the rules which govern it. Thus, if we are to study Mathematics we must divide it into sub-fields or, rather, sub-languages.

Given, therefore, that Mathematics is a language, we can say that a mathematical problem is nothing but a word (text) in the language of Mathematics. And because of the need here for restriction to certain strictly defined spaces, we are obliged to

---

* Received July 20, 1992; revised December, 9, 1992.

express problems in corresponding sub-languages. From the theory of languages we know that if we have one word, then we can use the rules of the language to produce equivalent words. Consequently, each problem has as equivalents all the words equivalent to that which was used to express the problem. We call one of these words the solution to the problem. So when we say that a problem has no solution, then that problem is the solution or rather that equivalent word (problem) which evidently proves that there is no solution.

The introduction of infinity into Mathematics was a decisive turning-point in the history of Mathematics. Without this concept a strict definition of the most fundamental concepts would be impossible. And whereas the finite appears accessible to us, in practice it often leads to an insuperable situation. Let us just try to conceive the finite number: $f(4, 4)$ where f is the function of *Ackermann*.

The tremendous development of computers enabled us to process a great deal of information, always however, in a finite space, the power of which is almost zero compared with the above number $f(4, 4)$. The very need for accessible finite spaces was the motivation behind the development and establishment of Interval Mathematics. In an attempt to use modern computing methods an indispensable prerequisite was that the input, processing and output of data through a computer system should be carried out automatically with the help of that computer. This attempt resulted in the modern field of Data Structures. Thus the algorithm became predominantly important. In my opinion, the Theory of Algorithms, as founded in our time, is the basic and principal achievement of the 20th century.

Let us say that we have a problem p which has a solution s and we are searching for an algorithm a, which when applied to the problem will lead to the solution. We symbolize this with the mapping:

$$p \xrightarrow{a} s$$

or simply: $ap = s$

In Mathematics, the probability of having a problem p with a solution s and having a finite algorithm a to solve it is almost zero. So usually, and not only because we are in finite spaces, the following model applies:

$$
\begin{array}{ccc}
p & \xrightarrow{a} & s \\
\downarrow & & \\
\bar{p} & \xrightarrow{\bar{a}} & \bar{s}
\end{array}
$$

where $\bar{p}$ is an "approximation" of p, $\bar{s}$ is an "approximation" of s and $\bar{a}$ is a feasible algorithm which, when applied to $\bar{p}$ gives the solution $\bar{s}$. Ultimately, whereas we are interested for s, we calculate $\bar{s}$, so that our final problem is to check how good $\bar{s}$ is. (N.B. Obviously some algorithm is also required for the mapping $p \rightarrow \bar{p}$)

The whole of the above model is incorporated into Mathematics. More specifically:

To define p, s, a we need what is known simply as *Mathematics*.

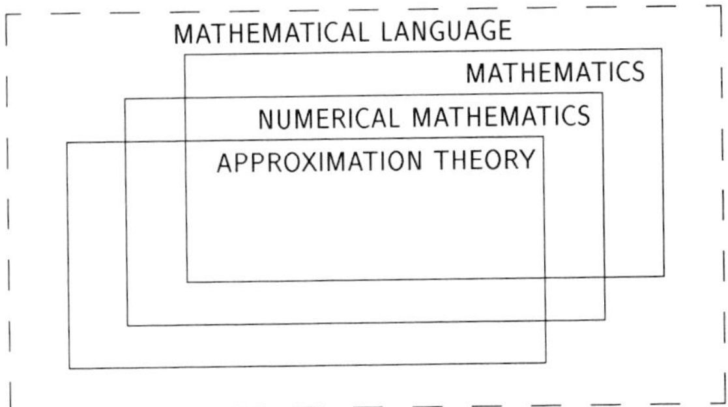

Figure 1. Mathematical language

To realize the mapping $\bar{p} \xrightarrow{\bar{a}} \bar{s}$ we need what is known as *Numerical Analysis* or better still, *Numerical Mathematics*.

Finally, for the comparison of p and $\bar{p}$, a and $\bar{a}$, s and $\bar{s}$ we need the *Approximation Theory*.

Thus, we essentially have the above overlapping scheme (Fig. 1):

In addition, however, all these must become accessible finite and here we see that we need:

a) The Finite Numerical Mathematics
b) The Finite Approximation Theory

The basic modern means for achieving the above is the computer.

2. The Iterative Method

In Numerical Mathematics we divide methods for solving problems into two large classes:

a) Direct methods
b) Iterative methods

The distinction between the two methods is made as follows:

Definition 2.1. A method (algorithm) M is called a *direct* method iff it is split into n methods $M_1, M_2, \ldots, M_n$, so that when applied to a problem p, which has the solution s, it transforms it equivalently into n successive problems p_i, $i = 1(1)n$, so that:

$$M_{i+1} p_i = p_{i+1}, \quad for \quad i = 0(1)n - 1, \quad where \quad p_0 = p, p_n = s$$

Definition 2.2. A method M is called *iterative* iff, when applied to a problem p, which has the solution s, it transforms it equivalently into a sequence of problems p_i, $i = 1$,

2, ... so that:

$$Mp_i = p_{i+1}, \quad i = 0, 1, 2, \ldots, \quad \text{where} \quad p_0 = p \quad \text{and} \quad \lim_{i \to \infty} p_i = s$$

Hence on the basis of the above two definitions 2.1 and 2.2 the *Elimination method* of *Gauss* is a direct method, while the *Gauss-Seidel method* is an iterative one.

However, in a direct method, if n is large and the methods M_i, $i = 1(1)n$ are independent, then it would be impossible to apply it, since a large number of independent algorithms would have to be created and put into the computer system. As we know, basically every direct method has a few independent steps (independent algorithms), each of which is probably repeated several times, each time, of course, with different data. Thus, here too, we actually have an "iterative" method. Moreover, in an iterative method too, we interrupt after a finite number of iterations.

As we know, each sequence of elements produced by an iterative method, since an finite automaton, such as the computer, is used to produce them, is a *cyclical sequence*. This cyclic property is not easy to ascertain, because apart from the finiteness of each computer system the number of its elements is too large. But even if we do ascertain it, this only leads to an impasse, since the only profit that we receive is that with this specific computer system we have, and with the method we applied, we cannot expect any further information. While the results do not provide us with elements capable of utilizing them reliably.

Let us consider a problem p, with the solution s which is an element of a topological space S. We use $\overline{T}$ to symbolize all the closed subsets of S.

Definition 2.3. Each element $t \in \overline{T}$ with the property $s \in t$ is called an *admissible* solution of the problem p.

Let A be a set of algorithms for the solution of the problem p, so that for every $a \in A$ we have $ap = t$, where t is an admissible solution.

Definition 2.4. An algorithm $a \in A$ is called *optimal* in relation to the problem p, iff $ap = t$, where $t = \{s\}$.

Definition 2.5. Two elements a, $b \in A$ are called *equal* ($a = b$), iff $t_a = t_b$, where $t_a = ap$ and $t_b = bp$.

Definition 2.6. For two elements a, $b \in A$ we introduce an *order relation* ($\leq$), as following:

$$a \leq b, \qquad \text{iff } t_a \subseteq t_b$$

Definition 2.7. For a, $b \in A$ we define *union* and *intersection* as following:

$$a \cup b = c \in A: t_c = t_a \cup t_b$$

$$a \cap b = c \in A: t_c = t_a \cap t_b$$

N.B. We assume that A is closed under union and intersection.

From the previous definitions for $a, b \in A$, we have:

$$a \cap b \leq a, b \leq a \cup b$$

We consider the iterative method E (algorithm) for the solution of the problem p: $Ep_i = p_{i+1}$, where $p_0 = p$ and p_i, $i = 1, 2, \ldots$ an admissible solution.

Definition 2.8. Let there be two iterative methods E and $\bar{E}$ for the solution of the problem p with the above property. We shall say that $\bar{E}$ is not *worse* than E and symbolize it with

$$\bar{E} \leq E, \quad \text{iff } Ep_i = p_{i+1},$$

$$\bar{E}\bar{p}_i = \bar{p}_{i+1}, \qquad \bar{p}_i \subseteq p_i, \quad i = 1, 2, \ldots, \text{where } p_0 = \bar{p}_0 = p.$$

Definition 2.9. An iterative method is called *monotone*, iff

$$Ep_i \subseteq p_i, \qquad i = 0, 1, 2, \ldots$$

We give the following theorem without proof.

Theorem 2.1. *For each iterative method E for the solution of the problem p there exists a monotone iterative method $\bar{E}$ for the solution of the same problem not worse than E.*

Definition 2.10. An iterative method E for the solution of the problem p is called *almost constant*, iff

$$Ep_i = p_i \qquad i = k, k + 1, k + 2, \ldots \qquad (k > 0)$$

So each direct method M for the solution of the problem p is split into $n + 1$ independent methods $M_0, M_1, \ldots, M_n$, so that M_n can be considered as an almost constant iterative method for the solution s of the problem p (see def. 2.1).

(For the above and some other theorems see [4]).

3. Interval Arithmetic—Software

There is no lack of possibilities for the application of all that was mentioned in the previous paragraph, since with the help of an appropriate Interval Arithmetic we can achieve a basis in different mathematical spaces. Interval Arithmetic was born of the need to replace infinite operators by finite ones, without running the risk of having a result and not knowing how it is related to the problem's solution. And whereas Interval Arithmetic includes the classical Arithmetic and converges towards it, when intervals converge towards one-element sets, it cannot be applied without the corresponding Arithmetic on the computer. Precisely this need was the basic reason for writing the paper [7]. Since then, so much has happened in this field, accompanied, of course, by the development of modern software, that today there are programming languages, such as PASCAL-XSC ([13]). Not only does this provide us with a fundamental tool in the field of Numerical Mathematics, but the variety of types, the possibility of determining new types and especially operators, greatly simplifies the realization of algorithms and contributes conclusively to the unification of programming languages. Thus the unified structure of Mathematics is also transferred to the field of programming languages, at the same time enabling a strict foundation of Numerical Mathematics and the creation of new methods which have no equivalent in classical Numerical Mathematics.

4. Structuregrams

Structuregrams like flowcharts, are an important tool for a comprehensive and unified form of expressing algorithms. Thus the creation of suitable software for an automatic design of structuregrams and their subsequent conversion in programs, into a language such as PASCAL, the structure of which is within the framework of the structure of structuregrams, will open new horizons for research, not only in Numerical Mathematics. It is precisely thanks to the above mentioned concerning iterative methods that we can appreciate the universality of structuregrams. The iterative factor is incorporated in the methods of programming languages definition (*Backus-Naur Form, Syntaxdiagrams*), since ultimately there is no difference between recursive and iterative methods. By using a general definition, we could have these two concepts as subcases. The structuregrams are based on iterative schemes, thus the structured form of algorithms is obtained. The blocks for questions with two or more possible answers, as they have been integrated into the structuregrams have a basic disadvantage. That is, they require an extension in width which can cause problems when they are presented using a medium such as a sheet paper or a computer screen, which are limited in width. In addition the slanted lines used only in the blocks for questions are also a weak point. Their difficulties are well known to those who have undertaken to drawing structuregrams. In the next paragraph we make a proposal to replace the blocks for questions by other blocks, so as to overcome the above problems. Of course, these new forms bring us closer to the structured programming languages. Thus, this improvement might appear to be particularly useful for an automatic design of structuregrams and in their subsequent automatic transferring into programs, and vice versa.

Before we present our proposal for the simplification of structuregrams, we wish to point out that: *Mathematics will be the field which, once unified itself, will constitute the model and the basic medium for a unified form of all human knowledge, at the same time setting aside that which is not useful.*

4.1 New Symbol-Forms for Questions

First of all we present a new symbol-form for questions in structuregrams in the case of two possible answers and then we generalize this when there are more than two possible answers. On the left, we have put the preferred symbol-form and on the right, the existing one.

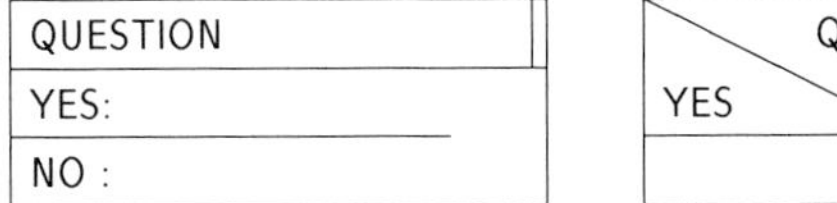
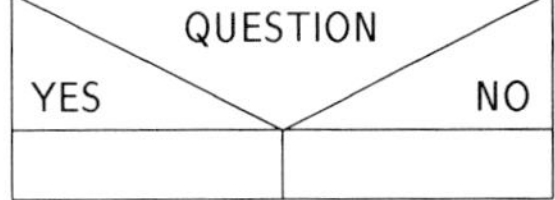

Figure 2. Structureblocks for questions with two answers

Remarks

1. The non-closing of rectangles (YES-NO) means that when the statements have been completed in the corresponding rectangle, the control is transferred through the vertical right column to the first horizontal closed line which follows, exactly as in the existing symbol-form.
2. The double line, which can also be placed on the left, can also be omitted. However, placing it is useful because in this way we see immediately that we are dealing with a question.

The generalization for the case with more than two answers is now evident.

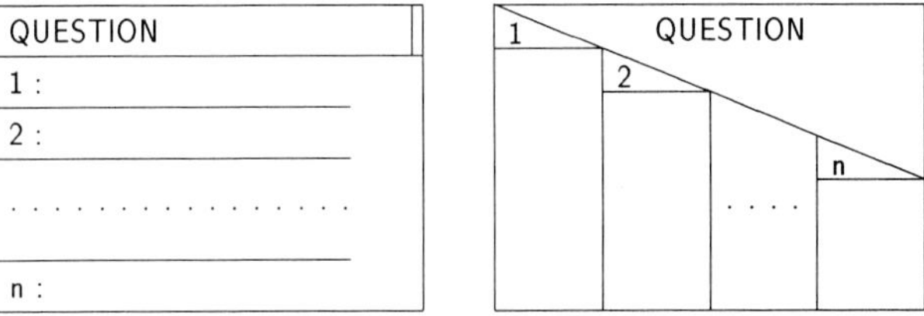

Figure 3. Structureblocks for questions with multiple answers

Remark
The problem becoming apparent here is that which the existing symbol-form can create when n is large and/or the statements in some answers are of great length.

Now we give the corresponding PASCAL statement for the above question with a multiple answer.

$$
\begin{aligned}
&CASE\ QUESTION\ OF\\
&\quad 1:\ BEGIN\quad END;\\
&\quad 2:\ BEGIN\quad END;\\
&\quad\quad \cdots\cdots\\
&\quad n:\ BEGIN\quad END\\
&END;
\end{aligned}
$$

The analogy with the symbol-form we propose is now quite evident.

CUBE (procedure)
$p = b - a^2/3$; $\quad q = c - ab/3 + 2a^3/27$
$R = (q/2)^2 + (p/3)^3$
$R < 0$
YES: $\quad q = 0$
$\quad\quad$ YES: $\quad f = \pi/2$
$\quad\quad$ NO : $\quad d = -q/2\sqrt{-(p/3)^3}$
$\quad\quad\quad\quad\quad t = \sqrt{1 - d^2}/d$; $\quad f = arctg(t)$
$\quad\quad L = 2\sqrt{-p/3}$
$\quad\quad R1 = Lcos(f/3) \quad\quad\quad\quad ; I1 = 0$
$\quad\quad R2 = Lcos((f + 2\pi)/3) \quad\quad ; I2 = 0$
$\quad\quad R3 = Lcos((f + 4\pi)/3) \quad\quad ; I3 = 0$
NO : $\quad q \neq 0$
$\quad\quad$ YES: $\quad R1 = \sqrt[3]{-q/2 + \sqrt{R}} + \sqrt[3]{-q/2 - \sqrt{R}} \quad ; I1 = 0$
$\quad\quad\quad\quad s = (R1/2)^2 + q/R1$
$\quad\quad\quad\quad s \geq 0$
$\quad\quad\quad\quad$ YES: $\quad R2 = -R1/2 + \sqrt{s} \quad ; I2 = 0$
$\quad\quad\quad\quad\quad\quad R3 = -R1/2 - \sqrt{s} \quad ; I3 = 0$
$\quad\quad\quad\quad$ NO : $\quad R2 = -R1/2 \quad\quad ; I2 = \sqrt{-s}$
$\quad\quad\quad\quad\quad\quad R3 = -R1/2 \quad\quad ; I3 = -\sqrt{-s}$
$\quad\quad$ NO : $\quad R1 = 0$; $I1 = 0$
$\quad\quad\quad\quad -p \leq 0$
$\quad\quad\quad\quad$ YES: $R2 = 0 \quad\quad\quad ; I2 = \sqrt{p}$
$\quad\quad\quad\quad\quad\quad R3 = 0 \quad\quad\quad ; I3 = -\sqrt{p}$
$\quad\quad\quad\quad$ NO : $R2 = \sqrt{-p} \quad ; I2 = 0$
$\quad\quad\quad\quad\quad\quad R3 = -\sqrt{-p} \quad ; I3 = 0$
$R1 = R1 - a/3$; $\quad R2 = R2 - a/3$; $\quad R3 = R3 - a/3$

Figure 4. Structuregram with the new symbol-forms

4.1.1 Example

For a better understanding of the advantages of this proposal, we give an example. The problems arise when we have a nest of such symbol-forms such as shown in this example, which of course, is not particularly complex.

We give the structuregrams for calculating the roots of a 3rd degree equation

$$x^3 + ax^2 + bx + c = 0$$

where a, b, c are real numbers and we assume that we do not have a complex arithmetic on the computer.

In the Figs. 4 and 5 we give the structuregrams for the above example with the new symbol-forms and with the existing ones, respectively.

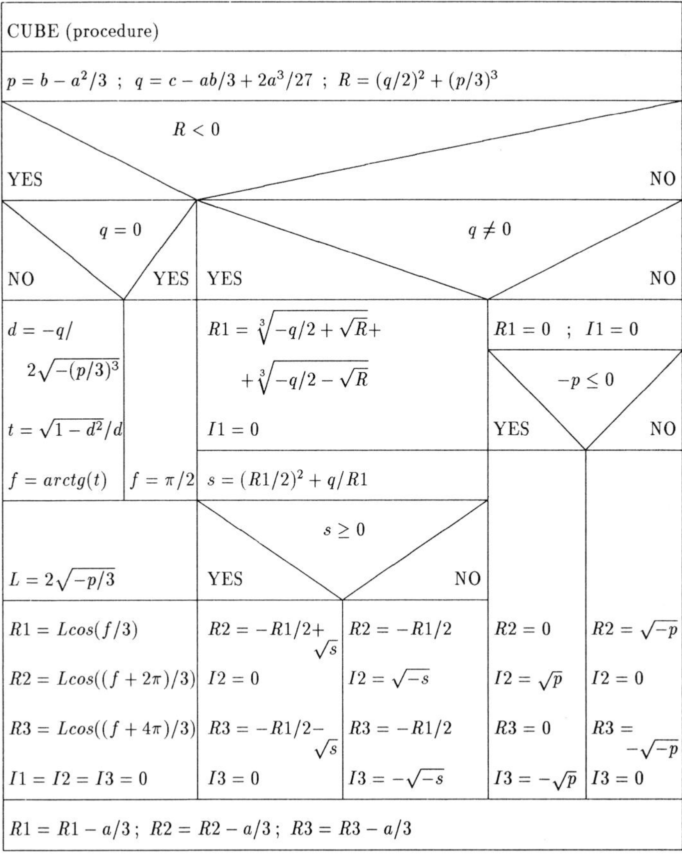

Figure 5. Structuregram with the existing symbol-forms

Remarks

1. The YES-NO answers in the internal questions have been placed slightly further to the right for better visual presentation. And here a pre- fixed FORMAT may be maintained, determining the width of the indentation based on the total width of its nest or will removing the indentation.

2. The structuregrams, thanks to the new symbol-forms for questions, have a shape/structure corresponding to that of a structured language such as PASCAL. Thus, for compound statements it is easy to see where the BEGIN-END will be placed. These new symbol-forms are a step towards the automatic conversion of a structuregram into the corresponding programme of a structured language.

References

[1] Aceto, L., Hennessy, M.: Termination, deadlock and divergence. J. ACM *39*, 147–187 (1992).
[2] Ackermann, W.: Solvable cases of the decision problem. Amsterdam: North Holland 1956.

[3] Alefeld, G., Herzberger, J.: Einführung in der Intervallrechnung. Bibl. Inst., Mannheim 1974.

[4] Apostolatos, N.: Eine Allgemeine Betrachtung von Numerischen Algorithmen. Beiträge zur Numerischen Mathematik 6, 11–25 (1977).

[5] Apostolatos, N.: An introduction to theory of algorithms. Book (Greek), Athens 1978.

[6] Apostolatos, N.: A proposal for the renaming of the scientific branch of Computer Science. The Computer Journal 32 (3), 282 (1989).

[7] Apostolatos, N., Kulisch, U.: Gundlagen einer Maschinenintervallarithmetik. Computing 2, 89–104 (1967).

[8] Borger, Egon: Berechenbarkeit, Komplexität, Logik, 2. Aufl. Braunschweig/Wiesbaden: Vieweg 1986.

[9] Cheney, E. W.: Introduction to approximation theory. New York: McGraw-Hill 1966.

[10] Collatz, L.: Funktionalanalysis und Numerische Mathematik. Berlin: Springer 1964.

[11] Hilbert, D.: Über das Unendliche. Mathematischen Annalen 95, 161–190 (1926).

[12] Kaucher, E., Kulisch, U., Ullrich, Ch. (eds.): Computer arithmetic—scientific computation and programming languages. Stuttgart: Teubner 1987.

[13] Klatte, R., Kulisch, U., Neaga, M. Ratz, D., Ullrich, Ch.: PASCAL-XSC. Berlin, Heidelberg: Springer 1991.

[14] Knuth, E. D.: The art of Computer programming Reading, Mass.: Addison-Wesley 1969.

[15] Kulisch, U.: Grundlagen des Numerischen Rechnens. Zürich: Bibl. Inst. 1976.

[16] Kulisch, U., Miranker, W. L.: Computer arithmetic in theory and practice. New York: Academic Press 1981.

[17] Kulisch, U., Stetter, H. J. (eds.): Scientific computation with automatic result verification. Wien: Springer 1988. (Computing Suppl. 6).

[18] Nassi, I., Shneiderman, B.: Flowchart techniques for structured programming. SIGPLAN Notices 8, 12–26 (1973).

Prof. Dr. N. Apostolatos
University of Athens
Department of Informatics, Section of Theoretical
  Informatics
Panepistemiopolis, G-15710 Athens, Greece
  and Technische Universität München
Institut für Angewandte Mathematik und Statistik
Arcisstrasse 21, D-W-8000 München 2,
  Federal Republic of Germany

Computing, Suppl. 9, 11–19 (1993)

A General Approach to a Class of Single-Step Methods for the Simultaneous Inclusion of Polynomial Zeros*

L. Atanassova, Bremen, and **J. Herzberger**, Oldenburg

Dedicated to Professor Dr. U. Kulisch on the occasion of his 60th birthday

Abstract — Zusammenfassung

A General Approach to a Class of Single-Step Methods for the Simultaneous Inclusion of Polynomial Zeros. A general model for the construction of certain classes of simultaneous single-step inclusion methods for polynomial roots is presented. This approach makes use of interval operations in order to guarantee the inclusion of the zeros. It is shown how the order of convergence of these methods can be determined. Some methods of the literature are special cases of this model and new methods can be constructed.

AMS Subject Classification: 65H05, 65G10

Key words: Polynomial roots, single-step methods, interval analysis.

Ein allgemeines Modell für eine Klasse von Einzelschrittverfahren zur simultanen Einschließung von Polynomwurzeln. Wir bringen ein allgemeines Modell für die Herleitung gewisser Klassen von simultanen Einschließungsverfahren von Polynomwurzeln. Zur Erzielung von garantierten Einschließungen macht es Gebrauch von der Intervallrechnung. Es wird eine Berechnung der Konvergenzordnung dieser Verfahren angegeben. Einige Verfahren in der Literatur sind spezielle Fälle dafür, und es können auch neue Verfahren damit hergeleitet werden.

1. Introduction

Given a polynomial of degree n with real coefficients $a_0, a_1, \ldots, a_{n-1}$

$$p(x) = x^n + a_{n-1}x^{n-1} + \cdots + a_0 = \prod_{i=1}^{n} (x - \xi_i) \tag{1}$$

with only simple real roots $\xi_1, \xi_2, \ldots, \xi_n$. If the initial inclusions

$$\xi_1 \in X_1, \xi_2 \in X_2, \ldots, \xi_n \in X_n$$

are given, then there exist many iterative methods for improving these inclusion intervals. They are converging, under certain conditions on the set of intervals $X_1, X_2, \ldots, X_n$, towards the roots $\xi_1, \xi_2, \ldots, \xi_n$. A survey on such methods can be found in the monograph of M. Petković [12].

---

* Received October 7, 1992; revised December 18, 1992.

More recently, some families of higher-order inclusion methods were developed by combining basic iteration formulas in a proper way (see [2], [8]). In the paper [3] the authors made a general approach for such total-step methods and showed how the order of convergence of these methods can easily be calculated. In this note we are proposing a general single-step approach. As far as the calculation of the order of convergence of these methods is concerned, this is not a trivial extension of the model presented in [3].

2. Basic Iteration Formula

Let $\varphi \colon \mathbb{R}^{n+k} \to \mathbb{R}$ be a real-valued function for $k \geq 1$ with the property

$$\varphi(x_1, x_2, \ldots, x_{n-1}, x_n, \ldots, x_{n+k})$$
$$= \varphi(x_{\pi(1)}, \ldots, x_{\pi(n-1)}, x_n, \ldots, x_{n+k}) \tag{2}$$
$$\text{for every permutation } \pi \text{ of } \{1, 2, \ldots, n-1\}.$$

With the help of this function φ we can derive a set of real-valued functions by

$$\varphi_i(x_1, \ldots, x_{n+k}) = \varphi(x_1, \ldots, x_{i-1}, x_{i+1}, \ldots, x_n, x_i, x_{n+1}, \ldots, x_{n+k}) \qquad 1 \leq i \leq n. \tag{3}$$

The functions φ_i shall be related to the zeros of the polynomial (1) by virtue of the following identities, which are kind of fixed-point relations.

$$\xi_i = \varphi_i(\xi_1, \ldots, \xi_{i-1}, x, \xi_{i+1}, \ldots, \xi_n, p(x), p'(x), \ldots, p^{(k-1)}(x)) \tag{4}$$
$$\text{for } 1 \leq i \leq n \text{ and } x \in \mathbb{R} \text{ (for a suitable } k\text{).}$$

Examples:

(a) $\displaystyle \varphi_i(x_1, \ldots, x_{n+1}) = x_i - \frac{p(x_i)}{\prod_{\substack{j=1 \\ j \neq i}}^{n} (x_i - x_j)}$ (Weierstrass [14])

(b) $\displaystyle \varphi_i(x_1, \ldots, x_{n+2}) = x_i - \frac{1}{\dfrac{p'(x_i)}{p(x_i)} - \sum_{\substack{j=1 \\ j \neq i}}^{n} \dfrac{1}{x_i - x_j}}$ (Ehrlich [7])

(c) A general family of formulas was given by Wang and Zheng in [13].

Now, we want to construct inclusion methods by means of interval arithmetic (see Moore [10], Alefeld and Herzberger [1] or Neumaier [11]). By doing this, we assume that for each such function φ_i there exists an interval extension

$$\tilde{\varphi}_i(X_1, \ldots, X_{i-1}, x_i, X_{i+1}, \ldots, X_n, p(x_i), \ldots, p^{(k-1)}(x_i)),$$
$$x_i \in X_i, \quad 1 \leq i \leq n \qquad \text{and} \qquad (X_1, \ldots, X_n) \subseteq (Y_1, \ldots, Y_n)$$

for some interval vector $(Y_1, \ldots, Y_n)$. In practice, this does not seem to be a restriction since all known iteration formulas solving this problem are rational functions for which such interval extensions exist in a trivial way.

Furthermore, these interval extensions shall have the property

$$d(\tilde{\varphi}_i(X_1,\ldots,X_{i-1},x_i,X_{i+1},\ldots,X_n,p(x_i),\ldots,p^{(k-1)}(x_i)))$$

$$\leq \gamma \cdot d(X_i)^\alpha \cdot \left(\sum_{\substack{j=1 \\ j \neq i}}^n d(X_j)^\beta \right), \quad \alpha, \beta \in \mathbb{N}\backslash\{0\}, \, 1 \leq i \leq n, \tag{5}$$

where $d(X) = d([x_1, x_2]) = x_2 - x_1$ is the width of the interval X.

Remark: The more general property

$$d(\tilde{\varphi}_i(X_1,\ldots,X_{i-1},x_i,X_{i+1},\ldots,X_n,p(x_i),\ldots,p^{(k-1)}(x_i)))$$

$$\leq \gamma \cdot d(X_i)^\alpha \cdot \left(\sum_{\substack{j=1 \\ j \neq i}}^n d(X_j) \right)^\beta, \quad \alpha, \beta \in \mathbb{N}\backslash\{0\}, \, 1 \leq i \leq n, \tag{5'}$$

is more difficult to be treated considering the speed of convergence due to the "mixed terms". But in view of the inequality

$$(*) \qquad (x_1 + x_2 + \cdots + x_n)^\beta$$

$$\leq \delta \cdot (x_1^\beta + x_2^\beta + \cdots + x_n^\beta), \quad \beta \in \mathbb{N}\backslash\{0\}, \, x_i \geq 0, \, 1 \leq i \leq n,$$

we always get a lower bound for the speed of convergence by replacing (5') by (5).

We are now able to define a fixed-point iteration method (simultaneous method) by

$$X_i^{(m+1)} = \tilde{\varphi}_i(X_1^{(m+1)},\ldots,X_{i-1}^{(m+1)},x_i^{(m)},X_{i+1}^{(m)},\ldots,X_n^{(m)},p(x_i^{(m)}),\ldots,p^{(k-1)}(x_i^{(m)}))$$

$$\cap X_i^{(m)}, \quad 1 \leq i \leq n, m \geq 0 \tag{6}$$

in the form of a single-step method.

There exist non-monotonic versions by omitting the intersection in (6). But in this case an additional assumption like "$\bigcup_{m>0} X_i^{(m)} \subseteq X_i$" would have to be introduced.

In practical computation, however, one usually prefers monotonic methods and for this reason we are considering only methods like (6).

Theorem 1: *Method* (6) *has the properties*

(i) $X_i^{(0)} \supseteq X_i^{(1)} \supseteq X_i^{(2)} \supseteq \cdots, 1 \leq i \leq n;$

(ii) $\xi_i \in X_i^{(m)}, 1 \leq i \leq n \Rightarrow \xi_i \in X_i^{(m+1)}, 1 \leq i \leq n$ *(inclusion property); this ensures in case* $\xi_i \in X_i^{(0)}, 1 \leq i \leq n$, *that the intersection can never become empty and thus method* (6) *is defined for* $m \geq 0;$

(iii) *method* (6) *is locally convergent if* $\xi_i \in X_i^{(0)}, 1 \leq i \leq n$, *i.e. for sufficiently small values* $d(X_i^{(0)}), 1 \leq i \leq n$, *we have* $X_i^{(m)} \to \xi_i$ *for* $m \to \infty, 1 \leq i \leq n;$

(iv) *the R-order of convergence of every sequence* $\{X_i^{(m)}\}, 1 \leq i \leq n$ *(which is the R-order of convergence of* $\{d(X_i^{(m)}\}, 1 \leq i \leq n)$ *is at least* τ^* *where*

$$\tau^* = \varepsilon^* \cdot \beta + \alpha$$

and ε^* *is the uniquely positive root of the polynomial equation*

$$\varepsilon^n - \beta\varepsilon - \alpha = 0$$

14   L. Atanassova and J. Herzberger

and $\varepsilon^* > 1$ (which means that $\tau^* > \alpha + \beta$ or greater than the R-order of the corresponding total-step method).

Proof: (i) is a trivial consequence of the definition of (6).

Of (ii): Assuming $\xi_j \in X_j^{(m)}$, $1 \leq j \leq n$, and that we have already proved $\xi_j \in X_j^{(m+1)}$, $1 \leq j < i$, for an arbitrary $m \geq 0$, from (4) and the inclusion property of the interval operations we get

$$\xi_i = \varphi_i(\xi_1, \ldots, \xi_{i-1}, x_i^{(m)}, \xi_{i+1}, \ldots, \xi_n, p(x_i^{(m)}), \ldots, p^{(k-1)}(x_i^{(m)}))$$

$$\in \tilde{\varphi}_i(X_1^{(m+1)}, \ldots, X_{i-1}^{(m+1)}, x_i^{(m)}, X_{i+1}^{(m)}, \ldots, X_n^{(m)}, p(x_i^{(m)}), \ldots, p^{(k-1)}(x_i^{(m)}))$$

and thus $\xi_i \in X_i^{(m+1)}$.

Of (iii): Applying the width operator d on formula (6) we get from (5)

$$d(X_i^{(m+1)}) \leq \gamma \cdot d(X_i^{(m)})^\alpha \cdot \left(\sum_{k=1}^{i-1} d(X_k^{(m+1)})^\beta + \sum_{k=i+1}^{n} d(X_k^{(m)})^\beta \right).$$

If we define $d(X^*) = \max_{1 \leq i \leq n} d(X_i)$ and by observing $d(X^{(m+1)*}) \leq d(X^{(m)*})$ we get

$$d(X^{(m+1)*}) \leq \tilde{\gamma} \cdot d(X^{(m)*})^{\alpha+\beta}.$$

Since $\alpha + \beta \geq 2$ this recursion is at least quadratically convergent and thus converges locally towards 0. But this means $d(X_i^{(m)}) \to 0$, $1 \leq i \leq n$ and with (ii) it follows the assertion.

Of (iv): As in (iii) after applying the width operator d on (6) we get the recursions

$$d(X_i^{(m+1)}) \leq \gamma \cdot d(X_i^{(m)})^\alpha \cdot \left(\sum_{k=1}^{i-1} d(X_k^{(m+1)})^\beta + \sum_{k=i+1}^{n} d(X_k^{(m)})^\beta \right), \qquad 1 \leq i \leq n. \quad (7)$$

According to the theory of Burmeister and J. W. Schmidt [6] (or [5]) the R-order of convergence of all the sequences $\{d(X_i^{(m)})\}$, $1 \leq i \leq n$, can be estimated by a lower bound which is a positive solution of the inequalities (8) with positive u_i, $1 \leq i \leq n$

$$\alpha \cdot u_i + \beta \cdot u_k \geq \tau \cdot u_i \qquad \text{for } k > i$$
$$\tau \cdot \beta \cdot u_k + \alpha \cdot u_i \geq \tau \cdot u_i \qquad \text{for } k < i \tag{8}$$

and the best estimation is obtained by the maximum of such values τ. On the other hand the structure of the recursions (7) suggests considering the matrix

$$M + \tau \cdot R = \begin{bmatrix} \alpha & \beta & 0 & \ldots & & 0 \\ 0 & \alpha & \beta & 0 & \ldots & 0 \\ & & 0 & & & 0 \\ & & & & & \beta \\ \tau \cdot \beta & & 0 & \ldots & 0 & \alpha \end{bmatrix}$$

Then we have to calculate the spectral radius of the matrix

$$(I - R)^{-1}M = (I + R)M = \begin{bmatrix} \alpha & \beta & 0 & \cdots & & 0 \\ 0 & \alpha & \beta & 0 & \cdots & 0 \\ & & & & & 0 \\ & 0 & & & & \\ & & & & & \beta \\ \alpha \cdot \beta & \beta^2 & 0 & \cdots & 0 & \alpha \end{bmatrix}$$

which is the unique positive root τ^* (according to the Perron-Frobenius Theorem and Descartes rule of signs) of the polynomial equation

$$\left(\frac{\tau - \alpha}{\beta}\right)^n - \beta \cdot \left(\frac{\tau - \alpha}{\beta}\right) - \alpha = 0. \tag{9}$$

Obviously, $\left(\dfrac{\tau^* - \alpha}{\beta}\right) > 1$ (or $\tau^* > \alpha + \beta$, where $\alpha + \beta$ is the order of convergence of the corresponding total-step method (see [3])). Thus we get an corresponding eigenvector for τ^* by $(u_1, u_2, \ldots, u_n)$ where

$$u_i = \left(\frac{\tau^* - \alpha}{\beta}\right)^{i-1} > 0, \qquad 1 \le i \le n,$$

with monotonic increasing components. Therefore, the inequalities (8) are fulfilled and by the theory of Burmeister and J. W. Schmidt we have

$$O_R(\{d(X_i^{(m)})\}, 0) \ge \tau^*, \qquad 1 \le i \le n. \qquad \blacksquare$$

Example: We consider the single-step interval version of Weierstrass method (see example (a)) by setting $\alpha = \beta = 1$ in our model and get the well-known result in [1].

Remarks: Immediately from the definition of the polynomial equation $\varepsilon^n - \beta\varepsilon - \alpha = 0$ for ε^* follows that $\lim_{n \to \infty} \varepsilon_n^* = 1$, where ε_n^* is the unique positive root of this polynomial as a function of the degree n of the original polynomial in (1). This means that τ^* tends to $\alpha + \beta$, the R-order of convergence of the corresponding total-step methods, as n tends to infinity.

Furthermore, the value of ε^* increases with increasing parameters α and β. This can be shown as follows:

Let ε^* be the positive root of the equation $\varepsilon^n - \beta\varepsilon - \alpha = 0$. Then we are choosing $\alpha' = \alpha + a$ and $\beta' = \beta + b$ with $a > 0$ and $b > 0$. Since $\varepsilon^{*n} - (\beta + b)\varepsilon^* - (\alpha + a) = \varepsilon^{*n} - \beta\varepsilon^* - \alpha - (b\varepsilon^* + a) = -(b\varepsilon^* + a) < 0$ it follows that for the positive root κ^* of the equation $\kappa^n - \beta'\kappa - \alpha' = 0$ the relation $\kappa^* > \varepsilon^*$ holds true.

Thus, with increasing parameters α and β the bound τ^* (given by $\beta\varepsilon^* + \alpha$) is increasing. Passing from total-step to single-step method one can improve the R-order of convergence by at least 1 if $\beta \ge \dfrac{1}{\varepsilon^* - 1}$ which might be fulfilled for reasonable parameters α, β and n especially if n is not too large. (This would surely be the case, for example, for a hypothetic method with $\alpha = 6$ and $\beta = 1$ for $n = 3$).

3. Combined Methods

Let us now consider a second function ψ with the defining properties (2)–(4) for which interval extensions $\tilde{\psi}_i$ exist with

$$d(\tilde{\psi}_i(X_1, \ldots, X_{i-1}, x_i, X_{i+1}, \ldots, X_n, p(x_i), \ldots, p^{(k-1)}(x_i)))$$

$$\leq \tilde{\gamma} \cdot d(X_i)^\sigma \cdot \left(\sum_{\substack{j=1 \\ j \neq i}}^{n} d(X_j)^\mu \right), \quad \sigma, \mu \in \mathbb{N} \setminus \{0\}, \; 1 \leq i \leq n. \tag{10}$$

Then we are able to define a "combined" single-step method by

$$Z_{i,j}^{(m)} = \begin{cases} \tilde{\psi}_i(X_1^{(m+1)}, \ldots, X_{j-1}^{(m+1)}, X_j^{(m)}, \ldots, X_{i-1}^{(m)}, x_i^{(m)}, X_{i+1}^{(m)}, \ldots, X_n^{(m)}, \\ \quad p(x_i^{(m)}), \ldots, p^{(k-1)}(x_i^{(m)})) \qquad \text{for } j < i, \\ \tilde{\psi}_i(X_1^{(m+1)}, \ldots, X_{i-1}^{(m+1)}, x_i^{(m+1)}, X_{i+1}^{(m+1)}, \ldots, X_{j-1}^{(m+1)}, X_j^{(m)}, \ldots, X_n^{(m)}, \\ \quad p(x_i^{(m+1)}), \ldots, p^{(k-1)}(x_i^{(m+1)})) \qquad \text{for } j > i \end{cases} \tag{11a}$$

$$X_j^{(m+1)} = \tilde{\varphi}_j(Z_{1,j}^{(m)}, \ldots, Z_{j-1,j}^{(m)}, x_j^{(m)}, Z_{j+1,j}^{(m)}, \ldots, Z_{n,j}^{(m)}, p(x_j^{(m)}), \ldots, p^{(k-1)}(x_j^{(m)}))$$

$$\cap X_j^{(m)}, \quad 1 \leq j \leq n, m \geq 0, \tag{11b}$$

(where j is the index of the outer loop).

For the methods combined in this way we can prove the following properties.

Theorem 2: *Method* (11a, b) *has the properties*:

(i) $X_i^{(0)} \supseteq X_i^{(1)} \supseteq X_i^{(2)} \supseteq \cdots, 1 \leq i \leq n$;

(ii) $\xi_i \in X_i^{(m)}, 1 \leq i \leq n \Rightarrow \xi_i \in X_i^{(m+1)}, 1 \leq i \leq n$; *this again ensures the feasibility of the method provided* $\xi_i \in X_i^{(0)}, 1 \leq i \leq n$;

(iii) *method* (11a, b) *is locally convergent if* $\xi_i \in X_i^{(0)}, 1 \leq i \leq n$;

(iv) *the R-order of convergence of every sequence* $\{X_i^{(m)}\}, 1 \leq i \leq n$ (*which is the R-order of convergence of* $\{d(X_i^{(m)})\}, 1 \leq i \leq n$) *is at least* τ^* *where*

$$\tau^* = \alpha + \beta(\mu + \varepsilon^*\sigma)$$

and ε^* *is the unique positive root of the polynomial equation*

$$\varepsilon^n - \varepsilon\sigma\beta - (\alpha + \mu\beta) = 0$$

and $\varepsilon^* > 1$ (*which means that* $\tau^* > \alpha + \beta(\mu + \sigma)$ *or greater than the R-order of the corresponding total-step method*).

Proof:

(i) $\rightarrow$ (iii) are analogously proved like for the corresponding statements in Theorem 1.

Of (iv): Applying the width operator d on (11a) we get

$$d(Z_{v,i}^{(m)}) \leq \begin{cases} \delta \cdot d(X_v^{(m)})^\sigma \cdot \left(\sum_{k=1}^{i-1} d(X_k^{(m+1)})^\mu + \sum_{\substack{k=i \\ k \neq v}}^{n} d(X_k^{(m+1)})^\mu \right) & \text{for } i < v, \\ \delta \cdot d(X_v^{(m+1)})^\sigma \cdot \left(\sum_{\substack{k=1 \\ k \neq v}}^{i-1} d(X_k^{(m+1)})^\mu + \sum_{k=i}^{n} d(X_k^{(m)})^\mu \right) & \text{for } i > v. \end{cases}$$

On the other hand (11b) gives

$$d(X_i^{(m+1)}) \leq \gamma \cdot d(X_i^{(m)})^\alpha \left(\sum_{v=1}^{i-1} d(Z_{v,i}^{(m)})^\beta + \sum_{v=i+1}^{n} d(Z_{v,i}^{(m)})^\beta \right).$$

A simple substitution and an application of inequality $(*)$ finally leads to the recursion

$$d(X_i^{(m+1)}) \leq \tilde{\gamma} \cdot d(X_i^{(m)})^\alpha \left(\sum_{v=1}^{i-1} d(X_v^{(m+1)})^{\sigma\beta} \left(\sum_{\substack{k=1 \\ k \neq v}}^{i-1} d(X_k^{(m+1)})^{\mu\beta} + \sum_{k=i}^{n} d(X_k^{(m)})^{\mu\beta} \right) \right.$$

$$\left. + \sum_{v=i+1}^{n} d(X_v^{(m)})^{\sigma\beta} \left(\sum_{k=1}^{i-1} d(X_k^{(m+1)})^{\mu\beta} + \sum_{\substack{k=i \\ k \neq v}}^{n} d(X_k^{(m)})^{\mu\beta} \right) \right), \quad 1 \leq i \leq n.$$

$$(12)$$

According to the theory of Burmeister and J. W. Schmidt [6] we now have to consider the inequalities

$$\begin{aligned} \alpha u_i + \beta\sigma u_k + \mu\beta u_\ell &\geq \tau u_i, & k &> i, \ell \geq i, \ell \neq k, \\ \tau\sigma\beta u_k + \alpha u_i + \mu\beta u_\ell &\geq \tau u_i, & k &< i \leq \ell, \\ \tau\mu\beta u_\ell + \alpha u_i + \sigma\beta u_k &\geq \tau u_i, & k &> i > \ell, \\ \tau\sigma\beta u_k + \tau\mu\beta u_\ell + \alpha u_i &\geq \tau u_i, & k &< i, \ell < i, \ell \neq k, \end{aligned}$$

$$(13)$$

and get for $\tau > 1$ and corresponding $u_i > 0$, $1 \leq i \leq n$ fulfilling (13) a lower bound for the R-order of convergence of all sequences $\{d(X_i^{(m)})\}$, $1 \leq i \leq n$ by the value of τ.

The structure of (12) suggests considering the matrix

$$M + \tau R = \begin{bmatrix} \alpha + \mu\beta & & \sigma\beta & & & \\ 0 & \alpha + \mu\beta & \sigma\beta & & & 0 \\ 0 & & & & & \\ & & & & & \sigma\beta \\ \tau\sigma\beta & & & \cdots & 0 & \alpha + \mu\beta \end{bmatrix}$$

and the spectral radius of $(I - R)^{-1} M = (I + R)M$ is the unique positive root τ^* of the polynomial equation

$$\left(\frac{\tau - (\alpha + \mu\beta)}{\sigma\beta} \right)^n - \left(\frac{\tau - (\alpha + \mu\beta)}{\sigma\beta} \right) \sigma\beta - (\alpha + \mu\beta) = 0.$$

Obviously, $\left(\dfrac{\tau^* - (\alpha + \mu\beta)}{\sigma\beta} \right) > 1$ (or $\tau^* > \alpha + \beta(\sigma + \mu)$ where the right-hand side is the order of convergence of the corresponding combined total-step method [3]), and therefore the corresponding eigenvector with the components $u_i = \left(\dfrac{\tau^* - (\alpha + \mu\beta)}{\sigma\beta} \right)^{i-1} > 0$, $1 \leq i \leq n$ has monotonically increasing components. From the definition of τ^* it also follows $\tau^* \cdot u_\ell \geq u_k$ for $k \geq \ell$. Using this property and with the help of the monotonicity of the u_i it is easily shown that τ^* and u_i, $1 \le i \le n$ are fulfilling the system (13). According to [6] this proves the assertion.   ∎

Similar remarks like in connection with Theorem 1 can be made.

4. Conclusions

(a) The principle of combination as described here can be applied iteratively and then leads to families of methods. This was done in a similar way with a special method in [9].

(b) In special cases the value of τ^* can be improved by a greater one when choosing another matrix out of the system (13). This was done for a special case in [2].

(c) Our model can also be extended for complex roots using complex interval arithmetic. The same can be done in case of multiple roots if their multiplicity is known in advance.

Acknowledgements

The authors would like to thank one of the referees for his helpful suggestions which made it possible to improve the presentation of this final version.

References

[1] Alefeld, G., Herzberger, J.: Introduction to interval computations. New York: Academic Press 1983.

[2] Atanassova, L.: On the R-order of a single-step Nourein type method. In: Atanassova, L., Herzberger, J. (eds.) Computer arithmetic and enclosure methods, pp. 179–187. Amsterdam: Elsevier (North-Holland) 1992.

[3] Atanassova, L., Herzberger, J.: A general approach to simultaneous inclusion methods for polynomial zeros. In: Atanassova, L., Herzberger J. (eds.) Computer arithmetic and enclosure methods, pp. 189–198. Amsterdam: Elsevier (North-Holland) 1992.

[4] Atanassova, L., Herzberger, J.: Bemerkungen zur allgemeinen Darstellung von simultanen Polynomwurzel-Einschließungsverfahren. ZAMM *73* (in press).

[5] Burmeister, W., Schmidt, J.W.: Determination of the cone radius for positive concave operators. Computing *33*, 37–49 (1984).

[6] Burmeister, W., Schmidt, J.W.: On the R-order of coupled sequences arising in single-step methods. Numer. Math. *53*, 653–661 (1988).

[7] Ehrlich, L. W.: A modified Newton method for polynomials. Comm. ACM *10*, 107–108 (1967).

[8] Kjurkchiev, N., Andreev, A.: A modification of Weierstrass-Dochev's method with rate of convergence $R + 2$ for simultaneous determination of zeros of a polynomial (in Russian). C.R. Acad. Bulg. Sc. *38*, 1461–1463 (1985).

[9] Kjurkchiev, N., Andreev, A.: On the generalization of the Alefeld-Herzberger's method. Computing *47*, 355–360 (1992).

[10] Moore, R. E.: Interval analysis. Englewood-Cliffs, N.J.: Prentice-Hall 1966.

[11] Neumaier, A.: Interval methods for systems of equations. Cambridge: Cambridge University Press 1990.

[12] Petkovic', M.: Iterative methods for simultaneous inclusion of polynomial zeros. Berlin: Springer 1989.

[13] Wang, X., Zheng, S.: A family of parallel and interval iterations for finding simultaneously all roots of a polynomial with rapid convergence. I.J. Comput. Math. *4*, 70–76 (1984).

[14] Weierstrass, K.: Neuer Beweis des Satzes, daß jede ganze rationale Funktion einer Veränderlichen dargestellt werden kann als ein Produkt aus linearen Funktionen derselben Veränderlichen. Ges. Werke *3*, (Johnson Reprint Corp., New York 1967), 251–269 (1903).

L. Atanassova
Institut für Dynamische Systeme
Universität Bremen
D-W-2800 Bremen 33
Federal Republic of Germany

J. Herzberger
Fachbereich Mathematik
Universität Oldenburg
D-W-2900 Oldenburg
Federal Republic of Germany

Computing, Suppl. 9, 21–32 (1993)

On Some Properties of an Interval Newton Type Method and its Modification[1],[*]

N. S. Dimitrova, Sofia

Dedicated to Professor U. Kulisch on the occasion of his 60th birthday

Abstract — Zusammenfassung

On some Properties of an Interval Newton Type Method and its Modification. Considered is an interval iterative Newton type method for enclosing a real simple root of the nonlinear equation $f(x) = 0$ in a given interval X. The method has a simple formulation in terms of extended interval arithmetic. Cubic convergence of the method is proved assuming that f possesses a Lipschitzian second derivative which vanishes at the root of the equation. A modification of the method with higher order of convergence is proposed. An algorithm with result verification is formulated and some numerical experiments are reported.

AMS Subject Classification: 65G10, 65H05, 65H10

Key words: Extended interval arithmetic, Newton's method, order of convergence.

Über einige Eigenschaften eines Newton—Intervallverfahrens und seiner Modifikation. Wir betrachten ein iteratives Newtonartiges Intervallverfahren zur Einschließung einer einfachen Lösung der Gleichung $f(x) = 0$ in einem vorgegebenen Intervall X. Das Verfahren hat eine einfache Darstellung in erweiterter Intervallarithmetik. Kubische Konvergenz des Verfahrens wird nachgewiesen falls die zweite Ableitung f'' in der Nullstelle verschwindet und eine Lipschitzbedingung befriedigt. Eine Modifikation des Verfahrens mit höherer Konvergenzordnung wird vorgeschlagen. Ein Algorithmus zur Ergebnisverifikation und numerische Ergebnisse werden vorgestellt.

1. Introduction

The present work contains further studies on the interval iterative Newton type procedure $X^{(n+1)} = X^{(n)} - {}^-f(X^{(n)})/{}^-F'(X^{(n)})$, $n = 0, 1, \ldots$. It is formulated in terms of extended interval arithmetic and does not involve intersection as the usual interval methods. This procedure was introduced in [9] and studied in some detail in [4]. In [5] global quadratic convergence of the method is proved and an algorithm with result verification using computer (interval) arithmetic is presented.

---

[1] This work has been partially supported by the Ministry of Education and Sciences—National Found Science Researches under contract No. MM-10/91.

[*] Received September 23, 1992; revised December 17, 1992.

Denote by IR the set of all real compact intervals. An interval $X \in IR$ with end-points $\underline{x}$ and $\bar{x}$, $\underline{x} \leq \bar{x}$ will be denoted by $X = [\underline{x}, \bar{x}]$. The width of X is defined by $\omega(X) = \bar{x} - \underline{x}$. An interval X with end-points x_1 and x_2 will be also denoted by $X = [x_1 \vee x_2]$, where $[x_1 \vee x_2] = \{[x_1, x_2],$ if $x_1 \leq x_2; [x_2, x_1],$ otherwise$\}$. The presentation $X = [x_1 \vee x_2]$ does not require $x_1 \leq x_2$. By x^{+0} and x^{-0} we mean the end-points $x^{+0} = \{\underline{x},$ if $|\underline{x}| \leq |\bar{x}|; \bar{x},$ otherwise$\}$; $x^{-0} = \{\bar{x},$ if $|\underline{x}| \leq |\bar{x}|; \underline{x},$ otherwise$\}$, which satisfy $|x^{+0}| \leq |x^{-0}|$. Thus we can write $X = [x^{+0} \vee x^{-0}]$. For $X = [x^{+0} \vee x^{-0}]$ the functional $\chi: IR \backslash [0,0] \to [-1,1]$ is defined by $\chi(X) = x^{+0}/x^{-0}$. In what follows by $+$, $-$, $\times$ and $/$ we mean the usual interval arithmetic operations as they are defined in [2], [10]. The extended interval arithmetic operations, denoted by $+^-$, $-^-$, $\times^-$ and $/^-$ are introduced as follows [4], [5], [6], [9]. For $X, Y \in IR$, $X = [\underline{x}, \bar{x}] = [x^{+0} \vee x^{-0}]$, $Y = [\underline{y}, \bar{y}] = [y^{+0} \vee y^{-0}]$ we define

$$X +^- Y = [\underline{x} + \bar{y} \vee \bar{x} + \underline{y}]$$

$$= \begin{cases} [\underline{x} + \bar{y}, \bar{x} + \underline{y}] & \text{if } \omega(X) \geq \omega(Y), \\ [\bar{x} + \underline{y}, \underline{x} + \bar{y}] & \text{if } \omega(X) < \omega(Y); \end{cases}$$

$$X -^- Y = [\underline{x} - \underline{y} \vee \bar{x} - \bar{y}]$$

$$= \begin{cases} [\underline{x} - \underline{y}, \bar{x} - \bar{y}] & \text{if } \omega(X) \geq \omega(Y), \\ [\bar{x} - \bar{y}, \underline{x} - \underline{y}] & \text{if } \omega(X) < \omega(Y); \end{cases}$$

$$X \times^- Y = \begin{cases} [x^{-0}y^{+0} \vee x^{+0}y^{-0}] & \text{if } 0 \notin X, Y, \\ y^{+0}X = [y^{+0}\underline{x} \vee y^{+0}\bar{x}] & \text{if } 0 \in X, 0 \notin Y; \end{cases}$$

$$X /^- Y = \begin{cases} [x^{+0}/y^{+0} \vee x^{-0}/y^{-0}] & \text{if } 0 \notin X, Y, \\ X/y^{-0} = [\underline{x}/y^{-0} \vee \bar{x}/y^{-0}] & \text{if } 0 \in X, 0 \notin Y. \end{cases}$$

It is known that the four usual interval-arithmetic operations $+$, $-$, $\times$, $/$ (see [2], [10]) are inclusion monotone in the sense that $X_1 \subseteq X$, $Y_1 \subseteq Y$ imply $X_1 * Y_1 \subseteq X * Y$ for any operation $* \in \{+, -, \times, /\}$. In [5] it is shown that the extended interval-arithmetic operations are quasi-inclusion monotone in the sense of the next two Propositions [5]:

Proposition 1. *Let* $X, X_1, Y, Y_1 \in IR$, $X \supseteq X_1$, $Y \subseteq Y_1$, $* \in \{+^-, -^-\}$. *Then*

(a) $\max\{\omega(X), \omega(X_1)\} \leq \min\{\omega(Y), \omega(Y_1)\}$ *implies* $X * Y \subseteq X_1 * Y_1$;
(b) $\min\{\omega(X), \omega(X_1)\} \geq \max\{\omega(Y), \omega(Y_1)\}$ *implies* $X * Y \supseteq X_1 * Y_1$. $\square$

Proposition 2. *Let* $X, X_1, Y, Y_1 \in IR, 0 \notin Y, Y_1, X \supseteq X_1, Y \subseteq Y_1, * \in \{\times^-, /^-\}$. *Then*

(a) $\min\{\chi(X), \chi(X_1)\} \geq \max\{\chi(Y), \chi(Y_1)\}$ *implies* $X * Y \subseteq X_1 * Y_1$;
(b) $\max\{\chi(X), \chi(X_1)\} \leq \min\{\chi(Y), \chi(Y_1)\}$ *implies* $X * Y \supseteq X_1 * Y_1$. $\square$

Denote by x^* the solution of the nonlinear equation $f(x) = 0$, where $f: D \to R$, $D \subseteq R$, is a continuously differentiable function. Suppose that the derivative f' of f has a constant sign in D, i.e. $f'(x) \neq 0$ for all $x \in D$. Let $f(X) = \{f(x): x \in X\}$ be the range of f on an interval $X \in ID$, $ID = \{X \in IR: X \subseteq D\}$. Since f is monotone on $X = [\underline{x}, \bar{x}]$, the presentation $f(X) = [f(\underline{x}) \vee f(\bar{x})]$ is valid. Obviously, $x^* \in X$

is equivalent to $0 \in f(X)$. Define an interval extension F' of f' as any interval function, $F': ID \to IR$, satisfying the conditions:

$$\begin{cases} \{f'(x): x \in X\} \subseteq F'(X) \text{ for } X \in ID; \\ F'(X) \subseteq F'(Y), \text{ whenever } X \subseteq Y, \quad X, Y \in ID; \\ 0 \notin F'(X) \text{ for all } X \in ID. \end{cases} \quad (1)$$

Assume further that F' satisfies a Lipschitz condition [10] with a constant $L, L > 0$, independent on X

$$\omega(F'(X)) \leq L\omega(X) \text{ for all } X \in ID. \quad (2)$$

Denote by $\mathcal{N}: ID \to IR$, the interval Newton type operator

$$\mathcal{N}(X) = X -^{-} f(X)/^{-} F'(X). \quad (3)$$

We present below some properties of (3) which we further need (see [5]).

Proposition 3. *Let $f: D \to R, D \subseteq R$, be a continuously differentiable function. Let $f(X)$ be the range of f on X and F' be an interval extension of the derivative f' with $0 \notin F'(X)$. Then $0 \in f(X)$ implies $\omega(X) \geq \omega(f(X)/^{-} F'(X))$.* $\square$

Theorem 1. *Let $f: D \to R, D \subseteq R$, be a continuously differentiable function. Let $f(X)$ be the range of f on X and F' be an interval extension of the derivative f', which satisfies the conditions (1). Then $\mathcal{N}(X) \subseteq X$ is a necessary and sufficient condition for existence of an unique solution of $f(x) = 0$ in the interval X, i.e. $\mathcal{N}(X) \subseteq X$ is equivalent to $0 \in f(X)$.* $\square$

Corollary 1. *Under the assumptions of Theorem 1 the necessary and sufficient condition for nonexistence of a solution of the equation $f(x) = 0$ in the interval X is $\mathcal{N}(X) \nsubseteq X$, i.e. $\mathcal{N}(X) \nsubseteq X$ is equivalent to $0 \notin f(X)$.* $\square$

Theorem 2. *Let the assumptions of Theorem 1 be fulfilled. Then the following assertions are valid:*

(a) *If $f(x^*) = 0$ and $x^* \in X$ then $x^* \in \mathcal{N}(X)$;*
(b) *If $f(x^*) = 0$ and $x^* \in X$ then $\mathcal{N}(\mathcal{N}(X)) \subseteq \mathcal{N}(X)$;*
(c) *$\mathcal{N}(X) = X$ if and only if $X = [x^*, x^*] = x^*$ and $f(x^*) = 0$.* $\square$

Using the interval Newton type operator $\mathcal{N}$ we define the following interval Newton type iterative procedure for enclosing the real simple root x^* of the nonlinear equation $f(x) = 0$:

$$\begin{cases} X^{(0)} \in ID, \\ X^{(n+1)} = \mathcal{N}(X^{(n)}), \quad n = 0, 1, \dots. \end{cases} \quad (4)$$

Theorem 3. *Let $f: D \to R, D \subseteq R$, be a continuously differentiable function. Let $f(X)$ be the range of f on X and F' be an interval extension of the derivative f', which satisfies the conditions (1)–(2). Then:*

(a) *If $\mathcal{N}(X^{(0)}) \nsubseteq X^{(0)}$, the equation has no solution in $X^{(0)}$ and the iteration procedure (4) terminates after the first step;*

(b) *If $\mathcal{N}(X^{(0)}) \subseteq X^{(0)}$, (4) produces a sequence of intervals $\{X^{(n)}\}$ with the following properties*:
    (b1) $X^{(0)} \supseteq X^{(1)} \supseteq \cdots \supseteq X^{(n)} \supseteq X^{(n+1)} \supseteq \cdots$;
    (b2) $x^* \in X^{(n)}$ *for* $n = 1, 2, \ldots$;
    (b3) $\lim_{n \to \infty} X^{(n)} = x^*$ *and* $\omega(X^{(n+1)}) \leq c\omega^2(X^{(n)})$, $c > 0$. $\square$

It is shown in [5] that if the computational costs for $f(X)$ and for $F'(X)$ are assumed equal then the efficiency index of (4) in the sense of Ostrowski (see [12]) is *eff* $\{(4)\} = \sqrt{2}$. A method with higher efficiency index is considered in section 3.

2. Cubic Convergence of the Method

It is known [11] that the classical real Newton's method $x^{(n+1)} = x^{(n)} - f(x^{(n)})/f'(x^{(n)})$, $n = 0, 1, \ldots$, for determining a real simple root x^* of the equation $f(x) = 0$ when f possesses a third continuous derivative and $f'(x^*) \neq 0$, $f''(x^*) = 0$ hold, is cubically convergent to x^* in the sense that $|x^{(n+1)} - x^*| \leq c|x^{(n)} - x^*|^3$, $c = \text{const}$, $c > 0$. It is shown in [1] that the interval Newton method of Moore (see [2], [10]),

$$\begin{cases} X^{(n+1)} = (m(X^{(n)}) - f(m(X^{(n)}))/F'(X^{(n)})) \cap X^{(n)}, \quad m(X^{(n)}) \in X^{(n)}, \\ n = 0, 1, 2, \ldots \end{cases} \tag{5}$$

does not preserve this property. To carry over the cubic convergence of the classical method to the interval Newton method (5) a modification of the last one is proposed in [1].

In this section we shall show that if the interval extension $F'(X)$ of the first derivative approximates quadratically the range $f'(X) = \{f'(x): x \in X\}$ and the second derivative of f is Lipschitzian, the interval sequence $\{X^{(n)}\}$ given by the interval Newton type procedure (4) is cubically convergent to the root.

Theorem 4. *Let* $f: D \to R$, $D \subseteq R$, *be twice continuously differentiable on* D. *Let* $f'(X) = \{f'(x): x \in X\}$ *be the range of* f' *and* F' *be an interval extension of* f', *which satisfies* (1) *and* $\omega(F'(X)) = \omega(f'(X)) + O(\omega^2(X))$. *Denote by* x^* *the root of the equation* $f(x) = 0$ *and suppose that* $f''(x^*) = 0$. *Let* F'' *be an interval extension of* f'', $F'': ID \to IR$ *satisfying the conditions*

$$\{f''(x): x \in X\} \subseteq F''(X) \text{ for } X \in ID;$$

$$F''(X) \subseteq F''(Y) \text{ whenever } X \subseteq Y, \quad X, Y \in ID;$$

$$\omega(F''(X)) \leq L\omega(X) \text{ for all } X \in ID, \tag{6}$$

where the constant $L > 0$ *does not depend on* X. *Then*:

(a) *If* $\mathcal{N}(X^{(0)}) \nsubseteq X^{(0)}$, *the equation has no solution in* $X^{(0)}$ *and the iterative procedure* (4) *terminates after the first step*;
(b) *If* $\mathcal{N}(X^{(0)}) \subseteq X^{(0)}$ *then* (4) *produces a sequence of intervals* $\{X^{(n)}\}$ *with the following properties*:

(b1) $X^{(0)} \supseteq X^{(1)} \supseteq \cdots \supseteq X^{(n)} \supseteq X^{(n+1)} \supseteq \cdots$;

(b2) $x^* \in X^{(n)}$ *for* $n = 1, 2, \ldots$;

(b3) $\lim_{n \to \infty} X^{(n)} = x^*$ *and* $\omega(X^{(n+1)}) \leq c\omega^3(X^{(n)}), c > 0$.

Proof: Statements (a) and (b1) resp. (b2) of the theorem follow immediately from Theorem 3(a), (b1) resp. (b2). We shall prove (b3). From the definitions of the operations $-^-, /^-$ and Proposition 3 we obtain

$$\omega(X^{(n+1)}) = \omega(X^{(n)}) - \omega(f(X^{(n)}) /^- F'(X^{(n)}))$$

$$= \omega(X^{(n)}) - \omega(f(X^{(n)})) / |F'^{-0}(X^{(n)})|$$

$$= \omega(X^{(n)}) - |f(\underline{x}^{(n)}) - f(\overline{x}^{(n)})| / |F'^{-0}(X^{(n)})|$$

$$= (1 - |f'(\xi)| / |F'^{-0}(X^{(n)})|)\omega(X^{(n)}),$$

wherein $\xi \in (\underline{x}^{(n)}, \overline{x}^{(n)}) \subseteq X^{(0)}$. Because of $f'(\xi) \in F'(X^{(0)})$, $|f'(\xi)| \geq |F'^{+0}(X^{(0)})|$ holds; further $|F'^{-0}(X^{(n)})| \leq |F'^{-0}(X^{(0)})|$ is valid. Therefore

$$\omega(X^{(n+1)}) \leq (1 - |F'^{+0}(X^{(0)})| / |F'^{-0}(X^{(0)})|)\omega(X^{(n)}).$$

With $q = 1 - |F'^{+0}(X^{(0)})| / |F'^{-0}(X^{(0)})|, 0 < q < 1$, we obtain $\omega(X^{(n+1)}) \leq q\omega(X^{(n)})$. This inequality means $\lim_{n \to \infty} X^{(n)} = x^*$.

The cubic convergence of the sequence $\{X^{(n)}\}$ remains to be shown. We have

$$\omega(X^{(n+1)}) \leq (1 - |F'^{+0}(X^{(n)})| / |F'^{-0}(X^{(n)})|)\omega(X^{(n)})$$

$$= (1/|F'^{-0}(X^{(n)})|)(|F'^{-0}(X^{(n)})| - |F'^{+0}(X^{(n)})|)\omega(X^{(n)})$$

$$= (1/|F'^{-0}(X^{(n)})|)\omega(F'(X^{(n)}))\omega(X^{(n)})$$

$$= (1/|F'^{-0}(X^{(n)})|)(\omega(f'(X^{(n)})) + O(\omega^2(X^{(n)})))\omega(X^{(n)})$$

$$\leq (1/|F'^{-0}(X^{(n)})|)(|f''(\eta)|\omega^2(X^{(n)}) + c_1\omega^3(X^{(n)})),$$

where $\underline{x}^{(0)} \leq \underline{x}^{(n)} < \eta < \overline{x}^{(n)} \leq \overline{x}^{(0)}$. Using the fact that $0 \in F''(X^{(n)})$ and $f''(\eta) \in F''(X^{(n)})$ we have $|f''(\eta)| \leq \omega(F''(X^{(n)}))$; condition (6) implies $\omega(F''(X^{(n)})) \leq L\omega(X^{(n)})$. Using the inclusion $F''(X^{(n)}) \subseteq F''(X^{(0)})$ we obtain

$$\omega(X^{(n+1)}) \leq (1/|F'^{-0}(X^{(n)})|)(L + c_1)\omega^3(X^{(n)})$$

$$\leq (1/|F'^{+0}(X^{(0)})|)(L + c_1)\omega^3(X^{(n)})$$

$$= c\omega^3(X^{(n)}),$$

where $c = (L + c_1)/|F'^{+0}(X^{(0)})|$. This proves the theorem. $\square$

Remark. The assumption $\omega(F'(X)) = \omega(f'(X)) + O(\omega^2(X))$ is satisfied using a centered form presentation for $F'(X)$ (see for example [2]).

3. Interval Iterative Methods with Higher Order of Convergence

Let $f: D \to R, D \subseteq R$, be a real valued function defined in D. Assume that f possesses a continuous derivative f' in D which has a constant sign in D, i.e. $f'(x) \neq 0$ for all

$x \in D$. Let $f(X) = \{f(x): x \in X\}$ be the range of f on an interval $X \in ID$. Denote by F' an interval extension of f', satisfying the conditions (1)–(2).

Consider the following modification of the interval iterative procedure (4):

$$\begin{cases} X^{(0)} \in ID; \\ X^{(k,0)} = X^{(k)}, \ k = 0, 1, 2, \ldots; \\ X^{(k,m)} = X^{(k,m-1)} \ -^{-} \ f(X^{(k,m-1)}) /^{-} \ F'(X^{(k)}), \\ m = 1, 2, \ldots, p; \\ X^{(k+1)} = X^{(k,p)}; \ k = 0, 1, 2, \ldots . \end{cases} \qquad (7)$$

Obviously, (7) reduces to (4) for $p = 1$.

Similar modifications of Moore's interval Newton method (5) are formulated in [3] and [8]. In [3] a theorem for order of convergence $(p + 1)$ of the sequence $\{X^{(k)}\}$ is formulated and in [8] a proof of the same statement is sketched.

Theorem 5. *Let $f: D \to R$, $D \subseteq R$, be a continuously differentiable function. Let $f(X)$ be the range of f on X and F' be an interval extension of the derivative f', which satisfies (1)–(2). Then:*

(a) *If $X^{(0,1)} \nsubseteq X^{(0)}$, the equation has no solution in $X^{(0)}$ and the iterative procedure (7) terminates after the first step;*
(b) *If $X^{(0,1)} \subseteq X^{(0)}$, then (7) produces a sequence of intervals $\{X^{(k)}\}$ with the following properties:*
    (b1) $X^{(0)} \supseteq X^{(1)} \supseteq \cdots \supseteq X^{(k)} \supseteq X^{(k+1)} \supseteq \cdots;$
    (b2) $x^* \in X^{(k)}$ *for* $k = 1, 2, \ldots;$
    (b3) $\lim_{k \to \infty} X^{(k)} = x^*$ *and* $\omega(X^{(k+1)}) \leq c(\omega(X^{(k)}))^{p+1}, \ c > 0.$

Proof: Statement (a) of our theorem follows from Corollary 1. Using the quasi-inclusion property of the operations $-^{-}$ and $/^{-}$ (Propositions 1–2) we obtain

$$X \ -^{-} \ f(X) /^{-} F'(X) \supseteq Y \ -^{-} \ f(Y) /^{-} F'(X) \ \text{for} \ X \supseteq Y, \ X, Y \in ID.$$

This leads to the inclusions

$$X^{(k)} = X^{(k,p)} \supseteq X^{(k,1)} \supseteq \cdots \supseteq X^{(k,p)} = X^{(k+1)}, \qquad k = 0, 1, 2, \ldots .$$

The proof of (b1) resp. (b2) follows then from Theorem 3(b1) resp. (b2).

We shall now prove the $(p + 1)$-st order of convergence of the sequence $\{X^{(k)}\}$. It follows from Theorem 3(b3) that there is a constant $c_1 > 0$, independent on k, such that the inequality

$$\omega(X^{(k,1)}) \leq c_1 \omega^2(X^{(k,0)}) = c_1 \omega^2(X^{(k)}) \qquad (8)$$

holds true. Let be $2 \leq m \leq p$. Then

$$\omega(X^{(k,m)}) = \omega(X^{(k,m-1)}) - \omega(f(X^{(k,m-1)}) /^{-} F'(X^{(k)}))$$

$$= \omega(X^{(k,m-1)}) - (1/|F'^{-0}(X^{(k)})|)\omega(f(X^{(k,m-1)}))$$

$$= (1 - |f'(\xi)|/|F'^{-0}(X^{(k)})|)\omega(X^{(k,m-1)}),$$

where $\xi \in (\underline{x}^{(k,m-1)}, \bar{x}^{(k,m-1)}) \subseteq X^{(k)}$. Since $f'(\xi) \in F'(X^{(k)})$ and $0 \notin F'(X^{(k)})$ (see (1)) it follows $|f'(\xi)| \geq |F'^{+0}(X^{(k)})|$. Thus we obtain

$$\omega(X^{(k,m)}) \leq (1 - |F'^{+0}(X^{(k)})|/|F'^{-0}(X^{(k)})|)\omega(X^{(k,m-1)})$$

$$= (1/|F'^{-0}(X^{(k)})|)(|F'^{-0}(X^{(k)})| - |F'^{+0}(X^{(k)})|)\omega(X^{(k,m-1)})$$

$$= (1/|F'^{-0}(X^{(k)})|)\omega(F'(X^{(k)}))\omega(X^{(k,m-1)}).$$

According to (1), $X^{(k)} \subseteq X^{(0)}$ implies $F'(X^{(k)}) \subseteq F'(X^{(0)})$ and therefore $|F'^{-0}(X^{(k)})| \geq |F'^{+0}(X^{(0)})|$. Using (2) we obtain

$$\omega(X^{(k,m)}) \leq (L/|F'^{+0}(X^{(0)})|)\omega(X^{(k)})\omega(X^{(k,m-1)})$$

$$= c_m \omega(X^{(k)})\omega(X^{(k,m-1)})$$

with $c_m = (L/|F'^{+0}(X^{(0)})|)$. Obviously, c_m does not depend on k. Applying the above inequality consecutively for $m = p, p - 1, \ldots, 2$ we obtain

$$\omega(X^{(k+1)}) = \omega(X^{(k,p)}) \leq c_p \omega(X^{(k)})\omega(X^{(k,p-1)})$$

$$\leq c_p c_{p-1}(\omega(X^{(k)}))^2 \omega(X^{(k,p-2)})$$

$$\leq \cdots$$

$$\leq c_p c_{p-1} \ldots c_2 (\omega(X^{(k)}))^{p-1} \omega(X^{(k,1)}),$$

where all constants c_i, $2 \leq i \leq p$, are independent on k. According to (8) we have

$$\omega(X^{(k+1)}) \leq c_p c_{p-1} \ldots c_2 c_1 (\omega(X^{(k)}))^{p+1}.$$

With $c = c_p c_{p-1} \ldots c_2 c_1$ we obtain finally $\omega(X^{(k+1)}) \leq c(\omega(X^{(k)}))^{p+1}$. $\square$

The method (7) requires p interval evaluations of $f(X)$ and one evaluation of $F'(X)$ per iteration step. Assuming that the computational costs for $f(X)$ and $F'(X)$ are about the same, we obtain for the efficiency index of (7) in the sense of Ostrowski [12] $\mathit{eff}\{(7)\} = (p + 1)^{1/(p+1)}$. The highest efficiency index is obtained for $p = 2$ (see [3]).

4. An Algorithm with Result Verification

Let S be a floating point system [7] and IS be the set of all intervals with end-points over S. The computer realization of algorithms using extended interval-arithmetic operations is discussed in detail in [4] and [5]. Two kinds of monotone roundings $\Diamond, \bigcirc: IR \to IS$ of intervals are used, defined for $A = [\underline{a}, \bar{a}]$ by:

$$\Diamond A = [\nabla \underline{a}, \Delta \bar{a}]; \qquad \bigcirc A = \begin{cases} [\Delta \underline{a}, \nabla \bar{a}] & \text{if } \Delta \underline{a} \leq \nabla \bar{a}, \\ \varnothing & \text{otherwise}, \end{cases}$$

where $\nabla a = \max\{x \in S: x \leq a\}$, $\Delta a = \min\{x \in S: x \geq a\}$. They generate the computer interval-arithmetic operations

$$A\langle * \rangle B = \Diamond(A * B), \qquad A(*)B = \bigcirc(A * B),$$

where $*$ can be any one of the interval-arithmetic operations defined above.

Using the quasi-inclusion properties of the operations $-^-$ and $/^-$ (Propositions 1–2) we obtain the following inclusions for $A, B \in IR$ (see [4], [5]).

$$\bigcirc A(-^-)\lozenge B \subseteq A -^- B \subseteq \lozenge A\langle -^-\rangle \bigcirc B \quad \text{if } \omega(A) \geq \omega(B),$$

$$\lozenge A(-^-)\bigcirc B \subseteq A -^- B \subseteq \bigcirc A\langle -^-\rangle \lozenge B \quad \text{if } \omega(A) < \omega(B);$$

$$\bigcirc A(/^-)\lozenge B \subseteq A /^- B \subseteq \lozenge A\langle /^-\rangle \bigcirc B \quad \text{if } \chi(A) \leq \chi(B),$$

$$\lozenge A(/^-)\bigcirc B \subseteq A /^- B \subseteq \bigcirc A\langle /^-\rangle \lozenge B \quad \text{if } \chi(A) > \chi(B).$$

Let $f: D \to R, D \subseteq R$, be a continuously differentiable function, whose derivative has an interval extension F' satisfying the conditions (1)–(2) for $X \in ID$. Denote by $ID_S = \{I \in IS: I \subseteq D\}$ the set of the computer intervals contained in D. Let $f(X) = [f(\underline{x}) \vee f(\bar{x})] = [\underline{f}, \bar{f}]$ be the range of f on $X = [\underline{x}, \bar{x}] \in ID_S$. According to the definitions of $\lozenge$ and $\bigcirc$ we have

$$\lozenge f(X) = [\nabla \underline{f}, \triangle \bar{f}]; \qquad \bigcirc f(X) = \begin{cases} [\triangle \underline{f}, \nabla \bar{f}] & \text{if } \triangle \underline{f} \leq \nabla \bar{f}, \\ \varnothing & \text{otherwise}. \end{cases}$$

The (optimal) machine intervals $\lozenge f(X)$ and $\bigcirc f(X)$ are difficult to be computed in the practice. The ideal situation would be to have $\nabla \underline{f} = \max\{f \in S: f \leq \underline{f}\}, \triangle \bar{f} = \min\{f \in S: f \geq \bar{f}\}$. In the general case this can not be achieved, since the computer realization of $\nabla \underline{f}$ and $\triangle \bar{f}$ depends on the particular expression for f. Because of the effect of rounding errors we can compute in general a lower bond for $\underline{f}$ instead of $\nabla \underline{f}$ and an upper bound for $\bar{f}$ instead of $\triangle \bar{f}$. In what follows by $\triangle a$ resp. ∇a we shall mean any two machine numbers such that $\triangle a \geq a$ resp. $\nabla a \leq a$.

We formulate now the following computer interval-arithmetic iteration method related to (7):

$$\begin{cases} X^{(0)} \in ID_S; \\ X^{(k,0)} = X^{(k)}, k = 0, 1, 2, \ldots; \\ X^{(k,m)} = X^{(k,m-1)}\langle -^-\rangle (\bigcirc f(X^{(k,m-1)})(/^-)\lozenge F'(X^{(k)})), \\ m = 1, 2, \ldots \textbf{ until } m = p \textbf{ or } \bigcirc f(X^{(k,m-1)}) = \varnothing \textbf{ or } X^{(k,m)} \supseteq X^{(k,m-1)}; \\ X^{(k+1)} = X^{(k,p)}; \\ k = 0, 1, 2, \ldots \textbf{ until } X^{(k+1)} \supseteq X^{(k)}. \end{cases} \tag{9}$$

According to the finite convergence principle [10] the iteration procedure (9) produces a finite sequence $\{X^{(k)}\}$ such that for some i, $X^{(i)} = X^{(i+l)}, l = 1, 2, \ldots,$ holds true. Since $\bigcirc f$ is inclusion isotone, i.e. $\bigcirc f(X) \subseteq \bigcirc f(Y)$ whenever $X \subseteq Y, X, Y \in ID_S$, statements (b1), (b2) of Theorem 5 are valid for (9); statement (b3) obtains the form $\lim_{k \to \infty} X^{(k)} \supset x^*$ and $\omega(X^{(k+1)}) \leq c(\omega(X^{(k)}))^{p+1}, k = 1, 2, \ldots, i - 1$. Statements (a) and (b) reduce to the test whether $0 \in \bigcirc f(X^{(0)})$ resp. $0 \in \lozenge f(X^{(0)})$. Because of the effect of rounding errors it can happen that $0 \notin \bigcirc f(X^{(0)})$. This does not necessarily mean that $f(x) = 0$ has no solution in $X^{(0)}$. The equation possesses no solution in $X^{(0)}$ iff $0 \notin \lozenge f(X^{(0)})$. But if $0 \notin \bigcirc f(X^{(0)})$ and $0 \in \lozenge f(X^{(0)})$ are simultaneously true then one can not claim existence/nonexistence of a solution in the initial interval. In this situation either another $X^{(0)}$ should be chosen or we should compute using higher precision. The same situation may occur on some step (k, m). Further iterations are then useless even when $X^{(k,m)}$ has not become sufficiently small.

The algorithm with result verification based on (9) is written in PASCAL-SC-like form and is displayed below under the name NHI.

Algorithm NHI

0. {Input data}
   Initial interval $X^{(0)}$; integer value p $(p \geq 1)$.

1. {Initial test for existence of a solution}
   Compute $\bigcirc f(X^{(0)})$;
   If $\bigcirc f(X^{(0)}) = \emptyset$ **then write** (message 1) and stop;
   If $0 \notin \bigcirc f(X^{(0)})$ **then**
       Compute $\diamondsuit f(X^{(0)})$;
       If $0 \notin \diamondsuit f(X^{(0)})$ **then**
          write (message 2) and stop;
       else
          write (message 3) and stop;
   else $\{0 \in \bigcirc f(X^{(0)})\}$
       Compute $\diamondsuit F'(X^{(0)})$;
       If $0 \in \diamondsuit F'(X^{(0)})$ **then**
          write (message 4) and stop;
       else goto 2.

2. {Iteration according (9)}
   $k := 0; m := 0$;
   $X^{(k,0)} := X^{(k)}$;
   repeat
       $m := m + 1$;
       $X^{(k,m)} := X^{(k,m-1)} \langle -^- \rangle \bigcirc f(X^{(k,m-1)})(/^-) \diamondsuit F'(X^{(k)})$;
       If $X^{(k,m)} \supseteq X^{(k,m-1)}$ **then**
          iwrite$(X^{(k,m-1)})$ and stop;
   until $m = p$;
   $X^{(k+1)} := X^{(k,p)}$;
   repeat
       $k := k + 1$;
       $X^{(k,0)} := X^{(k)}$;
       $m := 0$;
       repeat
          $m := m + 1$;
          Compute $\bigcirc f(X^{(k,m-1)})$;
          If $\bigcirc f(X^{(k,m-1)}) = \emptyset$ **or** $0 \notin \bigcirc f(X^{(k,m-1)})$ **then**
             write (message 5);
             iwrite $(X^{(k,m-1)})$ and stop;
          Compute $\diamondsuit F'(X^{(k)})$;

$$X^{(k,m)} := X^{(k,m-1)} \langle -\bar{} \rangle \bigcirc f(X^{(k,m-1)}) (/\bar{}) \diamond F'(X^{(k)});$$
    If $X^{(k,m)} \supseteq X^{(k,m-1)}$ **then**
    iwrite $(X^{(k,m-1)})$ and stop;
  until $m = p$;
    $X^{(k+1)} := X^{(k,p)}$;
until $X^{(k+1)} \supseteq X^{(k)}$;
iwrite $(X^{(k)})$.
{The resulting interval for the solution is $X^{(k,m-1)}$ or $X^{(k)}$}

Messages:

message 1 $=$ '$\bigcirc f(X^{(0)}) = \varnothing$.
         Restart the algorithm with another initial interval.'

message 2 $=$ 'The equation has no solution in the initial interval.'

message 3 $=$ 'The algorithm can not determine existence/nonexistence of a
         solution in the initial interval. Restart the algorithm with another
         initial interval.'

message 4 $=$ '$0 \in \diamond F'$ on the initial interval.
         Restart the algorithm with another initial interval or compute $\diamond F'$
         more accurately.'

message 5 $=$ '$\bigcirc f = \varnothing$. The enclosing interval can not be made smaller in this
         precision.'

The algorithm NHI displayed above reduces for $p = 1$ to the algorithm NEQ2, considered in [5].

5. Numerical Experiments

We consider two examples to which the algorithm NHI was applied with various values for p and various initial intervals for the solution of the equation $f(x) = 0$. A program was written in PASCAL-SC, where the extended (computer) interval-arithmetic operations $\langle -\bar{} \rangle$ and $(/\bar{})$ were simulated using the operator concept facilities of the program language.

Example 1. ([2], [3]) $f(x) = x^2(\tfrac{1}{3}x^2 + \sqrt{2}\sin x) - \dfrac{\sqrt{3}}{19}$.

The algorithm NHI was applied to the equation $f(x) = 0$ for $p = 1 \div 10$ for the initial interval $X^{(0)} = [0.1, 1]$. For all p the enclosing interval for the solution is

$$X^{(k,m)} = [0.392379507134, 0.392379507138].$$

The values of k and m $(1 \le m \le p)$ depend on the entered number p as follows:

$$p = 1 \quad \Rightarrow \quad m = 1, \quad k = 8; \qquad\qquad p = 6 \quad \Rightarrow \quad m = 4, \quad k = 2;$$

$$p = 2 \quad \Rightarrow \quad m = 2, \quad k = 4; \qquad\qquad p = 7 \quad \Rightarrow \quad m = 3, \quad k = 2;$$

$$p = 3 \quad \Rightarrow \quad m = 3, \quad k = 3; \qquad\qquad p = 8 \quad \Rightarrow \quad m = 3, \quad k = 2;$$

$$p = 4 \quad \Rightarrow \quad m = 1, \quad k = 3; \qquad\qquad p = 9 \quad \Rightarrow \quad m = 3, \quad k = 2;$$

$$p = 5 \quad \Rightarrow \quad m = 1, \quad k = 3; \qquad\qquad p = 10 \quad \Rightarrow \quad m = 1, \quad k = 2.$$

It should be noticed that for $p \geq 4$ the corresponding values of m do not achieve the upper bounds of p. We see from the above table that e.g. for $p = 4$ the resulting interval for the solution is $X^{(3,1)}$ instead of $X^{(3,4)} = X^{(4)}$. In this case the algorithm terminates according the criteria $X^{(k,m+1)} \not\subset X^{(k,m)}$.

Example 2. ([1]) $f(x) = \frac{1}{6}x^3 - \frac{1}{2}x^2 + \frac{3}{2}x - \frac{7}{6}$.

The equation $f(x) = 0$ has a root $x^* = 1$. The function f satisfies the hypotheses of Theorem 4, i.e. $f'(x^*) \neq 0$, $f''(x^*) = 0$. Algorithm NHI with $p = 1$ was applied to this example with various initial intervals and different computations of $\Diamond F'$. The following results were obtained.

(i) $X^{(0)} = [0.9, 1.4]$.
   The enclosing interval for the solution is

$$X^{(5)} = [0.999999999999, 1.00000000002].$$

(ii) $X^{(0)} = [0.2, 1.9]$.
   For this initial interval the program displays

$$\Diamond F'(X^{(0)}) = [-3.80000000000E - 01, 3.10500000000E + 00]$$

and the message:

> $0 \in \Diamond F'$ on the initial interval.
> Restart the algorithm with another initial interval or
> compute $\Diamond F'$ more accurately.

After a more accurate interval evaluation of the first derivative producing $\Diamond F'(X^{(0)}) = [1.0, 1.405]$, we obtain the following result:

$$X^{(4)} = [0.999999999997, 1.00000000002].$$

(iii) $X^{(0)} = [0.9, 0.99]$.
   We obtain the message:

> The equation has no solution in the initial interval.

This result is obtained without any iteration. It follows from the first part of the algorithm NHI, i.e. after computing $\bigcirc f$ and $\Diamond f$ on the given interval.

(iv) $X^{(0)} = [0.99, 1.0]$
   For this initial interval we obtain

$$\bigcirc f(X^{(0)}) = [-1.00001666000E - 02, -1.00000000000E - 11],$$

$$\Diamond f(X^{(0)}) = [-1.00166666570E - 01, 0.00000000000E + 00]$$

and the message

> The algorithm can not determine existence/nonexistence
> of a solution in the initial interval. Restart the
> algorithm with another initial interval.

(v) $X^{(0)} = [0.99, 1.00000000001]$.

After three iterations we obtain the resulting interval

$$X^{(3)} = [0.999999999997, 1.00000000001]$$

and the message:

> $\bigcirc f(X^{(3)}) = \emptyset$. The enclosing interval can not be made
> smaller in this precision.

References

[1] Alefeld, G.: Über die Konvergenzordnung des Intervall-Newton-Verfahrens. Computing *39*, 363–369 (1987).

[2] Alefeld, G., Herzberger, J.: Introduction to interval computations. New York: Academic 1983.

[3] Alefeld, G., Potra, F.: A new class of interval methods with higher order of convergence. Computing *42*, 69–80 (1989).

[4] Dimitrova, N., Markov, S. M.: Interval methods of Newton type for nonlinear equations. Pliska stud. math. bulg. *5*, 105–117 (1983).

[5] Dimitrova, N., Markov, S.: A validated Newton type method for nonlinear equations. Interval Computations (in press).

[6] Dimitrova, N. S., Markov, S. M., Popova, E. D.: Extended interval arithmetics: new results and applications. In: Atanassova, L., Herzberger, J., (eds.) Amsterdam: North-Holland, pp. 225–232, 1992.

[7] Kulisch, U., Miranker, W.: Computer arithmetic in theory and practice. New York: Academic Press 1980.

[8] Madan, V. P.: Interval contractions and rootfinding. Rivista di Mathematica ed Applicata *6*, 55–61 (1990).

[9] Markov, S.: Some applications of extended interval arithmetic to interval iterations. Computing [Suppl.] *2*, 69–84 (1980).

[10] Moore, R. E.: Interval analysis. Englewood Cliffs, N.J.: Prentice Hall 1966.

[11] Ortega, J. M., Rheinboldt, W. C.: Iterative solution of nonlinear equations in several variables. New York, London: Academic Press 1970.

[12] Ostrowski, A. M.: Solution of equations and systems of equations. New York: Academic Press 1960.

N. S. Dimitrova
Division for Mathematical Modelling
Institute of Biophysics
Bulgarian Academy of Sciences
Acad. G. Bonchev str., bldg. 21
BG-1113 Sofia, Bulgaria

Computing, Suppl. 9, 33–43 (1993)

© Springer-Verlag 1993
Printed in Austria

Verified Solution of the Integral Equations for the Two-Dimensional Dirichlet and Neumann Problem*

H.-J. Dobner, Karlsruhe

Dedicated to Professor U. Kulisch on the occasion of his 60th birthday

Abstract — Zusammenfassung

Verified Solution of the Integral Equations for the Two-Dimensional Dirichlet and Neumann Problem. In this article selfvalidating numerical methods for including the solution of the Dirichlet and Neumann problem in the plane are constructed. Here an additional error analysis to estimate roughly the quality of the computed solution is obsolete. The so-called verification or E-methods compute a mathematically guaranteed enclosure for the true solution of these problems.

AMS Subject Classification: 65D15, 65G10, 65R20, 45B05

Key words: Integral equations of potential theory, Dirichlet problem, Neumann problem, functoid, coefficient enclosure, function tube, mathematically guaranteed enclosure.

Die verifizierte Lösung der Integralgleichung für das zweidimensionale Dirichlet und Neumann Problem. In diesem Artikel werden selbstverifizierende numerische Verfahren hergeleitet, welche die Lösungen der ebenen Dirichlet- und Neumann Probleme einschließen, so daß eine zusätzliche Fehleranalyse zur Beurteilung der Güte der Näherungslösung überflüssig wird. Diese sogenannten Verifikations oder E-Verfahren berechnen jeweils mathematisch gesicherte Einschließungen für die exakte Lösung jener Probleme.

1. The Dirichlet and Neumann Problem in the Plane

In this section we consider the basic boundary value problems of potential theory. Let C be a continuously curved closed Jordancurve in the plane, which is parametrized by

$$C: x(s), y(s), \qquad \alpha \leq s \leq \beta,$$

in the counter clock sense. The interior of C is denoted by D_1, whereas the unbounded outer domain of C is denoted by D_2. Furthermore a real valued continuous function f is given on C. The following problems for the Laplace operator

$$\Delta u = u_{xx}(x, y) + u_{yy}(x, y) \tag{1}$$

are considered, where n denotes the outer normal on C:

---

* Received September 24, 1992; revised December 22, 1992.

- Interior Dirichlet problem

$$\Delta u = 0 \qquad \text{in } D_1,$$
$$u|_c = f.$$

- Outer Dirichlet problem

$$\Delta u = 0 \qquad \text{in } D_2,$$
$$u|_c = f.$$

- Interior Neumann Problem

$$\Delta u = 0 \qquad \text{in } D_1,$$
$$\left.\frac{\partial u}{\partial n}\right|_c = f.$$

- Outer Neumann Problem

$$\Delta u = 0 \qquad \text{in } D_2,$$
$$\left.\frac{\partial u}{\partial n}\right|_c = f.$$

The functions solving these boundary value problems are given as potential or logarithmic potential respectively, with density function μ_i (inner Dirichlet problem), ρ_o (outer Neumann problem), ρ_i (inner Neumann problem), μ_o (outer Dirichlet problem). From potential theory it is known that the unknown densities (cf. Martensen [16]) μ_i and ρ_o satisfy the Fredholm integral equation

$$v(s) + \frac{1}{2\pi} \int_\alpha^\beta k(s,t)v(t)\,dt = -\frac{1}{\pi}f(s), \qquad \alpha \le s \le \beta, \tag{2}$$

whereas the densities μ_0, ρ_i are solutions of

$$z(s) - \frac{1}{2\pi} \int_\alpha^\beta k(t,s)z(t)\,dt = \frac{1}{\pi}f(s), \qquad \alpha \le s \le \beta. \tag{3}$$

The kernel function $k(s,t)$ has the form

$$k(s,t) = -2\frac{(x(s) - x(t))y'(s) - (y(s) - y(t))x'(s)}{[x(s) - x(t)]^2 + [y(s) - y(t)]^2}, \qquad \alpha \le s, t \le \beta.$$

The first attempts of treating such boundary value problems have been made by Nyström [18], who used discretization methods for (2) together with quadrature formulae. Nyströms quadrature formulae method has been improved and refined by many authors (e.g. Fenyö/Stolle [8]), but all those methods hardly can give an effective, reliable and close error estimation for the computed solution. Therefore we develop for this kind of problems selfvalidating numerical methods, also called E-methods. In this case the numerically computed solution is guaranteed to exist within a function set of very small diameter. The foundations of such numerics are summarized subsequently.

2. On Selfvalidating Numerics

We give a brief summary of the foundations for inclusion theory. A more detailed description of inclusion theory is given in Dobner [7] and Kaucher/Miranker [10]. We list some important notations first. $I\mathbb{R}$ denotes the set of real intervals $[a] = [\underline{a}, \bar{a}]$, $\underline{a}$, $\bar{a} \in \mathbb{R}$, together with the usual arithmetic operations. $M = C[\alpha, \beta]$ equipped with the supremum norm and the order relation $\leq$ is a partially ordered Banach space; $\varphi_i \in M$, $i \in \mathbb{N}$, denotes a generating system of M. $I_n M = \{\sum_{i=1}^n [a_i]\varphi_i \,|\, [a_i] \in I\mathbb{R}\} = \{u \in M \,|\, \bigwedge_{s \in [\alpha, \beta]} u(s) \in \sum_{i=1}^n [a_i]\varphi_i(s)\} \subseteq M$ together with the pointwise explained standard operations will be called interval functoid or simply functoid.

Definition 1
The set valued function $X \in I_n M$ is an enclosure for the real valued function $x \in M$, if the relation $x \subseteq X$, that is

$$\bigwedge_{s \in [\alpha, \beta]} \{x(s)\} \subseteq X(s) \tag{4}$$

holds. Enclosures for operators are defined in the same way.

Convention: Throughout this paper real valued quantities are indicated with small letters: x, f, v, w, …, whereas the corresponding enclosures are indicated with capital letters: X, F, V, W, … .

Definition 2
For $X = \sum_{i=1}^n [x_i]\varphi_i$, $Y = \sum_{i=1}^n [y_i]\varphi_i \in I_n M$ the coefficient enclosure $\subseteq_n$ is defined as

$$X \subseteq_n Y: \quad \Leftrightarrow \quad \bigwedge_{i=1}^n [x_i] \subseteq [y_i].$$

Lemma 3
For X, $Y \in I_n M$ the implication

$$X \subseteq_n Y \quad \Rightarrow \quad \bigwedge_{s \in [\alpha, \beta]} X(s) \subseteq Y(s)$$

is true and moreover $(IM_n, \subseteq_n)$ is partially ordered.

Proof:
Follows immediately from definition 2. $\qquad\qquad\square$

Lemma 4
Each element X of $I_n M$ can be written as interval in the form

$$X = [\underline{x}, \bar{x}] \tag{5}$$

with continuous functions $\underline{x}$, $\bar{x} \in M$.

Proof:
X has the representation

$$X = \sum_{i=1}^n [x_i]\varphi_i, \qquad [x_i] = [\underline{x}_i, \bar{x}_i] \in I\mathbb{R}.$$

According to

$$\bigwedge_{i=1}^{n} \bigwedge_{s \in [\alpha, \beta]} \underline{x}_i(s) := \min\{\underline{x}_i \varphi_i(s), \overline{x}_i \varphi_i(s)\},$$

$$\bigwedge_{i=1}^{n} \bigwedge_{s \in [\alpha, \beta]} \overline{x}_i(s) := \max\{\underline{x}_i \varphi_i(s), \overline{x}_i \varphi_i(s)\},$$

there are defined continuous functions $\underline{x}_i(s)$, $\overline{x}_i(s)$, with

$$\bigwedge_{s \in [\alpha, \beta]} [\underline{x}_i(s), \overline{x}_i(s)] = [x_i] \varphi_i(s).$$

Hence $\underline{x}(s) := \sum_{i=1}^{n} \underline{x}_i(s)$, $\overline{x}(s) := \sum_{i=1}^{n} \overline{x}_i(s)$ are continuous functions such that (5) holds, this completing the proof. $\qquad \square$

Definition 5
In $I_n M$ we define $\|\cdot\|$ as follows

$$\bigwedge_{X \in I_n M} \|X\| := \sup_{x \in X} \|x\|.$$

Furthermore we have

Lemma 6
The elements of $I_n M$ are closed, bounded and convex sets.

Proof:
Let be $\varnothing \neq X \in I_n M$ then, following Lemma 4, X can be represented as interval in the partially ordered Banach space $M = C[\alpha, \beta]$, where each interval is closed, bounded and convex. $\qquad \square$

3. An Enclosure Method for the Integral Equation of the Inner Dirichlet and Outer Neumann Problem

Although the difference between the integral Eq. (2) arising from the inner Dirichlet (outer Neumann) problem and the integral Eq. (3) coming from the outer Dirichlet (inner Neumann) problem, is only a sign, their behaviour is quite different.

Equation (2) is a nonsingular one, whereas (3) is an algebraic singular equation, so that two different enclosure strategies have to be developed.

We consider Eq. (2) using the notation

$$v = f + k(v),$$

which can be decomposed in the form

$$v = f + k_1(v) + k_2(v),$$

here k_2 is an integral operator with a kernel of the form

$$k_2(s, t) = \sum_{i=1}^{m} f_i(s) g_i(t), \qquad f_i, g_i \in M, m \in \mathbb{N},$$

which can by assumption always be chosen so that $k_1 := k - k_2$ is a contractive operator. For the sake of convenience we define $f_0 := f$; we remind of the notation convention above and have

Theorem 7
Let $W_0, W_1, \ldots, W_m \in I_n M$ be function tubes containing the solutions of

$$\bigwedge_{i=0}^{m} w_i = f_i + k_1(w_i),$$

that is (cf. definition 1)

$$\bigwedge_{i=0}^{m} w_i \subseteq W_i. \tag{6}$$

If $U = (U_1, \ldots, U_m) \in I\mathbb{R}^m$ satisfies

$$AU \supseteq R \tag{7}$$

with the $m \times m$ matrix $A = (\delta_{ij} - \int_\alpha^\beta G_i(t) W_j(t)\, dt)_{i,j=1,\ldots,m}$, and the inhomogeneity $R = (\int_\alpha^\beta G_i(t) W_0(t)\, dt)_{i=1,\ldots,m}$, then the solution v of (2) is guaranteed to exist in the function tube

$$\bigwedge_{s \in [\alpha, \beta]} v(s) \in V_1(s) := W_0(s) + \sum_{i=1}^{m} U_i W_i(s). \tag{8}$$

Proof:
As usual in the theory of integral equations (see e.g. Delves/Mohamed [3], Tricomi [20]) let γ_1 denote the resolvent operator resp. the resolvent kernel of k_1, then (2) is equivalent to

$$\bigwedge_{s \in [\alpha, \beta]} v(s) = f_0(s) + \gamma_1(f_0(s)) + \int_\alpha^\beta \sum_{i=1}^{m} (f_i(s) + \gamma_1(f_i(s))) g_i(t) v(t)\, dt.$$

The quantities $f_i + \gamma_1(f_i)$, $i = 0, 1, \ldots, m$, solve the integral equations

$$w_i(s) = f_i(s) + \int_\alpha^\beta k_1(s, t) w_i(t)\, dt,$$

so that the solution v of (2) is given by

$$v(s) = w_0(s) + \sum_{i=1}^{m} u_i w_i(s),$$

where $u = (u_1, \ldots, u_m) \in \mathbb{R}^m$ solves the real linear system

$$au = r,$$

with $a = (\delta_{ij} - \int_\alpha^\beta g_i(t) w_j(t)\, dt)_{i,j=1,\ldots,m}$ and $r = (\int_\alpha^\beta g_i(t) w_0(t)\, dt)_{i=1,\ldots,m}$. If considering now the solution set

$$\tilde{U} := \{u \,|\, au = r, \mid a \in A, r \in R\}$$

we see that for the enclosure set U of (7) there holds $u \in \tilde{U} \subseteq U$, so that (8) follows.
$\qquad\square$

Note that the existence of a solution of (2) is proved by the algorithm in theorem 7 itself.

Remark 8

Function tubes $W_0, W_1, \ldots, W_m \in I_n M$ with (6) may be computed by the following scheme:

$$\text{Iterate } \bigwedge_{i=0}^{m} W_i^{(l+1)} := F_i + K_1(W_i^{(l)}), \qquad l = 0, 1, 2, \ldots$$

$$\text{until } \bigwedge_{i=0}^{m} W_i^{(l+1)} \subseteq W_i^{(l)},$$

with Schauder's fixed point theorem, we can conclude

$$\bigwedge_{i=0}^{m} \bigwedge_{s \in [\alpha, \beta]} w_i(s) \in W_i(s).$$

Remark 9

Instead of the E-method proposed in the foregoing theorem, in practice a modified version is often used: If a real constant $q \in [0, 1)$ is known such that

$$\bigwedge_{s,t \in [\alpha,\beta]} |k_1(s, t)| \leq q,$$

then the solution v of (2) is guaranteed to exist in the functoid element V_2:

$$\bigwedge_{s \in [\alpha,\beta]} v(s) \in V_2(s) := F_0(s) + [F_0][Q] + \sum_{i=1}^{m} U_i(F_i(s) + [F_i][Q]), \qquad (9)$$

where

$$\bigwedge_{i=0}^{m} \int_{\alpha}^{\beta} |f_i(t)| \, dt \in [F_i] \in I\mathbb{R},$$

$$[Q] := \left[\frac{-q}{1-q}, \frac{q}{1-q} \right] \in I\mathbb{R},$$

and $U = (U_1, \ldots, U_m) \in I\mathbb{R}^m$ satisfies

$$\sum_{j=1}^{m} \left(\delta_{ij} - \int_{\alpha}^{\beta} G_i(t)(F_j(t) + [F_j][Q]) \, dt \right) U_j$$

$$\supseteq \int_{\alpha}^{\beta} G_i(t)(F_0(t) + [F_0][Q]) \, dt, \qquad i = 1, \ldots, m.$$

In the next section the enclosure methods described above, are modified so that they are applicable to the singular problem (3).

4. An Enclosure Method for the Integral Equation of the Inner Neumann and Outer Dirichlet Problem

By a simple calculation we have

$$\bigwedge_{s \in [\alpha,\beta]} \frac{1}{2\pi} \int_{\alpha}^{\beta} k(t, s) \, dt = 1,$$

so that the homogeneous equation corresponding to (3) has nontrivial solutions. From potential theory (cf. Martensen [15]) it is well known that (3) has a unique solution z with prescribed total distribution

$$\int_\alpha^\beta z(t)\,dt = e, \qquad e \in \mathbb{R}. \tag{10}$$

Furthermore we know

Theorem 10
If and only if f is periodic with period length $\beta - \alpha$ satisfying additionally the condition

$$\int_\alpha^\beta f(t)\,dt = 0, \tag{11}$$

then the unique solution z_0 of

$$\bigwedge_{s \in [\alpha, \beta]} z_0(s) - \frac{1}{2\pi} \int_\alpha^\beta (k(t,s) - 1)z_0(t)\,dt = \frac{1}{\pi}\left(f(s) + \frac{e}{2}\right) \tag{12}$$

is a solution of (3) satisfying (10) and the converse is also true.

Proof:
By use of Wielandt's removal of an eigenvalue from the spectrum (cf. Wielandt [21]), we know that 1 is not any longer eigenvalue of (12); so the other statements of this theorem follow easily. $\qquad\square$

Remark 11
The function f prescribed on C has to fulfill the two conditions formulated in theorem 10. The second constraint (11) is a consequence of Gauß' theorem. If one of these conditions is not satisfied, then Eq. (12) is still nonsingular, but z_0 does not lead to a solution of the considered boundary value problem.

For the enclosure of (3) resp. (12) the selfvalidating algorithms of the previous chapter can be employed now, so that the following theorem (validation with constraints) can be formulated without proof.

Theorem 12
Let be $f \in F \in I_n M$ with

$$0 \in \int_\alpha^\beta F(t)\,dt \tag{13}$$

and

$$0 \in F(\beta) - F(\alpha). \tag{14}$$

If $Z_0 \in I_n M$ is an enclosure of

$$\bigwedge_{s \in [\alpha, \beta]} Z_0(s) - \frac{1}{2\pi} \int_\alpha^\beta (K(t,s) - 1)Z_0(t)\,dt \supseteq \frac{1}{\pi}\left(F(s) + \frac{e}{2}\right),$$

then there exists a pair

$$(\tilde{z}, \tilde{f}) \in (Z_0, F) \tag{15}$$

such that

$$\bigwedge_{s \in [\alpha, \beta]} \tilde{z}(s) - \frac{1}{2\pi} \int_\alpha^\beta k(s,t)\tilde{z}(t)\,dt = \frac{1}{\pi}\tilde{f}(s)$$

and at the same time

$$\bigwedge_{\omega \in \mathbb{Z}} \bigwedge_{s \in [\alpha, \beta]} \tilde{f}(s + \omega(\beta - \alpha)) = \tilde{f}(s),$$

$$\int_\alpha^\beta \tilde{f}(t)\,dt = 0.$$

Theorem 12 is an example for an enclosure with constraints. This new field is treated in a more general framework in Dobner [7], also methods for constructing set functions F fulfilling the constraints (13), (14) are described in detail there.

5. Numerical Experiments

First we describe some methods how for computational purposes a given kernel can be splitted into a sum of a degenerate and a contractive kernel.

- **Automatic differentiation**
  If the kernel k is sufficiently smooth, a truncated Taylor expansion of order m about (s_0, t_0) yields

$$k(s,t) = k_2(s,t) + k_1(s,t) = \sum_{i=1}^m f_i(s)g_i(t) + k_1(s,t)$$

$$:= \sum_{i=0}^m (s - s_0)^i \left(\sum_{j=0}^{m-i} \frac{1}{i!j!} \frac{\partial^{i+j}k(s_0,t_0)}{\partial s^i \partial t^j}(t - t_0)^j \right)$$

$$+ \sum_{i=0}^{m+1} (s - s_0)^i \frac{1}{(m+1-i)!i!} \frac{\partial^{m+1}f(\sigma,\tau)}{\partial s^i \partial t^{m+1-i}}(t - t_0)^{m+1-i},$$

(σ, τ) between (s,t) and (s_0, t_0).

a desired decomposition. The derivatives as well as enclosures for the remainder are computed with the technique of automatic differentiation (cf. Rall [19]).

Another possibility used for the examples of this paper is the method of

- **Fourier expansion**
  If k is continuous and $w_i(s)$, $i = 1, 2, \ldots$ a complete orthonormalsystem the kernel can be written as

$$k(s,t) = \sum_{i=1}^\infty \sum_{j=1}^\infty a_{ik} w_i(s) w_j(t),$$

with the Fourier coefficients

$$a_{ij} = \int_\alpha^\beta \int_\alpha^\beta k(s,t) w_i(s) w_j(t)\,ds\,dt, \qquad i, j = 1, 2, \ldots .$$

As degenerate part we take

$$k_2(s,t) = \sum_{i=1}^{m} \sum_{j=1}^{m} a_{ij} w_i(s) w_j(t).$$

By estimating the remainder $k_1 = k - k_2$ using Parsevals identity and interval computation (see Alefeld/Herzberger [1]) it is possible to choose the parameter m such that k_1 is contractive. The Fourier coefficients a_{ij} are enclosed with validation methods (see e.g. Corliss [2]).

We restrict ourselves to the integral Eqs. (2) (3); the solutions of the corresponding boundary value problems are given as potentials and can therefore be enclosed by evaluating the resulting definite integrals within small guaranteed bounds (cf. the integration schemes of Corliss [2]).

We report about some examples, having been treated with verification methods. The functoid $I_n M$ has always been chosen as

$$I_n M = \{[a_0] + [a_1]\cos s + \cdots + [a_n]\cos ns + [a_{n+1}]\sin s$$
$$+ \cdots [a_{2n}]\sin ns; \quad [a_0], \ldots, [a_{2n}] \in I\mathbb{R}\}. \tag{16}$$

Example 13

If the boundary curve C is a circle, the kernel reduces simply to

$$k(s,t) = 1$$

and last bit accuracy can be achieved in all cases.

Example 14

We consider the inner Dirichlet problem, where C is an ellipse with the semiaxis $a > b > 0$, and f a right hand side of the form

$$f(s) = \frac{1}{5 + 4\cos s}, \quad s \in [0, 2\pi].$$

In the table below we display the values of the maximal diameter of the coefficients $[a_i]$, $i = 0, 1, \ldots, 2n$, in (16) as well as the value of the maximal diameter of $V_2(s)$, $s \in [0, 2\pi]$. The computations have been performed for different values of a, b and n. Here the algorithm of remark 9 has been employed.

		$n = 8$		$n = 20$	
a	b	$\max\limits_{i=0,\ldots,2n} \mathrm{diam}[a_i]$	$\max\limits_{s\in[0,2\pi]} \mathrm{diam}\,V_2(s)$	$\max\limits_{i=0,\ldots,2n} \mathrm{diam}[a_i]$	$\max\limits_{s\in[0,2\pi]} \mathrm{diam}\,V_2(s)$
3	1	$4.1\cdot 10^{-2}$	$6.5\cdot 10^{-2}$	$9.0\cdot 10^{-6}$	$1.5\cdot 10^{-5}$
5	1	1.1	1.8	$6.0\cdot 10^{-3}$	$1.0\cdot 10^{-2}$
3	2	$6.7\cdot 10^{-6}$	$9.8\cdot 10^{-6}$	$8.0\cdot 10^{-12}$	$1.4\cdot 10^{-11}$
4	2	$7.9\cdot 10^{-4}$	$1.2\cdot 10^{-3}$	$1.5\cdot 10^{-9}$	$2.3\cdot 10^{-9}$
8	7	$2.9\cdot 10^{-10}$	$4.2\cdot 10^{-10}$	$2.9\cdot 10^{-10}$	$4.2\cdot 10^{-10}$

Between the average diameter and the maximal diameter of $V_2(s)$ the relation

$$\frac{\text{average diam } V_2(s)}{\text{max diam } V_2(s)} \approx 0.1$$

holds.

Example 15

Here the inner Neumann problem was solved, with the inhomogeneity

$$f(s) = \pi \frac{2\dfrac{b}{a}}{1 + \dfrac{b}{a}} \cos s, \qquad s \in [0, 2\pi],$$

where $a > b > 0$ are the semiaxis of the elliptic curve C. We used the E-method of theorem 12.

		$n = 20$		$n = 35$	
a	b	$\max\limits_{i=0,\ldots,2n} \text{diam}[a_i]$	$\max\limits_{s\in[0,2\pi]} \text{diam } Z_0(s)$	$\max\limits_{i=0,\ldots,2n} \text{diam}[a_i]$	$\max\limits_{s\in[0,2\pi]} \text{diam } Z_0(s)$
7	1	$3.7 \cdot 10^{-2}$	$6.8 \cdot 10^{-2}$	$4.8 \cdot 10^{-4}$	$8.8 \cdot 10^{-4}$
7	2	$5.8 \cdot 10^{-5}$	$1 \cdot 10^{-4}$	$8.6 \cdot 10^{-9}$	$1.6 \cdot 10^{-8}$
7	4	$1.6 \cdot 10^{-11}$	$4.2 \cdot 10^{-11}$	$4.0 \cdot 10^{-12}$	$1.7 \cdot 10^{-11}$
7	6	$2.9 \cdot 10^{-12}$	$1.7 \cdot 10^{-11}$	$2.9 \cdot 10^{-12}$	$1.7 \cdot 10^{-11}$
9	1	$1.7 \cdot 10^{-1}$	$3.1 \cdot 10^{-1}$	$5.0 \cdot 10^{-3}$	$9.2 \cdot 10^{-3}$

6. Concluding Remarks

We see that by increasing the functoid dimension, each desired accuracy can be achieved in principle. The accuracy is essentially governed by the ratio a/b of the semiaxis a, b of the elliptic curve. These E-computations needed about 3–4 times more than conventional schemes.

Acknowledgement

I wish to thank the referees for their constructive and helpful remarks.

References

[1] Alefeld, G., Herzberger, J.: Einführung in die Intervallrechnung. Bibliographisches Institut, Mannheim, 1974.

[2] Corliss, G. F.: Computing narrow inclusions for definite integrals. In: Kaucher, E., Kulisch, U., Ullrich, C. (eds.) Computer arithmetic-scientific computation and programming languages, pp. 150–169. Stuttgart: Teubner 1987.

[3] Delves, L. M., Mohamed, J. L.: Computational methods for integral equations. Cambridge: Cambridge University Press 1985.

[4] Dobner, H.-J.: Computing narrow inclusions for the solution of integral equations. Num. Funct. Anal. and Optimiz. *10* (9/10), 923–936 (1989).

[5] Dobner, H.-J., Kaucher, E.: Self-validating computations of linear and nonlinear integral equations of the second kind, In: Ullrich, C. (ed.) Contributions to computer arithmetic and self-validating numerical methods, J. C. Baltzer AG, Scientific Publishing Co, 273–290, 1990.

[6] Dobner, H.-J.: Contributions to computational analysis. Bull. Austral. Math. Soc. *41*, 231–235 (1990).

[7] Dobner, H.-J.: Numerische Methoden zur verifizierten Lösung von Integralgleichungen. Habilitationsschrift, Universität Karlsruhe 1992.

[8] Fenyö, S., Stolle, H. W.: Theorie und Praxis der linearen Integralgleichungen 4. Basel, Boston, Stuttgart: Birkhäuser 1984.

[9] Kaucher, E., Kulisch, U., Ullrich C. (eds.): Computer arithmetic-scientific computation and programming languages. Stuttgart: Teubner 1987.

[10] Kaucher, E., Miranker, W. L.: Self validating numerics for function space problems. New York: Academic Press 1984.

[11] Kulisch, U.: Grundlagen des numerischen Rechnens. Bibliographisches Institut, Mannheim 1976.

[12] Kulisch, U., Miranker, W. L.: Computer arithmetic in theory and practice. New York: Academic Press 1981.

[13] Kulisch U. (ed.): Wissenschaftliches Rechnen mit Ergebnisverifikation. Braunschweig, Wiesbaden: Vieweg 1989.

[14] Kulisch, U., Miranker W. L. (eds.): A new approach to scientific computation. New York: Academic Press 1983.

[15] Martensen, E.: Zur numerischen Behandlung des inneren Neumannschen und Robinschen Problems. ZAMM *39*, 377–380 (1959).

[16] Martensen, E.: Potentialtheorie. Stuttgart: Teubner 1968.

[17] Neumaier, A.: Interval methods for systems of equations. Cambridge: Cambridge University Press 1990.

[18] Nyström, E. J.: Über die praktische Auflösung von linearen Integralgleichungen mit Anwendungen auf Randwertaufgaben der Potentialtheorie. Soc. Scient. Fenn. Comm. Phys.-Math. IV. *15*, 1–52 (1928).

[19] Rall, L. B.: Automatic differentiation. Berlin, Heidelberg, New York: Springer 1981 (Lecture Notes in Computer Science 120).

[20] Tricomi, F. G.: Integral equations. New York: Dover Publ. 1985.

[21] Wielandt, H.: Das Iterationverfahren bei nicht selbstadjungierten linearen Eigenwertaufgaben. Math. Z. *50*, 93–143 (1943).

Hans-Jürgen Dobner
University of Karlsruhe
Kaiserstrasse 12
D-W-7500 Karlsruhe
Federal Republic of Germany

Computing, Suppl. 9, 45–65 (1993)

Two-Stage Interval Iterative Methods*

A. Frommer, Wuppertal and **G. Mayer,** Karlsruhe

Dedicated to Professor U. Kulisch on the occasion of his 60th birthday

Abstract — Zusammenfassung

Two-Stage Interval Iterative Methods. We present an interval version of the well-known two-stage iterative methods to approximate solutions of linear systems of equations. By using interval arithmetical tools we are able to verify such solutions within interval bounds. The method can also guarantee the non-singularity of the underlying coefficient matrices of the systems. We prove criteria of convergence for the method and we report on an optimality result for the enclosure.

AMS Subject Classification: 65F10, 65G10

Key words: Two-stage interval iterative methods, linear interval systems, H-splittings, H-compatible splittings, M-splittings.

Zweistufige Intervall-Iterationsverfahren. Wir stellen eine Intervall-Version von bekannten zweistufigen Iterationsverfahren zur Approximation linearer Gleichungssysteme vor. Mit intervallarithmetischen Hilfsmitteln werden Lösungen solcher Systeme innerhalb gewisser Intervallschranken nachgewiesen. Dabei kann mit dem Verfahren gleichzeitig die Regularität der zugrundeliegenden Koeffizientenmatrizen gezeigt werden. Wir beweisen Konvergenzkriterien für das Verfahren und machen eine Optimalitäts-aussage über die Einschließung.

1. Introduction

In this paper we generalize the classical two stage method to obtain lower and upper bounds for all solutions $\tilde{x}$ of linear systems

$$\tilde{A}\tilde{x} = \tilde{b} \tag{1}$$

with $\tilde{A} \in [A]$ and $\tilde{b} \in [b]$. Here, $[A]$ denotes an $n \times n$ interval matrix, and $[b]$ is an interval vector with n components. (For interval computations see [2], e.g.) We collect all these solutions into a set

$$L := \{\tilde{x} \,|\, \exists \tilde{A} \in [A], \tilde{b} \in [b]: \tilde{A}\tilde{x} = \tilde{b}\}. \tag{2}$$

Normally, L is not easy to describe, cf. [4], [5], [18] or [22]. The problem to enclose L occurs, e.g., when only one system (1) is to be solved with $\tilde{A}$ being regular but not representable on a computer or with $\tilde{x}$ consisting of non-machine numbers. In the first case one encloses $\tilde{A}$ by the nearest machine matrices which are not less and which are not greater, respectively, than $\tilde{A}$ to end up with an interval matrix $[A]$

---

* Received September 21, 1992; revised December 20, 1992.

of small width. This matrix forms the outset of methods to bound L. In the second case one uses interval arithmetic to get (hopefully tight) bounds which are representable on a computer. The necessity to enclose L also occurs when bounding solutions of nonlinear systems of equations by using, e.g., the interval Newton method (cf. [2]), or when dealing with practical problems which lead to linear systems with varying coefficients.

Interval methods to enclose L can already be found in books such as [2], [12], [22]. We add a new method based on two splittings

$$[A] = [M] - [N], \qquad [M] = [F] - [G]. \tag{3}$$

For $[F]$ we assume throughout the paper that the interval Gaussian algorithm (cf. [2]) is feasible without pivoting, producing for any righthand side $[c]$ an interval vector which we will denote by $IGA\,([F],[c])$. With this notation and with the splittings in (3) we define the interval function $f\colon \mathbf{IR}^n \times \mathbf{N} \to \mathbf{IR}^n$ ($\mathbf{IR}^n$: set of interval vectors with n components; $\mathbf{N}$: set of positive integers) in the following algorithmic way:

$$\left.\begin{aligned}
&[\hat{x}]^0 := [x]; \\
&\textbf{for } i := 1 \textbf{ to } s \textbf{ do} \\
&\qquad [\hat{x}]^i := IGA\,([F],[G][\hat{x}]^{i-1} + [N][x] + [b]); \\
&f([x],s) := [\hat{x}]^s.
\end{aligned}\right\} \tag{4}$$

Keeping s fixed yields the *stationary two-stage method*

$$[x]^{k+1} = f([x]^k, s), \quad k = 0, 1, \dots . \tag{5}$$

Allowing $s = s(k)$ to vary with k defines the *non-stationary two-stage method*

$$[x]^{k+1} = f([x]^k, s(k)), \quad k = 0, 1, \dots . \tag{6}$$

We point out that for $s = 1$ and $[F] = [M]$ the method (5) reduces to the iterative method $[x]^{k+1} = IGA\,([M], [N][x]^k + [b])$ which is based on the splitting $[A] = [M] - [N]$ alone, and which is considered in greater detail in [15], e.g.

Since $IGA\,([F],[c])$ can be identified with $\tilde{F}^{-1}\tilde{c}$ when $[F]$ and $[c]$ contain $\tilde{F}$ and $\tilde{c}$ as the only element, respectively, the methods (5) and (6) reduce to the classical two-stage methods mentioned at the beginning. For later use, we reformulate (6) for this non-interval case with

$$\left.\begin{aligned}
&\tilde{A} \in [A], \tilde{M} \in [M], \tilde{N} \in [N], \tilde{F} \in [F], \tilde{G} \in [G], \tilde{b} \in [b], \\
&\tilde{A} = \tilde{M} - \tilde{N}, \tilde{M} = \tilde{F} - \tilde{G}.
\end{aligned}\right\} \tag{7}$$

It reads

$$\tilde{x}^{k+1} = \tilde{f}(\tilde{x}^k, s(k)) \tag{8}$$

where $\tilde{f}\colon \mathbf{R}^n \times \mathbf{N} \to \mathbf{R}^n$ is defined by

$$\left.\begin{aligned}
&\hat{x}^0 := \tilde{x}; \\
&\text{for } i := 1 \text{ to } s \text{ do} \\
&\qquad \hat{x}^i := \tilde{F}^{-1}(\tilde{G}\hat{x}^{i-1} + \tilde{N}\tilde{x} + \tilde{b}); \\
&\tilde{f}(\tilde{x},s) := \hat{x}^s.
\end{aligned}\right\} \tag{9}$$

As can easily be seen, method (8) may be written as

$$\tilde{x}^{k+1} = \tilde{T}_{s(k)} \tilde{x}^k + \sum_{j=0}^{s(k)-1} \tilde{P}^j \tilde{F}^{-1} \tilde{b} \tag{10}$$

with

$$\tilde{P} := \tilde{F}^{-1} \tilde{G} \tag{11}$$

and

$$\tilde{T}_s := \tilde{P}^s + \sum_{j=0}^{s-1} \tilde{P}^j \tilde{F}^{-1} \tilde{N}. \tag{12}$$

Convergence results for (8) are given in [8], [13], [23], numerical results can be found in [10]. We will add such results for the interval method (5). We also include criteria which guarantee the non-singularity of each matrix $\tilde{A} \in [A]$ and which ensure the inclusion $L \subseteq [x]^0$.

We have arranged our paper as follows. In Section 2 we introduce some notation, in Section 3 we list several known lemmas needed later on. In Section 4 we consider the stationary two-stage method and in Section 5 we present the non-stationary case. In Section 6 we illustrate our theory by a numerical example.

2. Notation

By $\mathbf{R}^n$, $\mathbf{R}^{m \times n}$, $\mathbf{IR}$, $\mathbf{IR}^n$, $\mathbf{IR}^{m \times n}$ we denote the set of real vectors with n components, the set of real $m \times n$ matrices, the set of intervals, the set of interval vectors with n components and the set of $m \times n$ interval matrices, respectively. By 'interval' we always mean a real compact interval. We write interval quantities in brackets with the exception of point quantities (i.e., degenerate interval quantities) which we identify with the element which they contain. Examples are the null matrix 0 and the identity matrix I. We use the notation $[A] = [\underline{A}, \overline{A}] = ([a_{ij}]) = ([\underline{a}_{ij}, \overline{a}_{ij}]) \in \mathbf{IR}^{m \times n}$ simultaneously without further reference, and we proceed similarly for the elements of $\mathbf{R}^n$, $\mathbf{R}^{m \times n}$, $\mathbf{IR}$, $\mathbf{IR}^n$. If necessary, we identify the elements of $\mathbf{R}^{n \times 1}$ and $\mathbf{IR}^{n \times 1}$ in the usual way with those of $\mathbf{R}^n$ and $\mathbf{IR}^n$, respectively. We denote the topological interior of $[x] \in \mathbf{IR}^n$ by $int([x])$, and we write $\square S$ for the tightest interval enclosure of a given bounded subset $S \subseteq \mathbf{R}^n$ calling it the 'interval hull' of S.

By $A \geq 0$ we denote a non-negative $m \times n$ matrix, i.e., $a_{ij} \geq 0$, for $i = 1, \ldots, m$ and $j = 1, \ldots, n$. With $B \in \mathbf{R}^{m \times n}$ we define $A \leq B$ by $B - A \geq 0$. We call $x \in \mathbf{R}^n$ positive writing $x > 0$ if $x_i > 0$, $i = 1, \ldots, n$.

We also mention the standard notation from interval analysis ([2], [22])

$$|[a]| := \max\{|\tilde{a}| \,|\, \tilde{a} \in [a]\} = \max\{|\underline{a}|, |\overline{a}|\} \quad (absolute\ value)$$

$$\langle [a] \rangle := \min\{|\tilde{a}| \,|\, \tilde{a} \in [a]\} = \begin{cases} \min\{|\underline{a}|, |\overline{a}|\} & \text{if } 0 \notin [a] \\ 0 & \text{otherwise} \end{cases}$$
$$(minimal\ absolute\ value)$$

$$d([a]) := \bar{a} - \underline{a} \quad (width)$$

$$q([a], [b]) := \max\{|\underline{a} - \underline{b}|, |\bar{a} - \bar{b}|\} \quad (Hausdorff\ distance)$$

for intervals $[a]$, $[b]$. For $[A]$, $[B] \in \mathbf{IR}^{m \times n}$ we obtain $|[A]|$, $d([A])$, $q([A], [B]) \in \mathbf{R}^{m \times n}$ by applying $|\cdot|$, $d(\cdot)$, $q(\cdot, \cdot)$ entrywise, and for $m = n$ we define the comparison matrix $\langle [A] \rangle = (c_{ij}) \in \mathbf{R}^{n \times n}$ by setting

$$c_{ij} := \begin{cases} -|[a_{ij}]| & \text{if } i \neq j \\ \langle [a_{ij}] \rangle & \text{if } i = j \end{cases}.$$

Since real matrices can be viewed as degenerate interval matrices, $|\cdot|$ and $\langle \cdot \rangle$ can also be used for them. A collection of rules for $|\cdot|$, $\langle \cdot \rangle$, $d(\cdot)$, $q(\cdot, \cdot)$ can be found in [2], [22].

By $\mathbf{Z}^{n \times n}$ we denote the set of real $n \times n$ matrices with non-positive off-diagonal entries, by $\rho(A)$ we denote the spectral radius of a matrix $A \in \mathbf{R}^{n \times n}$.

We call $A \in \mathbf{R}^{n \times n}$ an *M-matrix* if A is non-singular, $A^{-1} \geq 0$ and $A \in \mathbf{Z}^{n \times n}$, and we define A to be an *H-matrix* if $\langle A \rangle$ is an M-matrix.

We call an interval function $f: \mathbf{IR}^n \to \mathbf{IR}^n$ a *P-contraction* if there is a non-negative matrix $P \in \mathbf{R}^{n \times n}$ such that

$$\rho(P) < 1 \quad \text{and} \quad q(f([x]), f([y])) \leq P q([x], [y])$$

for all vectors $[x]$, $[y] \in \mathbf{IR}^n$.

An interval matrix $[A] \in \mathbf{IR}^{n \times n}$ is termed an M-*matrix* if each element $\tilde{A} \in [A]$ is an M-matrix. In the same way the term 'H-*matrix*' can be extended to $\mathbf{IR}^{n \times n}$. It is easy to verify (use Lemma 4c) below) that

$[A]$ is an M-matrix if and only if $\underline{A}$ is an M-matrix and $\bar{a}_{ij} \leq 0$ for $i \neq j$, and that

$[A]$ is an H-matrix if and only if $\langle [A] \rangle$ is an M-matrix.

We call the representation

$$[A] = [M] - [N] \tag{13}$$

a *splitting* of $[A]$, specifying (13) as an

M-splitting, if $[M]$ is an M-matrix and $\underline{N} \geq 0$,
H-splitting, if $\langle [M] \rangle - |[N]|$ is an M-matrix,
H-compatible splitting, if $\langle [A] \rangle = \langle [M] \rangle - |[N]|$.

For real matrices these definitions were introduced in [8] and [27]. Note that the second splitting is called a *strong splitting* in [16] and [20].

For real matrices $A \in \mathbf{R}^{n \times n}$ we also recall *regular splittings* which are defined by

$$A = M - N, \quad M^{-1} \geq 0, \quad N \geq 0,$$

and *weak regular splittings* which are given by

$$A = M - N, \quad M^{-1} \geq 0, \quad M^{-1}N \geq 0$$

(cf. 2.4.15 in [24]). It is clear that each regular splitting is a weak regular one.

The interval vector $IGA([A],[b])$ (cf. Section 1) can be represented as

$$IGA([A],[b])$$
$$= [D]^1([U]^1([D]^2([U]^2(\ldots\cdot([U]^{n-1}([D]^n([L]^{n-1}\cdot\ldots\cdot([L]^2([L]^1[b]))\ldots)))\ldots)))) \tag{14}$$

with the matrices $[D]^i$, $[L]^i$, $[U]^i$ given explicitly in [1], [7] or [28]; we set

$$|[A]^G| := |[D]^1||[U]^1||[D]^2||[U]^2|\cdot\ldots\cdot|[U]^{n-1}||[D]^n||[L]^{n-1}|\cdot\ldots\cdot|[L]^2||[L]^1| \tag{15}$$

(see [7] or [22], p. 154). The significance of $|[A]^G|$ is best seen in Lemma 1 below.

We end this section by defining the R_1-factor $\sigma(\{[x]^k\})$ of a sequence $\{[x]^k\}$ in $\mathbf{IR}^n$ which converges towards some interval vector $[x]^*$. Choose any vector norm $\|\cdot\|$. Then

$$\sigma(\{[x]^k\}) := \limsup_{k\to\infty} \|q([x]^k,[x]^*)\|^{1/k}.$$

This factor is independent of the choice of $\|\cdot\|$ (cf. [2], p. 153). It can be considered as a measure for the speed of convergence of $\{[x]^k\}$.

3. Auxiliary Results

In this section we recall some known results which will be used in the Sections 4 and 5.

Lemma 1. *Let $[A] \in \mathbf{IR}^{n\times n}$, $[x]$, $[y] \in \mathbf{IR}^n$. Let the interval Gaussian algorithm be feasible (without pivoting) for $[A]$ and for any righthand side. Then*

a) $d(IGA([A],[x])) \ge |[A]^G|\cdot d([x])$.
b) $q(IGA([A],[x]), IGA([A],[y])) \le |[A]^G|\cdot q([x],[y])$.
c) A^{-1} *exists for any matrix $A \in [A]$. It satisfies $|A^{-1}| \le |[A]^G|$.*
d) $|[A]^G|^{-1}$ *exists.*
e) *If $[A]$ is an H-matrix then $|[A]^G| \le \langle[A]\rangle^{-1}$.*
f) *If $[A]$ is an M-matrix then $|[A]^G| = \langle[A]\rangle^{-1} = \underline{A}^{-1}$.*

Proof:

The parts a) and b) can easily be seen by (14) and by using standard rules for the diameter $d(\cdot)$ and for the distance $q(\cdot,\cdot)$; cf. [2]. Together with e) and f) they can also be found in [22], p. 100, p. 102 and p. 158. The non-singularity of $A \in [A]$ in part c) is clear by the assumption and by the definition of the interval Gaussian algorithm. The inequality in c) follows from $A^{-1}b \in IGA([A],b)$ for any $b \in \mathbf{R}^n$. (Choose for b the unit vectors pointing along the coordinate axes and use (14), (15).) Part d) of the lemma follows from the non-singularity of the diagonal matrices $|[D]^i|$ and of the triangular matrices $|[L]^i|$, $|[U]^i|$ in (15). $\qquad\square$

Lemma 2. *Let $g: \mathbf{IR}^n \to \mathbf{IR}^n$ be a P-contraction with contraction matrix P. Then for the iterative method*

$$[x]^{k+1} := g([x]^k), \qquad k = 0, 1\ldots \tag{16}$$

the following assertions hold.

a) *The method (16) converges for any starting vector $[x]^0$ to the unique fixed point $[x]^*$ of g.*
b) *If $g(\tilde{x})$ is a real vector for all $\tilde{x} \in \mathbf{R}^n$, then $d([x]^*) = 0$, i.e. $\underline{x}^* = \overline{x}^* =: \tilde{x}^*$, which is also the unique fixed point of g restricted to $\mathbf{R}^n$.*
c) *The R_1-factor of each sequence from (16) satisfies*

$$\sigma(\{[x]^k\}) \le \rho(P) < 1. \tag{17}$$

Proof:

a) is proved in [2], pp. 134–135.
b) follows from a) starting with any point vector $[x]^0 = [x^0, x^0]$ whence $d([x]^k) = 0, k = 0, 1, \ldots$.
c) is proved in [16], Lemma 4.   □

Although strict inequality can hold in (17) for the R_1-factor of any sequence of iterates generated by (16), one often considers $\rho(P)$ as an asymptotic measure for the speed of convergence of (16).

Lemma 3. *Let $g: \mathbf{IR}^n \to \mathbf{IR}^n$ be a P-contraction and let $[\delta], [\delta]^k \in \mathbf{IR}^n, k = 0, 1, \ldots$ satisfy*

$$\lim_{k \to \infty} [\delta]^k = [\delta] \quad \text{and} \quad 0 \in int([\delta]).$$

Then for the iteration

$$[y]^{k+1} := g([y]^k) + [\delta]^k$$

there is some integer $k_0 = k_0([y]^0)$ such that

$$g([y]^{k_0}) \subseteq int([y]^{k_0}). \tag{}$$ □

This lemma which justifies the use of the so-called ε-inflation ([25]) generalizes a result in [26] and is proved in [19].

We now turn over to some classical results for non-negative matrices and for M-matrices.

Lemma 4. *Let $A, B \in \mathbf{R}^{n \times n}, 0 < x \in \mathbf{R}^n, 0 \le \alpha, \beta \in \mathbf{R}$.*

a) *If $A \ge 0$ then $\rho(A)$ is an eigenvalue of A. To $\rho(A)$ there corresponds a non-negative eigenvector which we will call a Perron vector of A.*
b) *If $A \ge 0$ and $\alpha x \le Ax \le \beta x$ then $\alpha \le \rho(A) \le \beta$.*
c) *If $|A| \le B$ then $\rho(A) \le \rho(B)$.*
d) *If A is an M-matrix, $B \in \mathbf{Z}^{n \times n}$ and $A \le B$ then B is an M-matrix.*

Proof:

a) is a version of the Theorem of Perron and Frobenius. See [30], p. 46.
b) Use Exercise 2 in [30], p. 47.
c) and d) can be found in [24], 2.4.9 and 2.4.10, respectively. $\square$

Lemma 5. a) *If* $[A] = [M] - [N]$ *is an H-splitting, then* $[A]$ *and* $[M]$ *are H-matrices and* $\rho(\langle[M]\rangle^{-1}|[N]|) < 1$.
b) *If* $[A] = [M] - [N]$ *is an H-compatible splitting and if* $[A]$ *is an H-matrix, then* $[M] - [N]$ *is an H-splitting, in particular* $\rho(\langle[M]\rangle^{-1}|[N]|) < 1$.
c) *If* $[A] = [M] - [N]$ *is an M-splitting and if* $[A]$ *is an M-matrix, then* $[M] - [N]$ *is at the same time an H-splitting and an H-compatible splitting.*

Proof:

a) Since, trivially,

$$\langle[M]\rangle \geq \langle[M]\rangle - |[N]|,$$

Lemma 4d) shows that $\langle[M]\rangle$ is an M-matrix. The same holds for $\langle[A]\rangle$ since $\langle[A]\rangle \geq \langle[M]\rangle - |[N]|$ by Proposition 3.7.1 in [22]. Therefore, $B := \langle[M]\rangle - |[N]|$ is a regular splitting of the M-matrix B, hence $\rho(\langle[M]\rangle^{-1}|[N]|) < 1$ by 2.4.17 in [24].
b) follows from $\langle[A]\rangle = \langle[M]\rangle - |[N]|$ which is an M-matrix since $[A]$ is an H-matrix by the assumption.
c) follows from

$$\langle[A]\rangle = \underline{A} = \underline{M} - \overline{N} = \langle[M]\rangle - |[N]|. \qquad \square$$

Lemma 6. *Let* $\tilde{A} \in \mathbf{R}^{n \times n}$ *be non-singular with* $\tilde{A}^{-1} \geq 0$. *Let* $\tilde{A} = \tilde{M} - \tilde{N}$ *and* $\tilde{M} = \tilde{F} - \tilde{G}$ *be regular splittings. Then the matrix* $\tilde{T}_s$ *from* (12) *is non-negative, and there is a real number* $\theta \in (0, 1)$ *and a positive vector* $x \in \mathbf{R}^n$ *such that*

$$\tilde{T}_s x \leq \theta x \tag{18}$$

holds for all $s \in \mathbf{N}$. *In particular,* $\rho(\tilde{T}_s) \leq \theta < 1$ *and* $\lim_{k \to \infty} \prod_{j=0}^{k} \tilde{T}_{s(j)} = 0$ *for any sequence* $\{s(j)\}$ *of positive integers* $s(j)$. $\qquad \square$

This lemma follows immediately from the proof of Theorem 4.3 in [8] and Lemma 4.2 of that paper.

Lemma 7. *Let* $[A]$ *be an M-matrix and let* $[A] = [M] - [N]$ *be an M-splitting. Then for the iterative method*

$$[x]^{k+1} = IGA([M], [N][x]^k + [b]), \, k = 0, 1, \ldots, ([b] \in \mathbf{IR}^n) \tag{19}$$

the following assertions hold.

a) *The method* (19) *is feasible for any starting vector* $[x]^0 \in \mathbf{IR}^n$.
b) *All the sequences* $\{[x]^k\}$ *from* (19) *converge to the same limit* $[x]^* \supseteq L$; $[x]^*$ *depends on* $[M]$ *and* $[N]$.
c) *In each of the following three cases we get* $[x]^* = \square L$:

   (i) $[M]$ *is a triangular matrix;*
   (ii) $[M]$ *is a point matrix;*
  (iii) $0 \leq \underline{b}$ *or* $\underline{b} \leq 0 \leq \overline{b}$ *or* $\overline{b} \leq 0$.

d) $\sigma(\{[x]^k\}) \leq \rho(\underline{M}^{-1}\overline{N}) < 1$ *for any sequence* $\{[x]^k\}$ *produced by* (19).   (20)

e) *If* $[A] = [\hat{M}] - [\hat{N}]$ *is another M-splitting satisfying* $\underline{\hat{N}} \leq \underline{N}$, $\overline{\hat{N}} \leq \overline{N}$ *and* $d([\hat{N}]) \leq d([N])$ *then the iterative method*

$$[\hat{x}]^{k+1} = IGA([\hat{M}], [\hat{N}][\hat{x}]^k + [b]), \qquad k = 0, 1, \dots \qquad (21)$$

converges (for any starting vector) to a limit $[\hat{x}]^*$ *which satisfies*

$$[x]^* \subseteq [\hat{x}]^*. \qquad (22)$$

Moreover,

$$\rho(\underline{\hat{M}}^{-1}\overline{\hat{N}}) \leq \rho(\underline{M}^{-1}\overline{N}). \qquad (23)$$

$\square$

Lemma 7 is contained in the Theorems 1–4 in [17]. The proofs can be found in [4], [15], [16], [21].

By (20) we consider $\rho(\underline{M}^{-1}\overline{N})$ as an asymptotic measure for the speed of convergence of (19) although simple examples show that strict inequality can hold in (20) even if $\sigma(\{[x]^k\})$ is replaced by $\sup\{\sigma(\{[x]^k\})\,|\,\{[x]^k\}$ from (19)$\}$ (cf. [17] for details). Nevertheless there are many splittings in which equality holds in (20) for selected sequences of iterates. This justifies our interpretation of $\rho(\underline{M}^{-1}\overline{N})$ as a measure for the asymptotic rate of convergence.

In view of this interpretation, (23) means that the method (21) converges asymptotically at least as fast as method (19), at the expense of the quality of enclosure (cf. (22)).

Lemma 8. *Let* $A^{-1} \geq 0$ *and let* $A = M - N = \hat{M} - \hat{N}$ *be weak regular splittings. If*

$$\hat{M}^{-1} \geq M^{-1},$$

and if x is a Perron vector of $\hat{M}^{-1}\hat{N}$ *satisfying* $\hat{N}x \geq 0$, *then*

$$\rho(\hat{M}^{-1}\hat{N}) \leq \rho(M^{-1}N). \qquad \square$$

Lemma 8 is contained in Theorem 3.11 of [14]. See also Theorem 3.1 and Lemma 2.1(a) in [13].

4. The Stationary Two-Stage Method

We start this section with some estimates on $f([x], s)$. The notation in these estimates together with that of Section 1 will be used in the whole Sections 4 and 5 without further reference.

Lemma 9. *Let* $[x]$, $[y] \in \mathbf{IR}^n$ *and define f by* (4). *With*

$$P := |[F]^G| \cdot |[G]| \qquad (24)$$

and

$$T_s := P^s + \sum_{j=0}^{s-1} P^j |[F]^G| |[N]| \tag{25}$$

we get

$$q(f([x], s), f([y], s)) \le T_s \cdot q([x], [y]) \tag{26}$$

and

$$d(f([x], s)) \ge T_s \cdot d([x]). \tag{27}$$

Proof:

By elementary rules for q, and by Lemma 1 b) we get

$q(f([x], s), f([y], s))$

$\quad = q([\hat{x}]^s, [\hat{y}]^s)$

$\quad = q(IGA([F], [G][\hat{x}]^{s-1} + [N][x] + [b]), IGA([F], [G][\hat{y}]^{s-1} + [N][y] + [b]))$

$\quad \le |[F]^G| q([G][\hat{x}]^{s-1} + [N][x] + [b], [G][\hat{y}]^{s-1} + [N][y] + [b])$

$\quad \le |[F]^G| \{ q([G][\hat{x}]^{s-1}, [G][\hat{y}]^{s-1}) + q([N][x], [N][y]) \}$

$\quad \le |[F]^G| \{ |[G]| q([\hat{x}]^{s-1}, [\hat{y}]^{s-1}) + |[N]| q([x], [y]) \}.$

Repeating this process $s - 1$ times yields (26).

In an analogous way we obtain

$$d(f([x], s)) = d(IGA([F], [G][\hat{x}]^{s-1} + [N][x] + [b]))$$
$$\ge |[F]^G| \{ d([G][\hat{x}]^{s-1}) + d([N][x]) + d([b]) \}$$
$$\ge |[F]^G| \{ |[G]| d([\hat{x}]^{s-1}) + |[N]| d([x]) \}$$
$$\ge \cdots \ge T_s d([x]). \qquad \square$$

Lemma 10. *Let f and $\tilde{f}$ be given as in (4) and (9), respectively, where the matrices $\tilde{M}$, $[M]$ etc. satisfy the inclusion relation (7). Define $\tilde{P} \in \mathbf{R}^{n \times n}$ as in (11).*

a) *If $\tilde{A}\tilde{x} = \tilde{b}$ then $\tilde{x} = \tilde{f}(\tilde{x}, s)$.*
b) *Conversely, if $\tilde{x} = \tilde{f}(\tilde{x}, s)$ and if $\tilde{P}^s$ has not the eigenvalue one then $\tilde{A}\tilde{x} = \tilde{b}$.*
c) *If $\tilde{A}\tilde{x} = \tilde{b}$ and if $\tilde{x} \in [x]$ then $\tilde{x} \in f([x], s)$.*

Proof:

a) $\tilde{A}\tilde{x} = \tilde{b} \Leftrightarrow \tilde{F}\tilde{x} = \tilde{G}\tilde{x} + \tilde{N}\tilde{x} + b \Leftrightarrow \tilde{x} = \tilde{F}^{-1}(\tilde{G}\tilde{x} + \tilde{N}\tilde{x} + \tilde{b}) = \hat{x}^1.$

Repeating the arguments in an analogous way yields

$$\hat{x}^i = \tilde{x}, \qquad i = 1, \ldots, s,$$

hence $\tilde{x} = \tilde{f}(\tilde{x}, s)$.

b) By (10) with $\tilde{x}^{k+1} = \tilde{x}^k = \tilde{x}$ we get

$$\tilde{x} = f(\tilde{x}, s) = \tilde{P}^s \tilde{x} + \sum_{j=0}^{s-1} \tilde{P}^j \tilde{F}^{-1}(\tilde{N}\tilde{x} + \tilde{b}).$$

Multiplying with $I - \tilde{P}$ results in

$$(I - \tilde{P})(I - \tilde{P}^s)\tilde{x} = (I - \tilde{P})\sum_{j=0}^{s-1}\tilde{P}^j\tilde{F}^{-1}(\tilde{N}\tilde{x} + \tilde{b})$$

$$= (I - \tilde{P}^s)\tilde{F}^{-1}(\tilde{N}\tilde{x} + \tilde{b})$$

$$\Leftrightarrow \qquad 0 = (I - \tilde{P}^s)\{(I - \tilde{P})\tilde{x} - \tilde{F}^{-1}(\tilde{N}\tilde{x} + \tilde{b})\}$$

$$= (I - \tilde{P}^s)\tilde{F}^{-1}\{(\tilde{F} - \tilde{G} - \tilde{N})\tilde{x} - \tilde{b}\}$$

$$= (I - \tilde{P}^s)\tilde{F}^{-1}(\tilde{A}\tilde{x} - \tilde{b}).$$

By the assumption, $(I - \tilde{P}^s)^{-1}$ exists, hence $\tilde{A}\tilde{x} - \tilde{b} = 0$.

c) follows at once from a) and from the inclusion monotonicity of the interval arithmetic. $\qquad\square$

Theorem 1. *Let $s \in \mathbf{N}$ be fixed. If T_s in (25) satisfies*

$$\rho(T_s) < 1. \tag{28}$$

then the following assertions hold.

a) *The stationary two-stage method (5) is convergent to the same vector $[x]^*$ for all starting vectors $[x]^0 \in \mathbf{IR}^n$.*

b) *The R_1-factor of each sequence from (5) satisfies*

$$\sigma(\{[x]^k\}) \le \rho(T_s) < 1. \tag{29}$$

c) *If the iteration*

$$[y]^{k+1} = f([y]^k, 1) = IGA([F], [G][y]^k + [N][y]^k + [b]), \; k = 0, 1\ldots \tag{30}$$

is convergent for some starting vector $[y]^0$, then its limit $[y]^*$ equals $[x]^*$ from a).

d) *Any matrix $\tilde{A} \in [A]$ is non-singular.*

e) *The limit $[x]^*$ from a) contains the solution set L from (2).*

Proof:

a), b) By (26) and (28), f is a T_s-contraction, hence a) and b) are proved by Lemma 2.

c) Start (5) with $[x]^0 := [y]^*$. Then $[\hat{x}]^i = [y]^*$, $i = 0, 1, \ldots, s$, and $[x]^1 = [y]^*$. Repeating these conclusions yields $[x]^k = [y]^*$, $k = 0, 1, \ldots$, and a) finishes the proof.

d) If $\tilde{A} \in [A]$ were singular, there is a non-zero vector $\tilde{x}$ such that $\tilde{A}\tilde{x} = 0$. By Lemma 10 a) and (10) with $\tilde{b} = 0$ and $\tilde{x}^k = \tilde{x}$ we obtain $\tilde{x} = \tilde{f}(\tilde{x}, s) = \tilde{T}_s\tilde{x}$. Thus $\tilde{x}$ is an eigenvector belonging to the eigenvalue one, hence $\rho(\tilde{T}_s) \ge 1$. Now, (7), (11), (24) and Lemma 1 c) lead to $\tilde{P} \le P$. Therefore, $|\tilde{T}_s| \le T_s$ holds which implies the contradiction $\rho(\tilde{T}_s) \le \rho(T_s) < 1$ by Lemma 4 c) and (28).

e) Let $\tilde{x} \in L$, i.e. $\tilde{A}\tilde{x} = \tilde{b}$ for some $\tilde{A} \in [A]$, $\tilde{b} \in [b]$. By Lemma 10 a) we get

$$\tilde{x} = \tilde{f}(\tilde{x}, s) \in f(\tilde{x}, s). \tag{31}$$

Start (5) with $[x]^0 = [\tilde{x}, \tilde{x}]$. Then $\tilde{x} \in [x]^1$ by (31), and finally, $\tilde{x} \in [x]^k$, $k = 1, \dots$. Therefore, $\tilde{x} \in [x]^*$ which is unique by a). $\qquad\square$

By (29) we consider $\rho(T_s)$ as a measure for the speed of convergence of (5) although a similar remark holds as at the end of Lemma 7.

If $[A]$ is a point matrix, then $[M]$, $[N]$, $[F]$, $[G]$ are necessarily degenerate. If $[b]$ and $[x]^0$ are degenerate, too, then (5) produces a sequence in $\mathbf{R}^n$. Therefore, by Theorem 1 and Lemma 2 we get at once the following result.

Corollary 1. *If $[A] = [\tilde{A}, \tilde{A}]$, $[b] = [\tilde{b}, \tilde{b}]$ and if T_s satisfies $\rho(T_s) < 1$ then $\tilde{A}$ is non-singular, and the stationary two-stage method (5) converges for all starting vectors $[x]^0 \in \mathbf{IR}^n$ to the unique solution $\tilde{x}$ of $\tilde{A}x = \tilde{b}$.*

Since in most cases $[F]^G$ and thus T_s are unknown, it is hard to decide whether (28) holds. Fortunately, there are classes of matrices $[A]$ and splittings $[A] = [M] - [N]$, $[M] = [F] - [G]$ which guarantee the condition (28).

Theorem 2. *Let $[A] = [M] - [N]$ be an H-splitting and let $[M] = [F] - [G]$ be an H-compatible splitting. Then*

$$\rho(T_s) < 1, \qquad s = 1, 2, \dots, \tag{32}$$

hence the assertions of Theorem 1 and Corollary 1 hold. In addition, (30) converges for any starting vector to the limit $[x]^*$ of the stationary two-stage method (5). Thus $[x]^*$ is independent of s.

Proof:

By the definition of an H-splitting,

$$B := \langle [M] \rangle - |[N]| \tag{33}$$

is an M-matrix, hence $B^{-1} \geq 0$. By Lemma 5 a), $[M]$ is an H-matrix, therefore $\langle [M] \rangle^{-1} \geq 0$, and (33) is a regular splitting. By Lemma 5 b), $[M] = [F] - [G]$ is an H-splitting. Thus, $\langle [M] \rangle$ is an M-matrix and from

$$\langle [M] \rangle = \langle [F] \rangle - |[G]| \tag{34}$$

we see that $\langle [F] \rangle$ is an M-matrix, too. Hence (34) is again a regular splitting, and Lemma 1 e) yields $|[F]^G| \leq \langle [F] \rangle^{-1}$. With $\hat{P} := \langle [F] \rangle^{-1} |[G]|$ we get

$$\hat{T}_s := \hat{P}^s + \sum_{j=0}^{s-1} \hat{P}^j \langle [F] \rangle^{-1} |[N]| \geq T_s, \tag{35}$$

hence

$$\rho(T_s) \leq \rho(\hat{T}_s) \tag{36}$$

by Lemma 4 c). Now, the assertions of Lemma 6 are fulfilled for (33), (34) so that

$$\rho(\hat{T}_s) < 1. \tag{37}$$

Together with (36) this proves the first part of the theorem. The convergence of (30) follows from (32) and Theorem 1 applied with $s = 1$. The limit of (30) equals $[x]^*$ as a consequence of Theorem 1 c). $\qquad\square$

We next specialize Theorem 2 to M-splittings for which we are able to prove some optimality results. But we stress the fact that aside from part c) of the subsequent theorem, nothing is known about the amount of overestimation for $[x]^*$.

Theorem 3. *Let $[A]$ be an M-matrix and let $[A] = [M] - [N]$ and $[M] = [F] - [G]$ be M-splittings. Then the following results are true for fixed $s \in \mathbf{N}$.*

a) $\rho(T_s) < 1$, *hence the assertions of Theorem 1 and Corollary 1 hold. In particular,* $\{[x]^k\}$ *from (5) converges to the same vector $[x]^*$ independently of the choice of $[x]^0$, and it satisfies*

$$\sigma(\{[x]^k\}) \le \rho(T_s) < 1. \tag{38}$$

b) *(30) is convergent for any starting vector; the limit equals $[x]^*$, hence the limit $[x]^*$ in a) is independent of s.*

c) *In each of the following three cases we get $[x]^* = \square L$:*
  (i) *$[F]$ is a triangular matrix;*
  (ii) *$[F]$ is a point matrix;*
  (iii) *$0 \le \underline{b}$ or $\underline{b} \le 0 \le \bar{b}$ or $\bar{b} \le 0$.*

d) *The iteration*

$$[z]^{k+1} = IGA([M], [N][z]^k + [b]), \qquad k = 0, 1, \ldots \tag{39}$$

is convergent for any starting vector $[z]^0$ to the same limit $[z]^*$, which satisfies

$$[z]^* \supseteq [x]^* \tag{40}$$

and

$$\sigma(\{[z]^k\}) \le \rho(\langle [M] \rangle^{-1}|[N]|) = \rho(\underline{M}^{-1}\overline{N}) \le \rho(T_s) < 1 \tag{41}$$

e) $$\rho(T_s) \le \rho(T_r) \qquad \text{for } r \le s. \tag{42}$$

Proof:

a), b) By Lemma 5 c) $[M] - [N]$ is an H-splitting. Since $[M]$ is an M-matrix by the definition of an M-splitting, Lemma 5 c) can be applied once more to show that $[M] = [F] - [G]$ is an H-compatible splitting. Thus Theorem 2 proves a) and b).

c) Since $\underline{G} \ge 0$, $\underline{N} \ge 0$ we have $[G][y] + [N][y] = ([G] + [N])[y]$, so that the iteration (30) can be written as

$$[y]^{k+1} = IGA([F], ([G] + [N])[y]^k + [b]).$$

Lemma 7 c) can be applied to the M-splitting $[A] = [F] - ([G] + [N])$ which proves c) by using b).

d) $[A] = [M] - [N]$ and $[A] = [F] - ([G] + [N])$ are two M-splittings of $[A]$ satisfying $\underline{N} \le \underline{G} + \underline{N}$, $\overline{N} \le \overline{G} + \overline{N}$, $d([N]) \le d([G] + [N])$. Therefore, using b) and Lemma 7 d), e) yields (40) and

$$\sigma(\{[z]^k\}) \le \rho(\underline{M}^{-1}\overline{N}) \le \rho(\underline{F}^{-1}(\overline{G} + \overline{N})) < 1. \tag{43}$$

By Lemma 1 f) we get $P = \underline{F}^{-1}\overline{G}$ with P from (24). Since

$$\rho(P) < 1 \tag{44}$$

which can be seen from (43) and Lemma 4 c), e.g., we get

$$0 \leq T_s = P^s + \sum_{j=0}^{s-1} P^j \underline{F}^{-1}\overline{N} = P^s + (I - P^s)(I - P)^{-1}\underline{F}^{-1}\overline{N}$$

$$= P^s + (I - P^s)\underline{M}^{-1}\overline{N} = \underline{M}^{-1}\overline{N} + P^s\underline{M}^{-1}\underline{A}. \tag{45}$$

Define

$$C_\varepsilon := \underline{M}^{-1}(\overline{N} + \varepsilon ee^T)$$

for $\varepsilon \in \mathbf{R}$ and $e := (1, 1, \ldots, 1)^T \in \mathbf{R}^n$. If $\varepsilon > 0$ then C_ε is positive, hence irreducible. By the theorem of Perron and Frobenius C_ε has a positive eigenvector u associated with the eigenvalue $\rho_\varepsilon := \rho(C_\varepsilon) > 0$. By (43) and by the continuity of the spectral radius, $\varepsilon > 0$ can be chosen so small that $0 < \rho_\varepsilon < 1$ holds. Now $C_\varepsilon u = \rho_\varepsilon u$ implies $\rho_\varepsilon \underline{M}u = \overline{N}u + \varepsilon ee^T u$, hence

$$\rho_\varepsilon \underline{A}u = \rho_\varepsilon(\underline{M} - \overline{N})u = (1 - \rho_\varepsilon)\overline{N}u + \varepsilon ee^T u > 0. \tag{46}$$

Thus (46) shows that $\underline{A}u > 0$ is true. Defining

$$T_s(\varepsilon) := C_\varepsilon + P^s\underline{M}^{-1}\underline{A} = T_s + \underline{M}^{-1}\varepsilon ee^T \geq 0$$

we get

$$T_s(\varepsilon)u = \rho_\varepsilon u + P^s\underline{M}^{-1}\underline{A}u \geq \rho_\varepsilon u$$

whence

$$\rho(T_s(\varepsilon)) \geq \rho_\varepsilon$$

by Lemma 4 a). Letting ε tend to zero proves $\rho(T_s) \geq \rho(\underline{M}^{-1}\overline{N})$.

e) To prove e) we follow essentially the lines in [13], Theorem 5.1 and Corollary 5.2. We represent $\underline{A}$ by means of two weak regular splittings

$$\underline{A} = M_s - N_s = M_r - N_r$$

for which

$$M_s^{-1}N_s = T_s, \qquad M_r^{-1}N_r = T_r, \qquad M_s^{-1} \geq M_r^{-1}$$

hold. Then we show that $N_s x \geq 0$ holds for a Perron vector x of T_s. Hence the assertion follows from Lemma 8, applied to the M-matrix $\underline{A}$.

To start we remark that (44) implies $\rho(P^s) = \rho(P)^s < 1$, hence $(I - P^s)^{-1}$ exists.

Define

$$M_s := \underline{M}(I - P^s)^{-1}, \qquad N_s := M_s - \underline{A}.$$

Then

$$M_s^{-1}N_s = I - M_s^{-1}\underline{A} = I - (I - P^s)\underline{M}^{-1}(\underline{M} - \overline{N})$$

$$= P^s + (I - P^s)\underline{M}^{-1}\overline{N} \stackrel{(45)}{=} T_s \geq 0$$

and

$$M_s^{-1} = (I - P^s)\underline{M}^{-1} = (I - P^s)(I - P)^{-1}\underline{F}^{-1}$$

$$= \sum_{j=0}^{s-1} P^j\underline{F}^{-1} \geq 0, \tag{47}$$

which proves $\underline{A} = M_s - N_s$ to be a weak regular splitting.

Analogously,

$$M_r := \underline{M}(I - P^r)^{-1}, \qquad N_r := M_r - \underline{A}$$

imply a weak regular splitting $\underline{A} = M_r - N_r$ satisfying $T_r = M_r^{-1} N_r$.

Since $s \geq r$, (47) yields

$$M_s^{-1} \geq \sum_{j=0}^{r-1} P^j\underline{F}^{-1} = M_r^{-1}.$$

Let x be a Perron vector of T_s, and let $\rho := \rho(T_s)$. Then

$$T_s x = \rho x \tag{48}$$

holds. By (45) we get $T_s \geq P^s$, hence

$$\rho \geq \rho(P^s). \tag{49}$$

If $\rho = \rho(P_s)$ then (44) implies $\rho = \rho(P^s) \leq \rho(P^r) \leq \rho(T_r)$ where we used (49) with s replaced by r. Thus (42) is proved.

Assume now $\rho > \rho(P^s)$. Equation (48) yields

$$N_s x = \rho M_s x = \rho\underline{M}(I - P^s)^{-1}x. \tag{50}$$

The definition of N_s implies

$$N_s x = \underline{M}((I - P^s)^{-1} - I)x + \overline{N}x = \underline{M}P^s(I - P^s)^{-1}x + \overline{N}x$$

$$= \underline{F}(I - P)P^s(I - P^s)^{-1}x + \overline{N}x = \underline{F}P^s\underline{F}^{-1}\underline{M}(I - P^s)^{-1}x + \underline{N}x$$

$$= (\overline{G}\underline{F}^{-1})^s\frac{1}{\rho}N_s x + \overline{N}x, \tag{51}$$

where we used (50) for the last equality. Since

$$\rho((\overline{G}\underline{F}^{-1})^s) = \rho((\overline{G}P^{s-1})\cdot\underline{F}^{-1}) = \rho(\underline{F}^{-1}\cdot(\overline{G}P^{s-1})) = \rho(P^s) < \rho$$

by assumption, the Neumann series of $\left(I - \frac{1}{\rho}(\overline{G}\underline{F}^{-1})^s\right)^{-1}$ is convergent and apparently non-negative, therefore (51) implies

$$N_s x = \left(I - \frac{1}{\rho}(\overline{G}\underline{F}^{-1})^s\right)^{-1}\overline{N}x \geq 0,$$

and the theorem is proved. $\qquad\square$

Note that (41) lets expect an asymptotically faster convergence for the iteration (39) than for the method (5). At a first glance this seems to be a drawback for the stationary two-stage method. A closer look, however, reveals that (39) may be computed as costly as the enclosure $IGA([A],[b])$. This is particularly true if $[M]$ is chosen badly. Furthermore, due to (40) the enclosure of L can be worse when using (39). Inequality (42) indicates that the two-stage method converges the faster the more steps in the loop of (5) are done. Note, however, that this is an asymptotic statement which needs not be true for a particular starting vector $[x]^0$, and which can also fail if the iteration is stopped after a finite number of steps due to some stopping criterion.

We add now another criterion which guarantees (28). This criterion will be very useful in practice as will be shown later.

Theorem 4. *Let $s \in \mathbf{N}$ be fixed and let $f([x], s) =: ([f]_i)$ satisfy*

$$[f]_i \subsetneqq [x]_i, \qquad i = 1, \ldots, n, \tag{52}$$

for some vector $[x] \in \mathbf{IR}^n$. Then $\rho(T_s) < 1$, hence the assertions of Theorem 1 and Corollary 1 hold. Starting the stationary two-stage method (5) with $[x]^0 := [x]$ from (52) yields

$$[x]^k \subseteq [x]^{k-1} \subseteq \cdots \subseteq [x]^1 \subseteq [x]^0. \tag{53}$$

In addition, all these iterates contain the solution set L from (2).

Proof:

By (52) and (27) we get

$$d([x]) > d(f([x], s)) \geq T_s d([x]).$$

Hence $d([x]) > 0$, and $\rho(T_s) < 1$ follows from Lemma 4 b). Thus, Theorem 1 and Corollary 1 apply. The inclusion (53) is a direct consequence of (52) and of the inclusion monotonicity of the interval arithmetic; $L \subseteq [x]^k$ follows from Lemma 10 c).   □

We remark that the slightly weakened condition

$$f([x], s) \subseteq [x]$$

guarantees at least a fixed point $\tilde{x} \in [x]$ of every function $\tilde{f}(x, s)$. This can be seen using Brouwer's fixed point theorem. But as Lemma 10 b) and the one dimensional example $A = 3$, $F = -G = -N = 1$, $b = 0$, $\tilde{x} = 1$, $s = 2$, $[x] = [1, 1]$ show, this fixed point needs not be a solution of the corresponding system $\tilde{A}x = \tilde{b}$.

The condition (52) in Theorem 4 guarantees that f is a T_s-contraction. Let, conversely, f be a T_s-contraction and modify (5) by iterating according to

$$\left.\begin{array}{l} [y]^k = blow([x]^k, \varepsilon) \\ [x]^{k+1} = f([y]^k, s) \end{array}\right\} \quad k = 0, 1, \ldots. \tag{54}$$

Here, $blow([x], \varepsilon)$ is defined by $blow([x], \varepsilon) := [x] + [-\varepsilon, \varepsilon] \cdot (1, \ldots, 1)^T$, where $\varepsilon > 0$ is a small real number. (This blowing up of the iterates $[x]^k$ is known as ε-*inflation* in the literature; cf. [25], e.g.) Lemma 3 guarantees that condition (52) is fulfilled in a finite number of steps, at least if (28) holds. Thus L is contained in all subsequent iterates built according to (5). This shows that

i) we have verified the non-singularity of each matrix $\tilde{A} \in [A]$, i.e., we have verified a unique solution $\tilde{x}$ for each linear system $\tilde{A}x = \tilde{b}$ with $\tilde{A} \in [A]$, $\tilde{b} \in [b]$;
ii) we have obtained lower and upper bounds for these solutions;
iii) we can improve these bounds by using (5).

We remark that the function *blow* can be modified in several ways; cf. [11] for a practical realization in the scientific programming language PASCAL-XSC.

5. The Non-Stationary Two-Stage Method

In this section we consider the non-stationary two-stage method (6). We start with a theorem which is partly an analogue of Theorem 2 in the stationary case.

Theorem 5. *Let* $[A] = [M] - [N]$ *be an H-splitting and let* $[M] = [F] - [G]$ *be an H-compatible splitting. Then the following assertions hold for any sequence* $\{s(k)\}_{k=0}^{\infty}$ *of positive integers.*

a) *The particular stationary two-stage method*

$$[y]^{k+1} = f([y]^k, 1) = IGA([F], [G][y]^k + [N][y]^k + [b]), \tag{55}$$

is convergent to the same limit $[x]^*$ *for any starting vector* $[y]^0$.

b) *The non-stationary two-stage method (6) is convergent to the vector* $[x]^*$ *from a) for any starting vector* $[x]^0$. *In particular,* $[x]^*$ *is independent of the special sequence* $\{s(k)\}_{k=0}^{\infty}$ *being chosen.*
c) *Any matrix* $\tilde{A} \in [A]$ *is non-singular.*
d) *The limit* $[x]^*$ *from a) contains the solution set L from (2).*
e) *If* $\lim_{k \to \infty} s(k) = \infty$ *then the R_1-factor of any sequence* $\{[x]^k\}$ *from (6) satisfies*

$$\sigma(\{[x]^k\}) \leq \rho(\langle[M]\rangle^{-1}|[N]|) \tag{56}$$

Proof:

a) is already stated and proved in Theorem 2.
b) Start (6) with $[x]^0 = [x]^*$. As in the proof of Theorem 1 c) one sees at once $[x]^k = [x]^*$, $k = 0, 1, \ldots$. Thus

$$f([x]^*, s(k)) = [x]^*.$$

Let now $\{[x]^k\}$ be any sequence constructed by (6). Then by (26) we get

$$q([x]^{k+1}, [x]^*) = q(f([x]^k, s(k)), f([x]^*, s(k)))$$

$$\leq T_{s(k)} q([x]^k, [x]^*)$$

$$\leq \cdots \leq \left(\prod_{j=0}^{k} T_{s(j)}\right) q([x]^0, [x]^*). \tag{57}$$

Repeating the arguments in the proof for Theorem 2 and using the matrix $\hat{T}_s \geq T_s$ of (35) we get

$$q([x]^{k+1}, [x]^*) \leq \left(\prod_{j=0}^{k} \hat{T}_{s(j)} \right) q([x]^0, [x]^*). \tag{58}$$

As in Theorem 2 we apply Lemma 6 with (33), (34). This proves

$$\lim_{k \to \infty} \prod_{j=0}^{k} \hat{T}_{s(j)} = 0,$$

hence $\lim_{k \to \infty} q([x]^k, [x]^*) = 0$ and $\lim_{k \to \infty} [x]^k = [x]^*$.

c), d) follow from a) and from Theorem 2 applied with $s = 1$.

e) Starting with (35) we get $\hat{P}^s \leq \hat{T}_s$, hence $\rho(\hat{P}^s) \leq \rho(\hat{T}_s) < 1$ by (37) and Lemma 4 c). This implies $\rho(\hat{P}) < 1$. Therefore $\hat{T}_s$ can be represented as

$$\hat{T}_s = \hat{P}^s + (I - \hat{P}^s)(I - \hat{P})^{-1}\langle[F]\rangle^{-1}|[N]|$$

$$= \hat{P}^s + (I - \hat{P}^s)(\langle[F]\rangle - |[G]|)^{-1}|[N]|)$$

$$= \hat{P}^s + (I - \hat{P}^s)\langle[M]\rangle^{-1}|[N]|,$$

where we used that $[M] = [F] - [G]$ is an H-compatible splitting.

By the assumption and by Lemma 5 a) we have

$$\rho := \rho(\langle[M]\rangle^{-1}|[N]|) < 1.$$

Let $\varepsilon > 0$ and choose a monotone vector norm $\|\cdot\|$ on $\mathbf{R}^n$ such that the induced operator norm satisfies

$$\rho \leq \|\langle[M]\rangle^{-1}|[N]|\| \leq \rho + \varepsilon.$$

(Cf. [2], p. 154 for the definition and for the existence of such a norm. Note that $\langle[M]\rangle^{-1} \geq 0$.) Since $\rho(\hat{P}) < 1$ we have $\lim_{i \to \infty} \hat{P}^i = 0$. Using $\lim_{k \to \infty} s(k) = \infty$ we can find an integer k_0 such that $\|\hat{P}^{s(k)}\| \leq \varepsilon$ for $k \geq k_0$. For these k we thus get

$$\|\hat{T}_{s(k)}\| \leq \|\hat{P}^{s(k)}\| + (\|I\| + \|\hat{P}^{s(k)}\|) \cdot \|\langle[M]\rangle^{-1}|[N]|\|$$

$$\leq \varepsilon + (1 + \varepsilon)(\rho + \varepsilon) =: \rho(\varepsilon).$$

Choose ε so small that $\rho(\varepsilon) < 1$. Then, by (58),

$$\|q([x]^{k+1}, [x]^*)\| \leq \left\| \prod_{j=0}^{k_0} \hat{T}_{s(j)} \right\| \cdot \left\| \prod_{j=k_0+1}^{k} \hat{T}_{s(j)} \right\| \cdot \|q([x]^0, [x]^*)\|$$

$$\leq \left\| \prod_{j=0}^{k_0} \hat{T}_{s(j)} \right\| \cdot \rho(\varepsilon)^{k-k_0} \cdot \|q([x]^0, [x]^*)\| \tag{59}$$

whence $\sigma(\{[x]^k\}) \leq \rho(\varepsilon)$. Letting ε tend to zero proves e). $\qquad\square$

For a point matrix $[A]$ and a point vector $[b]$ one can state and prove for the method (6) an analogue of Corollary 1 nearly literally. Therefore we will omit it.

It is clear that Theorem 5 applies for M-splittings $[A] = [M] - [N]$ and $[M] = [F] - [G]$ when $[A]$ is an M-matrix. An analogue of Theorem 3 can therefore easily be proved. By the same reasons as above we do not formulate it. We only remark that (56) now reads

$$\sigma(\{[x]^k\}) \leq \rho(\underline{M}^{-1}\overline{N}). \tag{60}$$

Since there are matrices $[A]$, vectors $[b]$ and sequences $\{[z]^k\}$ in Theorem 3 d) for which $\sigma(\{[z]^k\}) = \rho(\underline{M}^{-1}\overline{N})$ (cf. [15]), we can expect a faster convergence for the non-stationary two-stage method than for the 'outer iteration' (39), provided $\lim_{k\to\infty} s(k) = \infty$. In addition, the superset relation (40) combined with Theorem 5 a), b) guarantees $[x]^* \subseteq [z]^*$ for the limits $[x]^*$ of (6) and $[z]^*$ of (39). Therefore the non-stationary method (6) seems to be superior to the method (39). Note, however, that the amount of work done in the loop of (4) ('inner iterations') does not enter into (60)—at least not directly. Therefore, due to (60), the *number of iterations* needed to fulfill some stopping criterion may be less for (6) than for (39). But the *computing time* may behave conversely if the loop of (4) is too costly.

6. A Numerical Example

In this section we illustrate some of our theoretical results by an example in which the interval hull of L is known. We restrict ourselves to the stationary case. Numerical experiments with the non-stationary case and numerical comparisons between both cases, and with other iterative methods, will be reported in a future paper.

Example. Define $[A] \in \mathbf{IR}^{30 \times 30}$ and $[b] \in \mathbf{IR}^{30}$ by

$$[a]_{ij} := \begin{cases} [2,3] & , \quad \text{if } i = j \\ [-2^{-|i-j|}, 0] & , \quad \text{otherwise} \end{cases}$$

$$[b]_i := \begin{cases} [1,2] & , \quad \text{if } i \text{ is odd} \\ [-2,-1] & , \quad \text{if } i \text{ is even} \end{cases}$$

and let $[F]$ be the lower triangular matrix which coincides in its lower triangular part with the entries of $[A]$. Let $[G]_p$ coincide with $-[A]$ in the first p super-diagonals, otherwise set $[G]_p$ equal to zero. Set $[M]_p := [F] - [G]_p$ and choose $[N]_p$ in such a way that $[A] = [M]_p - [N]_p$ holds. Then the assumptions of Theorem 3 are fulfilled since

$$\underline{A}e \geq 2 - 2\sum_{i=1}^{29} \frac{1}{2^i} = 2 - \frac{1 - \dfrac{1}{2^{29}}}{1 - \dfrac{1}{2}} = \frac{1}{2^{28}} > 0$$

where $e := (1, \ldots, 1)^T \in \mathbf{R}^{30}$, hence by a well-known Theorem of Fan (cf. [6]) $\underline{A}$ is an M-matrix and therefore, by Lemma 4 d), $[A]$ is an M-matrix, too. Using Lemma 4 d) once more, one can easily show that $[M]_p$ and $[F]$ are M-matrices, hence

Theorem 3 a) and c) show that the stationary two-stage method (5) is convergent to $[x]^* = \Box L$. For $s = 1$ and $p = 0$ one gets the Gauß-Seidel method which we want to compare with several non-trivial stationary two-stage methods, for which $p \in \{1,\ldots,28\}$. From our numerical results done on a HP Workstation 720 in PASCAL-XSC, the entries of $[x]^* = \Box L$ can be guessed to be

$$[x]_1^* = [\, -1.\overline{93} \;, \quad 3.\overline{39}\,] \qquad [x]_{16}^* = [-11.9\overline{84} \;, \quad 10.\overline{84}\,]$$

$$[x]_2^* = [\, -4.\overline{24} \;, \quad 3.\overline{42}\,] \qquad [x]_{17}^* = [-10.\overline{78} \;, \quad 11.\overline{87}\,]$$

$$[x]_3^* = [\, -4.\overline{21} \;, \quad 5.\overline{621}\,] \qquad [x]_{18}^* = [-11.\overline{75} \;, \quad 10.\overline{57}\,]$$

$$[x]_4^* = [\, -6.3\overline{48} \;, \quad 5.\overline{48}\,] \qquad [x]_{19}^* = [-10.\overline{39} \;, \quad 11.4\overline{39}\,]$$

$$[x]_5^* = [\, -6.\overline{15} \;, \quad 7.\overline{51}\,] \qquad [x]_{20}^* = [-11.1\overline{96} \;, \quad 9.\overline{96}\,]$$

$$[x]_6^* = [\, -8.\overline{12} \;, \quad 7.\overline{21}\,] \qquad [x]_{21}^* = [\, -9.\overline{6} \;, \quad 10.\overline{6}\,]$$

$$[x]_7^* = [\, -7.\overline{75} \;, \quad 9.0\overline{75}\,] \qquad [x]_{22}^* = [-10.\overline{30} \;, \quad 9.\overline{03}\,]$$

$$[x]_8^* = [\, -9.5\overline{60} \;, \quad 8.\overline{60}\,] \qquad [x]_{23}^* = [\, -8.\overline{60} \;, \quad 9.5\overline{60}\,]$$

$$[x]_9^* = [\, -9.\overline{03} \;, \quad 10.\overline{30}\,] \qquad [x]_{24}^* = [\, -9.0\overline{75} \;, \quad 7.\overline{75}\,]$$

$$[x]_{10}^* = [-10.\overline{6} \;, \quad 9.\overline{6}\,] \qquad [x]_{25}^* = [\, -7.\overline{21} \;, \quad 8.\overline{12}\,]$$

$$[x]_{11}^* = [\, -9.\overline{96} \;, \quad 11.1\overline{96}\,] \qquad [x]_{26}^* = [\, -7.\overline{51} \;, \quad 6.\overline{15}\,]$$

$$[x]_{12}^* = [-11.4\overline{39} \;, \quad 10.\overline{39}\,] \qquad [x]_{27}^* = [\, -5.\overline{48} \;, \quad 6.3\overline{48}\,]$$

$$[x]_{13}^* = [-10.\overline{57} \;, \quad 11.\overline{75}\,] \qquad [x]_{28}^* = [\, -5.\overline{621} \;, \quad 4.\overline{21}\,]$$

$$[x]_{14}^* = [-11.\overline{87} \;, \quad 10.\overline{78}\,] \qquad [x]_{29}^* = [\, -3.\overline{42} \;, \quad 4.\overline{24}\,]$$

$$[x]_{15}^* = [-10.\overline{84} \;, \quad 11.9\overline{84}\,] \qquad [x]_{30}^* = [\, -3.\overline{39} \;, \quad 1.\overline{93}\,]$$

Here the upper bars denote the periodicity of the decimal number. We stopped our iterations whenever $[x]^{k+1} = [x]^k$ occured on the machine for the first time, denoting the number of iterates by k_{max}. For the Gauß-Seidel method we got $k_{max} = 639$; for $s = 1$ and $p > 0$ we sometimes got $k_{max} > 639$ due to a different way of programming and therefore due to a change in the effects of rounding, although this case theoretically coincides with the Gauß-Seidel method for our matrices and splittings. (Note that $[G]_p[x]^k + [N]_p[x]^k + [b] = ([G]_p + [N]_p)[x]^k + [b]$ due to $\underline{G}_p \geq 0$ and $\underline{N}_p \geq 0$.) We varied p systematically from 1 to 28 and s from 1 to 8 (at least). The most significant parts of our results are listed in the following tables.

$p = 1$	k_{max}	t_{rel}
$s = 1$	645	1.05
$s = 2$	428	1.09
$s = 4$	344	1.51
$s = 8$	326	2.63

$p = 3$	k_{max}	t_{rel}
$s = 1$	642	1.05
$s = 2$	341	0.90
$s = 4$	192	0.89
$s = 8$	123	1.07

$p = 6$	k_{max}	t_{rel}
$s = 1$	649	1.06
$s = 2$	327	0.90
$s = 3$	220	0.85
$s = 4$	166	0.83
$s = 5$	134	0.82
$s = 6$	113	0.82
$s = 7$	97	0.81
$s = 8$	86	0.82

$p = 10$	k_{max}	t_{rel}
$s = 1$	645	1.05
$s = 2$	323	0.93
$s = 4$	162	0.88
$s = 8$	82	0.86

$p = 20$	k_{max}	t_{rel}
$s = 1$	640	1.03
$s = 2$	321	1.00
$s = 4$	161	0.98
$s = 8$	81	0.97

$p = 28$	k_{max}	t_{rel}
$s = 1$	639	1.02
$s = 2$	320	1.01
$s = 4$	161	1.01
$s = 8$	81	1.01

Here, t_{rel} is the quotient of the time used for the stationary two-stage method just being considered and of the time used for the Gauß-Seidel method. Therefore, if t_{rel} is less than one, the corresponding two-stage method is faster than the Gauß-Seidel method. One sees that the greatest gain of time occurs for $p = 6$, $s = 7$ where the two-stage method is about 20% faster than the Gauß-Seidel method. For all choices of p the values of k_{max} decrease with increasing s. Considering k_{max} as a measure for the speed of convergence, this phenomenon agrees with part e) of Theorem 3. That the times do not always behave monotonically is due to the increasing amount of work to be done in the loop of (4) when increasing s. Note that k_{max} is almost constant for s being fixed and $p \geq 6$.

Acknowledgement

We would like to thank the anonymous referees for several valuable suggestions.

References

[1] Alefeld, G.: On the convergence of some interval-arithmetic modifications of Newton's method. SIAM J. Numer. Anal. *21*, 363–372 (1984).

[2] Alefeld, G., Herzberger, J.: Introduction to interval computations. New York: Academic Press 1983.

[3] Alefeld, G., Platzöder, L.: A quadratically convergent Krawczyk-like algorithm. SIAM J. Numer. Anal. *20*, 210–219 (1983).

[4] Barth, W., Nuding, E.: Optimale Lösung von Intervallgleichungssystemen. Computing *12*, 117–125 (1974).

[5] Beeck, H.: Über die Struktur und Abschätzung der Lösungsmenge von linearen Gleichungs-systemen mit Intervallkoeffizienten. Computing *10*, 231–244 (1972).

[6] Fan, K.: Topological proofs for certain theorems on matrices with nonnegative elements. Monatsh. Math. *62*, 219–237 (1958).

[7] Frommer, A., Mayer, G.: Efficient methods for enclosing solutions of systems of nonlinear equations. Computing *44*, 221–235 (1990).

[8] Frommer, A., Szyld, D. B.: H-splittings and two-stage iterative methods. Numer. Math. 1993 (in press).

[9] Golub, G. H., Overton, M. L.: Convergence of a two-stage Richardson iterative procedure for solving systems of linear equations. In: Watson, G.A. (ed.) Numerical analysis (Proceedings of the Ninth Biennial Conference, Dundee, Scotland, 1981), pp. 128–139 (Lecture Notes in Mathematics 912) New York: Springer 1982.

[10] Golub, G. H., Overton, M. L.: The convergence of inexact Chebyshev and Richardson iterative methods for solving linear systems. Numer. Math. 53, 571–593 (1988).

[11] Klatte, R., Kulisch, U., Neaga, M., Ratz, D., Ullrich, Chr.: PASCAL-XSC. Sprachbeschreibung mit Beispielen. Berlin: Springer 1991.

[12] Kulisch, U. W., Miranker, W. L. (eds.): A new approach to scientific computation. New York: Academic Press 1983.

[13] Lanzkron, P. J., Rose, D. J., Szyld, D. B.: Convergence of nested classical iterative methods for linear systems. Numer. Math. 58, 685–702 (1991).

[14] Marek, I., Szyld, D. B.: Comparison theorems for weak splittings of bounded operators. Numer. Math. 58, 387–397 (1990).

[15] Mayer, G.: Reguläre Zerlegungen und der Satz von Stein und Rosenberg für Intervallmatrizen. Habilitationsschrift, Universität Karlsruhe, 1986.

[16] Mayer, G.: Comparison theorems for iterative methods based on strong splittings. SIAM J. Numer. Anal. 24, 215–227 (1987).

[17] Mayer, G.: On the speed of convergence of some iterative processes. In: Greenspan, D., Rósza, P. (eds.) Colloquia Mathematica Societatis János Bolyai, 50, Numerical Methods, Miskolc, 1986, pp. 207–228. Amsterdam: North Holland 1988.

[18] Mayer, G.: Old and new aspects for the interval Gaussian algorithm. In: Kaucher, E., Markov, S. M., Mayer, G. (eds.) Computer arithmetic, scientific computation and mathematical modelling. IMACS Annals on Computing and Applied Mathematics, Baltzer, Basel, 1991, 329–349.

[19] Mayer, G.: On ε-inflation in verification algorithms. (Submitted for publication).

[20] Neumaier, A.: New techniques for the analysis of linear interval equations. Linear Algebra Appl. 58, 273–325 (1984).

[21] Neumaier, A.: Further results on linear interval equations. Linear Algebra Appl. 87, 155–179 (1987).

[22] Neumaier, A.: Interval methods for systems of equations. Cambridge: Cambridge University Press 1990.

[23] Nichols, N. K.: On the convergence of two-stage iterative processes for solving linear equations. SIAM J. Numer. Anal. 10, 460–469 (1973).

[24] Ortega, J. M., Rheinboldt W. C.: Iterative solution of nonlinear equations in several variables. New York: Academic Press 1970.

[25] Rump, S. M.: Solving algebraic problems with high accuracy. In: Kulisch, U. W., Miranker, W. L. (eds.) A new approach to scientific computation, pp. 53–120. New York: Academic Press 1983.

[26] Rump, S. M.: On the solution of interval linear systems. Computing 47, 337–353 (1992).

[27] Schneider, H.: Theorems on M-splittings of a singular M-matrix which depend on graph structure. Linear Algebra Appl. 58, 407–424 (1984).

[28] Schwandt, H.: Schnelle fast global konvergente Verfahren für die Fünf-Punkt-Diskretisierung der Poissongleichung mit Dirichletschen Randbedingungen auf Rechteckgebieten. Dissertation, Techn. Univ. Berlin, 1981.

[29] Szyld, D. B., Jones M. T.: Two-stage and multi-splitting methods for the parallel solution of linear systems. Technical Report Preprint MCSP 165-0790, Mathematics and Computer Science Division, Argonne National Laboratory, July 1990. SIAM J. Matrix Anal. Appl. 13, 671–679 (1992).

[30] Varga, R. S.: Matrix iterative analysis. Englewood Cliffs, New Jersey: Prentice-Hall 1962.

A. Frommer
Fachbereich Mathematik
Bergische Universitat GH Wuppertal
Gaußstrasse 20
D-W-5600 Wuppertal
Federal Republic of Germany

G. Mayer
Institut für Angewandte Mathematik
Universitat Karlsruhe
Kaiserstrasse 12
D-W-7500 Karlsruhe
Federal Republic of Germany

Computing, Suppl. 9, 67–78 (1993)

Computing
© Springer-Verlag 1993
Printed in Austria

Convergence Acceleration for Some Rootfinding Methods*

W. Han and **F. A. Potra**, Iowa City

Dedicated to Professor U. Kulisch on the occasion of his 60th birthday

Abstract — Zusammenfassung

Convergence Acceleration for Some Rootfinding Methods. We present simple, efficient extrapolation formulas to accelerate the convergence of super-linearly convergent sequences. Applications are given for some rootfinding methods such as Newton's method and the secant method. Numerical examples are given showing the effectiveness of the extrapolation formulas.

AMS Subject Classification: 65B99, 65H05

Key words: Convergence acceleration, rootfinding, secant method, Newton's method.

Konvergenzbeschleunigung einiger Verfahren zur Nullstellenberechnung. Wir geben einfache und effektive Extrapolationsformeln zur Beschleunigung der Konvergenz überlinear konvergierender Folgen an. Diese Formeln werden angewandt auf Verfahren zur Nullstellenbestimmung wie zum Beispiel dem Newtonverfahren oder dem Sekantenverfahren. Numerische Beispiele zeigen die Effektivität der Extrapolationsformeln.

1. Introduction

The acceleration of convergence has been an active field of research in numerical analysis. The most important results obtained before 1970 in this field are summarized in the excellent survey paper of Joyce [3] (see also [2]). A more recent survey is contained in the dissertation of Walz [8]. We note that the vast majority of the results obtained so far apply to rather slow convergent sequences, and especially on sequences that admit logarithmic error expansions. There are relatively fewer results for faster convergent sequences. Of course, if a sequence has a fast convergence rate, one cannot expect that acceleration techniques could bring too much improvement. However, even a modest improvement may be useful because, as we show in the present paper this can lead to saving one iteration for some popular rootfinding methods, such as the secant method or Newton's method.

The problem of accelerating sequences produced by iteration procedures of the form

$$x_{n+1} = T(x_n), \qquad n = 0, 1, \dots \tag{1}$$

was studied by Meinardus [4] in case

---

* Received September 11, 1992; revised October 22, 1992.

68   W. Han and F. A. Potra

$$0 < |T'(x^*)| < 1, \tag{2}$$

where $x^* = \lim_{n \to \infty} x_n$, and by Walz [8] in case

$$T'(x^*) = \cdots = T^{(p-1)}(x^*) = 0, \qquad T^{(p)}(x^*) \neq 0. \tag{3}$$

The acceleration procedures proposed in the above mentioned papers depend on an unknown quantity λ. If condition (2) is satisfied, then $\lambda = T'(x^*)$ and eventually it can be properly approximated by using the available information contained in the sequence $\{x_n\}$. In case (3) with $p \geq 2$, no explicit formula for λ is known and, unfortunately, the approximation procedure for λ proposed in [9] does not seem to work adequately. This will be discussed at the end of this paper.

In the present paper, we consider the problem of accelerating sequences $\{x_n\}$ that are super-linearly convergent to a root x^* and satisfy an error relation of the form

$$x_n - x^* = c_n \prod_{j=1}^{p} (x_{n-j} - x^*), \tag{4}$$

or

$$x_n - x^* = c_n(x_{n-1} - x^*)^p, \tag{5}$$

or more generally,

$$x_n - x^* = c_n \prod_{j=1}^{p} (x_{n-j} - x^*)^{\alpha_j}, \tag{6}$$

where, $p > 1$, $\alpha_j \geq 0$, $1 \leq j \leq p$, and $\{c_n\}$ is a sequence of constants such that

$$\lim_{n \to \infty} c_n = c \neq 0. \tag{7}$$

If condition (3) is satisfied, and $T^{(p)}$ is continuous at x^*, then (5) and (7) are clearly satisfied with $c = T^{(p)}(x^*)/p!$.

We say that a sequence $\{\hat{x}_n\}$ accelerates $\{x_n\}$, if

$$\hat{x}_n - x^* = \delta_n(x_n - x^*), \qquad \delta_n \to 0 \text{ as } n \to \infty. \tag{8}$$

Our acceleration schemes depend only on information that is available at the nth iteration. For sequences satisfying (4), we define

$$\hat{x}_n = x_n - \frac{(x_n - x_{n-1})^2}{x_n + x_{n-(p+1)} - 2x_{n-1}}, \tag{9}$$

for sequences satisfying (5), we set

$$\hat{x}_n = x_n - \frac{(x_{n-1} - x_n)^{p+1}}{(x_{n-2} - x_{n-1})^p}, \tag{10}$$

and for sequences satisfying (6), we use

$$\hat{x}_n = x_n - \frac{(x_{n-1} - x_n)^{\alpha_1+1}}{(x_{n-p-1} - x_{n-p})^{\alpha_p}} \prod_{j=2}^{p} (x_{n-j} - x_{n-j+1})^{\alpha_j - \alpha_{j-1}}. \tag{11}$$

We note that (9) reduces to Aitken's method (see [1] or [2]) in the particular case when $p = 1$. In any case, we prove that (8) is true, i.e., the accelerated sequence is (super-linearly) faster than the original sequence. We apply our schemes to the secant method, Newton's method, as well as to a method of order $1.839\ldots$ considered in [7] and [5].

In our analysis, we will use the notion of divided difference. A divided difference of a function f is symmetric with respect to its arguments. The first-order and second-order divided differences can be defined by

$$f[x, y] = \begin{cases} \dfrac{f(y) - f(x)}{y - x} & \text{if } y \neq x, \\[2mm] f'(x) & \text{if } y = x; \end{cases}$$

$$f[x, y, z] = \begin{cases} \dfrac{f[y, z] - f[x, y]}{z - x} & \text{if } y \neq x, \\[2mm] \dfrac{1}{2} f''(x) & \text{if } x = y = z. \end{cases}$$

Divided differences of higher orders are defined similarly. We will need the following relation between divided differences and derivatives of C^k-functions:

$$f[x_0, x_1, \ldots, x_k] = \frac{1}{k!} f^{(k)}(\xi) \qquad \text{for some } \xi \in [\min\{x_0, \ldots, x_k\}, \max\{x_0, \ldots, x_k\}].$$

Throughout the paper, we will use the convention that $\prod_{j=l}^{k} a_j = 1$, if $l > k$.

2. Acceleration of Super-Linearly Convergent Methods

Consider a sequence of iterates $\{x_n\}$ produced by a super-linearly convergent method to approximate a solution x^* of a nonlinear scalar equation

$$f(x) = 0. \tag{12}$$

First we consider the case when the error relation (4) holds, together with (7). From Theorem 3.2 of [6], it follows that the sequence $\{x_n\}$ has the exact Q-order of convergence τ, where $\tau \in [1, 2)$ is the largest root of

$$t^k - \sum_{j=0}^{p-1} t^j = 0.$$

Denote the iteration errors

$$\varepsilon_{n-j} = x_{n-j} - x^*, \qquad 0 \leq j \leq p.$$

Then, the error relation (4) can be rewritten in the form

$$\varepsilon_n = c_n \prod_{j=1}^{p} \varepsilon_{n-j}.$$

Let us define a new iterate

$$\hat{x}_n = x_n - \frac{(x_n - x_{n-1})^2}{x_n + x_{n-(p+1)} - 2x_{n-1}}. \tag{13}$$

For its error, we have, using (4),

$$\hat{\varepsilon}_n \equiv \hat{x}_n - x^*$$

$$= \varepsilon_n - \frac{(\varepsilon_n - \varepsilon_{n-1})^2}{\varepsilon_n + \varepsilon_{n-(p+1)} - 2\varepsilon_{n-1}}$$

$$= \frac{\varepsilon_n \varepsilon_{n-(p+1)} - \varepsilon_{n-1}^2}{\varepsilon_n + \varepsilon_{n-(p+1)} - 2\varepsilon_{n-1}}$$

$$= \delta_n \varepsilon_n, \tag{14}$$

where

$$\delta_n = \frac{c_n - c_{n-1}}{c_n(1 + c_n c_{n-1} \prod_{j=2}^{p} \varepsilon_{n-j}^2 - 2c_{n-1} \prod_{j=2}^{p} \varepsilon_{n-j})} = O(|c_n - c_{n-1}|). \tag{15}$$

By (7), $\delta_n \to 0$ as $n \to \infty$. Hence the iterates $\{\hat{x}_n\}$ defined by (13) accelerate the convergence of $\{x_n\}$.

Next, we consider the case when the errors satisfy (5) together with (7). Let us define

$$\hat{x}_n = x_n - \frac{(x_{n-1} - x_n)^{p+1}}{(x_{n-2} - x_{n-1})^p} \tag{16}$$

and show that $\{\hat{x}_n\}$ accelerates the convergence of $\{x_n\}$. Equations (5) and (16) imply

$$\hat{\varepsilon}_n = \varepsilon_n - \frac{(\varepsilon_{n-1} - \varepsilon_n)^{p+1}}{(\varepsilon_{n-2} - \varepsilon_{n-1})^p} = \delta_n \varepsilon_n, \tag{17}$$

where

$$\delta_n = 1 - \frac{c_{n-1}(1 - c_n \varepsilon_{n-1}^{p-1})^{p+1}}{c_n(1 - c_{n-1} \varepsilon_{n-2}^{p-1})^p}. \tag{18}$$

By (7), we have

$$|\delta_n| = O(|c_n - c_{n-1}|) \to 0 \qquad \text{as } n \to \infty. \tag{19}$$

Finally, we consider the general case (6). We define

$$\hat{x}_n = x_n - \frac{(x_{n-1} - x_n)^{\alpha_1 + 1}}{(x_{n-p-1} - x_{n-p})^{\alpha_p}} \prod_{j=2}^{p} (x_{n-j} - x_{n-j+1})^{\alpha_j - \alpha_{j-1}}. \tag{20}$$

For an error analysis, we use the following two relations implied by (6):

$$\varepsilon_{n-1}^{\alpha_1} = \frac{\varepsilon_n}{c_n} \prod_{j=2}^{p} \varepsilon_{n-j}^{-\alpha_j},$$

$$\varepsilon_{n-1} = c_{n-1} \prod_{j=1}^{p} \varepsilon_{n-j-1}^{\alpha_j}.$$

We have

$$\hat{\varepsilon}_n = \varepsilon_n - \frac{(\varepsilon_{n-1} - \varepsilon_n)^{\alpha_1+1}}{(\varepsilon_{n-p-1} - \varepsilon_{n-p})^{\alpha_p}} \prod_{j=2}^{p-1} (\varepsilon_{n-j} - \varepsilon_{n-j+1})^{\alpha_j - \alpha_{j-1}}$$

$$= \delta_n \varepsilon_n ,$$

where

$$\delta_n = 1 - \frac{c_{n-1}}{c_n} \frac{\left(1 - \dfrac{\varepsilon_n}{\varepsilon_{n-1}}\right)^{\alpha_1+1}}{\left(1 - \dfrac{\varepsilon_{n-p}}{\varepsilon_{n-p-1}}\right)^{\alpha_p}} \prod_{j=2}^{p} \left(1 - \frac{\varepsilon_{n-j+1}}{\varepsilon_{n-j}}\right)^{\alpha_j - \alpha_{j-1}} \to 0 \qquad \text{as } n \to \infty .$$

2.1 Application to the Secant Method

The secant method is defined by the recursion formula

$$x_n = x_{n-1} - f(x_{n-1}) \frac{x_{n-1} - x_{n-2}}{f(x_{n-1}) - f(x_{n-2})} \qquad n \geq 2. \tag{21}$$

When the initial guesses x_0 and x_1 are sufficiently close to a root x^* of $f(x)$, the secant method converges. For the sake of completeness, we include a derivation of the error relation for the secant method.

$$x_n - x^* = x_{n-1} - x^* - f[x_{n-2}, x_{n-1}]^{-1} f(x_{n-1})$$

$$= f[x_{n-2}, x_{n-1}]^{-1} \{ f[x_{n-2}, x_{n-1}](x_{n-1} - x^*) - (f(x_{n-1}) - f(x^*)) \}$$

$$= f[x_{n-2}, x_{n-1}]^{-1} \{ f[x_{n-2}, x_{n-1}] - f[x_{n-1}, x^*] \} (x_{n-1} - x^*)$$

$$= f[x_{n-2}, x_{n-1}]^{-1} f[x_{n-2}, x_{n-1}, x^*] (x_{n-1} - x^*)(x_{n-2} - x^*). \tag{22}$$

Hence, for the secant method, we have the error relation (4) with $p = 2$, and

$$c_n = \frac{f[x_{n-2}, x_{n-1}, x^*]}{f[x_{n-2}, x_{n-1}]} .$$

Assuming that x^* is a simple root, we have

$$c_n \to c = \frac{f''(x^*)}{2f'(x^*)} \qquad \text{as } n \to \infty .$$

Thus, when we apply the secant method to compute a simple root x^* with $f''(x^*) \neq 0$, we can use the extrapolation formula,

$$\hat{x}_n = x_n - \frac{(x_n - x_{n-1})^2}{x_n + x_{n-3} - 2x_{n-1}} , \tag{23}$$

to obtain a more accurate approximation.

If $f''(x^*) = 0$, then from (22), we have

$$x_n - x^* = c_n (x_{n-1} - x^*)(x_{n-2} - x^*)^2 ,$$

72 W. Han and F. A. Potra

where

$$c_n = \frac{f[x_{n-2}, x_{n-1}, x^*, x^*]}{f[x_{n-2}, x_{n-1}]} + \frac{x_{n-1} - x^*}{x_{n-2} - x^*} \frac{f[x_{n-1}, x^*, x^*, x^*]}{f[x_{n-2}, x_{n-1}]} \to \frac{f^{(3)}(x^*)}{6f'(x^*)} \quad \text{as } n \to \infty.$$

So, when $f''(x^*) = 0$ and $f^{(3)}(x^*) \neq 0$, we can use the extrapolation formula (20) with $p = 2$, $\alpha_1 = 1$ and $\alpha_2 = 2$:

$$\hat{x}_n = x_n - \frac{(x_{n-1} - x_n)^2(x_{n-2} - x_{n-1})}{(x_{n-3} - x_{n-2})^2}. \tag{24}$$

In general, if $f''(x^*) = \cdots = f^{(l)}(x^*) = 0$ and $f^{(l+1)}(x^*) \neq 0$ for some $l \geq 2$, then according to (20), the extrapolation formula is

$$\hat{x}_n = x_n - \frac{(x_{n-1} - x_n)^2(x_{n-2} - x_{n-1})^{l-1}}{(x_{n-3} - x_{n-2})^l}. \tag{25}$$

2.2 Application to a Method of Order 1.839 ...

The following method was proposed in [7] for solving scalar nonlinear equations and generalized in [5] to nonlinear operator equations:

$$x_n = x_{n-1} - \frac{f(x_{n-1})}{f[x_{n-2}, x_{n-1}] + f[x_{n-3}, x_{n-1}] - f[x_{n-3}, x_{n-2}]}. \tag{26}$$

Compared to the secant method, which requires the same amount of work (a function evaluation per iteration step), the algorithm (26) has a higher convergence order (1.839 vs. 1.618).

An error relation of the form (4) for (26) can be derived as follows.

$$x_n - x^* = x_{n-1} - x^* - \frac{f(x_{n-1}) - f(x^*)}{f[x_{n-2}, x_{n-1}] + f[x_{n-3}, x_{n-1}] - f[x_{n-3}, x_{n-2}]}$$

$$= (x_{n-1} - x^*)\frac{f[x_{n-2}, x_{n-1}] + f[x_{n-3}, x_{n-1}] - f[x_{n-3}, x_{n-2}] - f[x_{n-1}, x^*]}{f[x_{n-2}, x_{n-1}] + f[x_{n-3}, x_{n-1}] - f[x_{n-3}, x_{n-2}]}$$

$$= c_n(x_{n-1} - x^*)(x_{n-2} - x^*)(x_{n-3} - x^*), \tag{27}$$

where

$$c_n = \frac{f[x_{n-3}, x_{n-2}, x_{n-1}, x^*] + \dfrac{x_{n-1} - x^*}{(x_{n-2} - x^*)(x_{n-3} - x^*)} f[x_{n-3}, x_{n-2}, x_{n-1}]}{f[x_{n-2}, x_{n-1}] + f[x_{n-3}, x_{n-1}] - f[x_{n-3}, x_{n-2}]}. \tag{28}$$

Thus, from (28) (see also [5]), we obtain

$$c_n \to c = \frac{f^{(3)}(x^*)}{6f'(x^*)} \quad \text{as } n \to \infty.$$

Therefore, if $f^{(3)}(x^*) \neq 0$, we have the error relation (4) with $p = 3$, and the extrapolation formula is

$$\hat{x}_n = x_n - \frac{(x_n - x_{n-1})^2}{x_n + x_{n-4} - 2x_{n-1}} . \tag{29}$$

If $f^{(3)}(x^*) = \cdots = f^{(l)}(x^*) = 0$ and $f^{(l+1)}(x^*) \neq 0$ for some $l \geq 3$, then as in the discussion for the secant method, we can use the following extrapolation formula

$$\hat{x}_n = x_n - \frac{(x_{n-1} - x_n)^2 (x_{n-3} - x_{n-2})^{l-2}}{(x_{n-4} - x_{n-3})^{l-2}} . \tag{30}$$

2.3 Application to Newton's Method

Newton's method generates a sequence $\{x_n\}$ by the recursive formula

$$x_n = x_{n-1} - \frac{f(x_{n-1})}{f'(x_{n-1})} \qquad n \geq 1. \tag{31}$$

If the initial guess x_0 is sufficiently close to a root x^*, then x_n converges to x^* quadratically.

We have

$$x_n - x^* = x_{n-1} - x^* - \frac{f(x_{n-1}) - f(x^*)}{f[x_{n-1}, x_{n-1}]}$$

$$= \frac{f[x_{n-1}, x_{n-1}] - f[x_{n-1}, x^*]}{f[x_{n-1}, x_{n-1}]} (x_{n-1} - x^*)$$

$$= \frac{f[x_{n-1}, x_{n-1}, x^*]}{f[x_{n-1}, x_{n-1}]} (x_{n-1} - x^*)^2 ,$$

so that

$$\varepsilon_n = c_n \varepsilon_{n-1}^2 , \tag{32}$$

where

$$c_n = \frac{f[x_{n-1}, x_{n-1}, x^*]}{f[x_{n-1}, x_{n-1}]} . \tag{33}$$

If the root x^* is simple, then

$$c_n \to c = \frac{f''(x^*)}{2f'(x^*)} .$$

If $f''(x^*) \neq 0$, then $p = 2$, and the extrapolation formula (10) becomes

$$\hat{x}_n = x_n - \frac{(x_{n-1} - x_n)^3}{(x_{n-2} - x_{n-1})^2} . \tag{34}$$

When $f''(x^*) = 0$ and $f^{(3)}(x^*) \neq 0$, we can write the error relation in the form

$$x_n - x^*$$

$$= \frac{f[x_{n-1}, x_{n-1}, x^*]}{f[x_{n-1}, x_{n-1}]}(x_{n-1} - x^*)^2$$

$$= \frac{f[x_{n-1}, x_{n-1}, x^*] - f[x_{n-1}, x^*, x^*] + f[x_{n-1}, x^*, x^*] - f[x^*, x^*, x^*]}{f[x_{n-1}, x_{n-1}]}(x_{n-1} - x^*)^2$$

$$= c_n(x_{n-1} - x^*)^3,$$

where

$$c_n = \frac{f[x_{n-1}, x_{n-1}, x^*, x^*] + f[x_{n-1}, x^*, x^*, x^*]}{f[x_{n-1}, x_{n-1}]} \to c = \frac{f^{(3)}(x^*)}{3f'(x^*)} \quad \text{as } n \to \infty.$$

Therefore, in this case, the extrapolation formula is

$$\hat{x}_n = x_n - \frac{(x_{n-1} - x_n)^4}{(x_{n-2} - x_{n-1})^3}. \tag{35}$$

More generally, if $f''(x^*) = \cdots = f^{(l)}(x^*) = 0$ and $f^{(l+1)}(x^*) \neq 0$, $l \geq 2$, then the extrapolation formula is

$$\hat{x}_n = x_n - \frac{(x_{n-1} - x_n)^{l+2}}{(x_{n-2} - x_{n-1})^{l+1}}. \tag{36}$$

3. Numerical Examples

We present some experiments with extrapolation formulas described in the previous section.

Example 3.1. In the first example, we solve the equation

$$e^x - \frac{1}{0.1 + x^2} = 0$$

which has a root $x^* = 0.64975068180006853$.

For the secant method, we choose the initial guesses

$$x_0 = 1, \qquad x_1 = 0.9$$

Let us use $\hat{x}_n^{(1)}$ to denote the accelerating iterate computed from (23), and $\hat{x}_n^{(2)}$ the accelerating iterate computed from (20) with $p = 2$, $\alpha_1 = \alpha_2 = 1$. The following table contains the errors of the secant method iterates, $x_n - x^*$, and those of the extrapolated ones, $\hat{x}_n^{(1)} - x^*$ and $\hat{x}_n^{(2)} - x^*$.

n	$x_n - x^*$	$\hat{x}_n^{(1)} - x^*$	$\hat{x}_n^{(2)} - x^*$
2	-0.5315×10^{-1}		
3	0.1272×10^{-1}	0.3477×10^{-2}	-0.3067×10^{-1}
4	0.7469×10^{-3}	0.1109×10^{-3}	0.2741×10^{-3}
5	-0.1022×10^{-4}	0.2691×10^{-6}	-0.1517×10^{-5}
6	0.8263×10^{-8}	0.5528×10^{-10}	-0.4706×10^{-9}
7	0.9148×10^{-13}	-0.4730×10^{-13}	0.1332×10^{-14}

For the method (26), we take

$$x_0 = 1.1, \qquad x_1 = 1, \qquad x_2 = 0.9$$

We use $\hat{x}_n^{(1)}$ to denote the accelerating iterate computed from (29), and $\hat{x}_n^{(2)}$ the accelerating iterate computed from (20) with $p = 3, \alpha_1 = \alpha_2 = \alpha_3 = 1$. Then we have the following numerical results:

n	$x_n - x^*$	$\hat{x}_n^{(1)} - x^*$	$\hat{x}_n^{(2)} - x^*$
3	-0.4755×10^{-1}		
4	0.3249×10^{-2}	-0.1454×10^{-2}	-0.2255×10^{-1}
5	0.4608×10^{-4}	0.1623×10^{-4}	-0.5654×10^{-4}
6	0.1047×10^{-7}	0.1983×10^{-8}	0.3340×10^{-8}
7	-0.2776×10^{-14}	-0.1776×10^{-14}	-0.6661×10^{-15}

For Newton's method, we take

$$x_0 = 1$$

The numerical results are the following:

n	$x_n - x^*$	$\hat{x}_n - x^*$
1	-0.6364×10^{-1}	
2	-0.4631×10^{-2}	-0.3431×10^{-2}
3	-0.2334×10^{-4}	0.4754×10^{-5}
4	-0.5898×10^{-9}	0.8755×10^{-11}

One must be cautious in using the extrapolation formulas (23) and (29). When the iterates are very close to a root of the function, the denominators in (23) and (29) are close to zero, and the loss-of-significance error will dominate the iterate error. Therefore, the numerical results of the extrapolation will deteriorate when the iterates are very close to the root.

Example 3.2. We examine an example where $f''(x^*) = 0$. The equation

$$f(x) = x - 1 + 0.1(x - 1)^3 = 0$$

has a root $x^* = 1$. Let us use the secant method to compute the root. Note that

$f''(x^*) = 0$ and $f^{(3)}(x^*) \neq 0$, so the correct extrapolation formula is (24). We denote by x_n the iterate given by the secant method, $\hat{x}_n^{(1)}$ the extrapolated iterate computed from the "wrong" formula (23), and $\hat{x}_n^{(2)}$ the extrapolated iterate computed from (24). With the initial guesses $x_0 = 0$, $x_1 = 0.1$, we have the following numerical table.

n	$x_n - x^*$	$\hat{x}_n^{(1)} - x^*$	$\hat{x}_n^{(2)} - x^*$
2	-0.1345		
3	-0.1144×10^{-1}	0.8972×10^{-2}	0.1149×10^{1}
4	-0.2243×10^{-4}	0.1262×10^{-3}	0.4966×10^{-5}
5	-0.2941×10^{-9}	0.3445×10^{-8}	0.8484×10^{-10}

We notice that the "wrong" extrapolation formula produces worse results than the secant iterates. We also notice that the error in the first extrapolated iterate $\hat{x}_3^{(2)}$ from the correct formula (24) is large, because the values x_0 and x_1 are far away from the root.

Example 3.3. We end this paper by considering an example presented in [9] as an illustration of the acceleration method proposed there. As we mentioned in the introduction, that acceleration method depends on an unknown quantity λ. If

$$f(x^*) = 0, \quad f'(x^*) \neq 0 \quad \text{and} \quad f''(x^*) \neq 0, \tag{37}$$

then the Newton iteration mapping

$$T(x) = x - \frac{f(x)}{f'(x)}$$

satisfies (3) with $p = 2$. Therefore, in this case, the acceleration formula proposed in [9] reduces to

$$y_n^{(1)} = \frac{x_{n+1} - \lambda^{2^n} x_n}{1 - \lambda^{2^n}}.$$

Because λ is unknown, it is suggested in [9] that it should be replaced by

$$\tilde{\lambda}_{n+2} = \frac{x_{n+2} - x_{n+1}}{x_{n+1} - x_n}. \tag{38}$$

In this case, $y_n^{(1)}$ would depend on x_{n+2} so that we will denote

$$\tilde{x}_{n+2} = \tilde{y}_n^{(1)} = \frac{x_{n+1} - \tilde{\lambda}_{n+2} x_n}{1 - \tilde{\lambda}_{n+2}}. \tag{39}$$

The example presented in [9] consists in applying Newton's method for finding $x^* = \sqrt{3}$, starting from $x_0 = 3$. Here, $f(x) = x^2 - 3$, so that (37) is clearly satisfied, and Newton's method becomes

$$x_n = \frac{1}{2}\left(x_{n-1} + \frac{3}{x_{n-1}}\right) \qquad n \geq 1.$$

The following table contains the results for $\{\hat{x}_n\}$ given by our acceleration method (34) and for $\{\tilde{x}_n\}$ given by (38) and (39).

n	$x_n - x^*$	$\hat{x}_n - x^*$	$\tilde{x}_n - x^*$
1	0.2680×10^0		
2	0.1795×10^{-1}	-0.6538×10^{-1}	0.2324×10^{-2}
3	0.9205×10^{-4}	-0.1282×10^{-2}	0.9417×10^{-6}
4	0.2446×10^{-8}	-0.4745×10^{-6}	0.1301×10^{-12}

While (34) works reasonably well, it appears as if (38)–(39) were a "deceleration" method. The explanation is that by substituting (38) in (39), we obtain, with n instead of $n + 2$,

$$\tilde{x}_n = x_n - \frac{(x_n - x_{n-1})^2}{x_n + x_{n-2} - 2x_{n-1}} \tag{40}$$

which is exactly Aitken's method. Or, it is known that this method cannot be applied to super-linearly convergent sequences. In fact, proceeding as in (14), (15), we deduce that $\tilde{x}_n - x^* = \tilde{\delta}_n(x_n - x^*)$, $\tilde{\delta}_n = \dfrac{\tilde{c}_n - \tilde{c}_{n-1}}{\tilde{c}_n(1 + \tilde{c}_n\tilde{c}_{n-1} - 2\tilde{c}_{n-1})}$, $\tilde{c}_n = \dfrac{x_n - x^*}{x_{n-1} - x^*}$. For Newton's method, we have $\tilde{c}_n = c_n\varepsilon_{n-1}$, where $\varepsilon_n = x_n - x^*$ and c_n is given by (33). Thus,

$$\tilde{\delta}_n = \frac{c_n\varepsilon_{n-1} - c_{n-1}\varepsilon_{n-2}}{c_n\varepsilon_{n-1}(1 + c_nc_{n-1}\varepsilon_{n-1}\varepsilon_{n-2} - 2c_{n-1}\varepsilon_{n-2})}$$

$$= \frac{c_nc_{n-1}\varepsilon_{n-2}^2 - c_{n-1}\varepsilon_{n-2}}{c_{n-1}c_n\varepsilon_{n-2}^2(1 + c_nc_{n-1}\varepsilon_{n-1}\varepsilon_{n-2} - 2c_{n-1}\varepsilon_{n-2})}$$

$$= \frac{1 - \dfrac{1}{c_n\varepsilon_{n-2}}}{1 + c_nc_{n-1}\varepsilon_{n-1}\varepsilon_{n-2} - 2c_{n-1}\varepsilon_{n-2}},$$

so that, because $\lim_{n\to\infty} c_n = c < \infty$,

$$\lim_{n\to\infty} |\tilde{\delta}_n| = \infty.$$

Therefore, $\{\tilde{x}_n\}$ is "super-linearly slower" than the original Newton sequence.

References

[1] Atkinson, K. E.: An introduction to numerical analysis, 2nd ed. New York: John Wiley & Sons 1988.
[2] Brezinski, C.: Accélération de la Convergence en Analyse Numérique. Berlin: Springer 1977 (Lecture Notes in Mathematics, 584).
[3] Joyce, D. C.: Survey of extrapolation processes in numerical analysis. SIAM Review *13*, 435–488 (1971).
[4] Meinardus, G.: Über das asymptotische Verhalten von Iterationsfolgen. Z. Ang. Math. Mech. *63*, 70–72 (1983).

[5] Potra, F. A.: On an iterative algorithm of order 1.839 ... for solving nonlinear operator equations. Numer. Funct. Anal. and Optimiz. 7, 75–106 (1984–85).
[6] Potra, F. A.: On Q-order and R-order of convergence. J. Optimization Theory and Applications 63, 415–431 (1989).
[7] Traub, J.: Iterative methods for the solution of equations. New Jersey: Prentice-Hall, Englewood Cliffs 1964.
[8] Walz, G.: Approximation von Funktionen durch asymptotische Entwicklungen und Eliminations-prozeduren. Dissertation, Universität Mannheim, 1987.
[9] Walz, G.: Asymptotic expansion and acceleration of convergence for higher order iteration process. Numer. Math. 59, 529–540 (1991).

Weimin Han
F. A. Potra
Department of Mathematics
University of Iowa
Iowa City, IA 52242, U.S.A.

Computing, Suppl. 9, 79–92 (1993)

Computing
© Springer-Verlag 1993
Printed in Austria

A Program for Enclosing Cumulative Binomial Probabilities*

G. Heindl, Wuppertal

Dedicated to Prof. Dr. U. Kulisch on the occasion of his 60th birthday

Abstract — Zusammenfassung

A Program for Enclosing Cumulative Binomial Probabilities. Based on a simple theoretical algorithm for computing cumulative binomial probabilities, a PASCAL-SC program for enclosing such probabilities is derived. In several real life problems the high quality of the computed enclosures is demonstrated.

AMS Subject Classification: 60-04, 62-04, 65-04, 65G10, 65U05

Key words: Inclusion algorithms, cumulative binomial probabilities, PASCAL-SC.

Ein Programm zur Einschließung kumulativer binomischer Wahrscheinlichkeiten. Aus einem einfachen Algorithmus zur Berechnung kumulativer binomischer Wahrscheinlichkeiten wird ein PASCAL-SC Programm zur Einschließung solcher Wahrscheinlichkeiten entwickelt. An mehreren praktischen Beispielen wird die hohe Güte der berechneten Einschließungen demonstriert.

1. Introduction

If in any trial of a sequence of mutually independent trials the event A appears with probability p, then

$$b(i, n, p) := (n!/(i! * (n - i)!)) * p^i * (1 - p)^{n-i} \tag{1}$$

is the (binomial) probability for "i times A in n trials".

We are interested in the computation of sufficiently good approximating lower and upper bounds for (cumulative binomial) probabilities

$$w := \sum_{i=il}^{iu} b(i, n, p) \tag{2}$$

"for at least il times A and at most iu times A in n trials", especially in cases where $n, p, il, iu,$ are not suited for deriving a sufficiently precise inclusion of w from statistical tables or from one of the wellknown limit theorems.

In the following a suitable PASCAL-SC program, using only the most elementary features of PASCAL-SC, is developed. The program rests on a theoretical algorithm which will be described first.

---

* Received September 30, 1992; revised December 4, 1992.

2. A Theoretical Algorithm for Computing Cumulative Binomial Probabilities

Since $p = 0$ and $p = 1$ are trivial cases we assume $0 < p < 1$.

Then $b(i, n, p) > 0$, $i = 0, \ldots, n$, and, writing m for $n + 1$, we have the wellknown relations

$$b(i - 1, n, p)/b(i, n, p) = (i/(m - i)) * ((1 - p)/p) =: q(i), \quad i = 1, \ldots, n. \tag{3}$$

Since $q(i) < 1$ iff $i < m * p$, $q(i) = 1$ iff $i = m * p$ and $q(i) > 1$ iff $i > m * p$, it follows that $b(i, n, p)$ is maximal for $i = km$, where

$$km := [m * p] \tag{4}$$

(the largest integer not greater than $m * p$). $km \in \{0, \ldots, n\}$ is a consequence of $p < 1$. In addition, if $m * p$ is not an integer, then

$$b(i - 1, n, p) < b(i, n, p) \qquad \text{whenever } i \geq 1 \text{ and } i \leq km,$$

$$b(i - 1, n, p) > b(i, n, p) \qquad \text{whenever } i \geq km + 1 \text{ and } i \leq n.$$

If $m * p$ is an integer, then $km = m * p \geq 1$,

$$b(i - 1, n, p) < b(i, n, p) \qquad \text{whenever } i \geq 1 \text{ and } i \leq km - 1,$$

$$b(km - 1, n, p) = b(km, n, p) \qquad \text{and}$$

$$b(i - 1, n, p) > b(i, n, p) \qquad \text{whenever } i \geq km + 1 \text{ and } i \leq n.$$

Clearly these estimates remain valid when the probabilities $b(i, n, p)$ are replaced by the scaled quantities

$$f(i) := b(i, n, p)/b(km, n, p), \qquad i = 0, \ldots, n, \tag{5}$$

which can be computed from the initial value $f(km) = 1$ and the recurrences

$$\begin{aligned} f(i - 1) &= q(i) * f(i) \\ &= (i/(m - i)) * ((1 - p)/p) * f(i), \qquad i = km, \ldots, 1, \quad \text{if } km > 0, \end{aligned} \tag{6}$$

$$\begin{aligned} f(i) &= q(i)^{-1} * f(i - 1) \\ &= ((m - i)/i) * (p/(1 - p)) * f(i - 1), \qquad i = km + 1, \ldots, n, \quad \text{if } km < n, \end{aligned} \tag{7}$$

derived from (3).

The main advantage of the chosen scaling is that it causes $q(i) \leq 1$ (even $q(i) < 1$ for $i < km$) in (6) and $q(i)^{-1} < 1$ in (7).

Let us show now how the $f(i)$ can be used for computing

$$w := \sum_{i=il}^{iu} b(i, n, p)$$

for given $il, iu \in \mathbb{N}_0$: $il \leq iu \leq n$. From

$$\sum_{i=0}^{n} b(i, n, p) = 1 \text{ we conclude } \sum_{i=0}^{n} f(i) = b(km, n, p)^{-1}. \text{ Hence}$$

$$w = b(km, n, p) * \sum_{i=il}^{iu} f(i) = \left(\sum_{i=il}^{iu} f(i) \right) \Big/ \left(\sum_{i=0}^{n} f(i) \right). \tag{8}$$

In order to derive w from this formula and the recurrences (6) and (7) in an economical way, three cases are distinguished:

Case 1: $iu < km$

In this case we compute $s1 := \sum_{i=iu+1}^{km} f(i)$, $s2 := \sum_{i=il}^{iu} f(i)$, $s3 := \sum_{i=0}^{il-1} f(i)$ ($:= 0$ if $il = 0$), $s := \sum_{i=km+1}^{n} f(i)$ ($:= 0$ if $km = n$) and w from $w = s2/(s1 + s2 + s3 + s)$.

Case 2: $il > km$

In this case we compute $s1 := \sum_{i=km}^{il-1} f(i)$, $s2 := \sum_{i=il}^{iu} f(i)$, $s3 := \sum_{i=iu+1}^{n} f(i)$ ($:= 0$ if $iu = n$), $s := \sum_{i=0}^{km-1} f(i)$ ($:= 0$ if $km = 0$) and w from $w = s2/(s1 + s2 + s3 + s)$.

Case 3: $il \le km \le iu$

In this last case we compute $s1 := \sum_{i=il}^{km} f(i)$, $s2 := \sum_{i=0}^{il-1} f(i)$ ($:= 0$ if $il = 0$), $s3 := \sum_{i=km+1}^{iu} f(i)$ ($:= 0$ if $iu = km$), $s4 := \sum_{i=iu+1}^{n} f(i)$ ($:= 0$ if $iu = n$) and w from $w = (s1 + s3)/(s1 + s2 + s3 + s4)$.

Let us now use this algorithm for constructing

3. A PASCAL-SC Program for Enclosing w

The program will be developed in the ATARI ST version of PASCAL-SC [3]. The underlying floating point system is $S := S(10, 13, -98, 100)$. maxint $= 2^{31} - 1$. Although the author tried to make the listing as selfexplaining as possible, a separate description of the main features will be given first.

The input data are specified as follows:
il, iu, n:integer:$0 \le il \le iu \le n \le$ maxint $- 1$.
p:real:$0.0 < p < 1.0$ (dec $:= 1$)   or $p = p1/p2$, $p1$, $p2$:integer:$1 \le p1 \le p2$ (dec $:= 2$). The second case is considered in order to avoid the necessity of entering p as an interval.

Note that if e.g. $p = 1/6$ (the probability for an ace when casting a fair die) then p is not a member of S!

The first step of the program consists in an (exact) computation of

$$m := n + 1 \quad (\le \text{maxint}) \quad \text{and} \quad km := [m * p].$$

If dec $= 1$ then obviously km is derived most easily from the PASCAL-SC statement

$$km := \text{trunc}(m * <p);.$$

If dec $= 2$ we use a similar statement

$$km := \text{etrunc}(m, p1, p2);$$

where etrunc is a suitable PASCAL-SC function based on scalp.

The comments following the crucial statements of etrunc (in the listing) can serve for a correctness proof.

Next there are constructed the procedures itoiminus1 and iminus1toi, the first for deriving an inclusion of $f(i-1)$ from an inclusion of $f(i)$ and (6), the second for deriving an inclusion of $f(i)$ from an inclusion of $f(i-1)$ and (7).
itoiminus1 needs an inclusion $0.0 \leq bu \leq (1-p)/p \leq bo$,
iminus1toi needs an inclusion $0.0 \leq ub \leq p/(1-p) \leq ob$.

Both inclusions are computed in the main program taking into account the value of dec.

In order to derive inclusions of the sums from which w is derived, the procedures sumdown and sumup are introduced.

Given jl, ju:integer:$0 \leq jl \leq ju \leq km$ and var $tl, tr, sl\ sr$:real:$0.0 \leq tl \leq f(ju) \leq tr$, then after the call

$$\text{sumdown}(jl, ju, tl, tr, sl, sr);$$

$0.0 \leq tl \leq f(jl) \leq tr$ and $0.0 \leq sl \leq \sum_{i=jl}^{ju} f(i) \leq sr$ hold.

Given jl, ju:integer:$km \leq jl \leq ju \leq n$ and var tl, tr, sl, sr:real:$0.0 \leq tl \leq f(jl) \leq tr$, then after the call

$$\text{sumup}(jl, ju, tl, tr, sl, sr);$$

$0.0 \leq tl \leq f(ju) \leq tr$ and $0.0 \leq sl \leq \sum_{i=jl}^{ju} f(i) \leq sr$ hold.

In both procedures summation is stopped after $tr \leq$ minpos for the first time. The rest of the sum is estimated above by the number of remaining terms times minpos. The validity of this estimate is assured by $ju \leq km$ in sumdown and by $jl \geq km$ in sumup.

By stopping the summations in accordance with a lower demand for precision, it is possible to speed up the procedures considerably.

Concerning this improvement it should be remarked that the entries of the well-known Tables of the Binomial Probability Distribution of the National Bureau of Standards. Washington 1952, are given with seven decimal places.

But, although the resulting precision seems to be high enough for all practical purposes, it is not sufficient e.g. for finding the smaller of the two probabilities $b(24, 32, 0.28)$ and $b(32, 41, 0.36)$ (given both by 0.0000000 in the tables).

Now, with a suitable case statement, an inclusion

$$0.0 \leq w1 \leq w \leq w2 \leq 1.0$$

of w is easily computed in any of the three cases considered in the theoretical algorithm.

The following listing shows a possible realization of the described steps and their embedding into a complete interactive PASCAL-SC program for testing the quality of the suggested approach.

```
program binomial (input,output);
{ ===========================================
  PASCAL-SC program for computing inclusions of cumulative binomial
  probabilities
        w = b(il,n,p) + ··· + b(i,n,p) + ··· + b(iu,n,p)                    +)
  where
        b(i,n,p) := (n!/(i!*(n − i)!)*p**i*(1 − p)**(n − i),
        il,i,iu,n : integer : 0 <= il <= i <= iu <= n <= maxint − 1
        p : real : 0.0 < p < 1.0 or
        p = p1/p2, p1,p2 : integer : 1 <= p1 < p2.
After the start of the program the user is asked to type in n, p (or p1,p2), il
and iu.
The program is written in the ATARTI ST version of PASCAL-SC. The
underlying floating point system is S(10,13,−98,100) =: S. maxint = 2**31 − 1.
----------
+) Arithmetic operators appearing in comments denote the corresponding
   unrounded operations!
  ===========================================}
const minpos = 1.0E-99; {smallest positive number in S}
var we,n,m,dec,km,p1,p2,del,il,iu : integer;
    p,cpl,cpr,ub,ob,bu,bo,tl,tr,sl,sr : real;
    s1l,s1r,s2l,s2r,s3l,s3r,s4l,s4r,wl,wr : real;
    location : (left,middle,right);
{}
function etrunc(m,p1,p2 : integer) : integer;
{etrunc(m,p1,p2) = [m*(p1/p2)] if m,p1 >= 0, p2 > 0 and p1 <= p2}
var v1,v2 : array[1..2] of real;
    h : integer;
begin
  h := trunc(m*>(p1 /> p2));
  v1[1] := h; {>= [m*(p1/p2)]} v1[2] := −m;
  v2[1] := p2;                 v2[2] := p1;
  while scalp(v1,v2,1) > 0.0
  {i.e. v1[1]*p2 − m*p1 rounded up is > 0.0
   i.e. v1[1]*p2 − m*p1 > 0.0
   i.e. v1[1] > m*(p1/p2)
   i.e. v1[1] > [m*(p1/p2)] (since v1[1] is integer valued)}
  do
     begin
       h := h − 1 ; v1[1] := h
     end;
  {For the actual system v1[1] must be reduced at most once.
   (m,p1,p2) = (9999999,10000001,100000000) is an example where a
   reduction is necessary.}
  etrunc := h
end{etrunc};
```

```
procedure itoiminus1(i : integer; var tl,tr : real);
{Input: i,tl,tr : 1 <= i <= n and
                  0.0 <= tl <= f(i) := b(i,n,p)/b(km,n,p) <= tr
                  (km = [m*p], m = n + 1).
          bu,bo : (global) real :
                  0.0 <= bu <= (1 − p)/p <= bo.
  Output:tl,tr : 0.0 <= tl <= f(i − 1) = (i/(m − i))*((1 − p)/p)*f(i) <= tr.}
var d : integer;
begin
  d := m − i;   {d is a positive integer}
  tl := (i/< d)*<bu*<tl;
  tr := (i/> d)*>bo*>tr
end{itoiminus1};
{}
procedure iminus1toi(i : integer ; var tl,tr : real);
{Input: i,tl,tr : 1 <= i <= n and
                0.0 <= tl <= f(i − 1) := b(i − 1,n,p)/b(km,n,p) <= tr
                (km = [m*p], m = n + 1).
          ub,ob : (global) real :
                0.0 <= ub <= p/(1 − p) <= ob.
  Output:tl,tr : 0.0 <= tl <= f(i) = ((m − i)/i)*(p/(1 − p))*f(i − 1) <= tr.}
var d : integer;
begin
  d := m − i;   {d is a positive integer}
  tl := (d/< i)*<ub*<tl;
  tr := (d/> i)*>ob*>tr
end{iminus1toi};
{}
procedure sumdown(jl,ju : integer ; var tl,tr,sl,sr : real);
{Input: jl,ju,tl,tr :   0 <= jl <= ju <= km (= [m*p], m = n + 1) and
                        0.0 <= tl <= f(ju) := b(ju,n,p)/b(km,n,p) <= tr.
  Output: tl,tr,sl,sr : 0.0 <= tl <= f(jl) := b(jl,n,p)/b(km,n,p) <= tr and
                        0.0 <= sl <= f(jl) + ··· + f(ju) <= sr.
  Used procedures : itoiminus1.}
var i : integer;
begin
  sl := tl; sr := tr;   {0.0 <= sl <= f(ju) <= sr}
  i := ju;
  {0.0 <= tl <= f(i) <= tr, 0.0 <= sl <= f(i) + ··· + f(ju) <= sr for i = ju}
  while (i > jl) and (tr > minpos) do
    begin
      itoiminus1(i,tl,tr);
      {0.0 <= tl <= f(i − 1) <= tr}
      sl := sl +< tl;
      sr := sr +> tr;
      {0.0 <= sl <= f(i − 1) + ··· + f(ju) <= sr}
```

```
        i := i − 1
          {0.0 <= tl <= f(i) <= tr, 0.0 <= sl <= f(i) + ··· + f(ju) <= sr}
        end;
    if i > jl then {tr <= minpos and therefore
                    0.0 < f(l) <= minpos, l = 0, ..., i}
      sr := sr +> ((i − jl)* > minpos)
end{sumdown};
procedure sumup(jl,ju : integer ; var tl,tr,sl,sr : real);
{Input: jl,ju,tl,tr :   km <= jl <= ju <= n (km = [m∗p], m = n + 1) and
                        0.0 <= tl <= f(jl) := b(jl,n,p)/b(km,n,p) <= tr.
  Output: tl,tr,sl,sr : 0.0 <= tl <= f(ju) := b(ju,n,p)/b(km,n,p) <= tr and
                        0.0 <= sl <= f(jl) + ··· + f(ju) <= sr.
  Used procedures : iminus1toi.}
var i : integer;
begin
    sl := tl; sr := tr;   {0.0 <= sl <= f(jl) <= sr}
    i := jl + 1;
    {0.0 <= tl <= f(i − 1) <= tr, 0.0 <= sl <= f(jl) + ··· + f(i − 1) <= sr
     for i = jl − 1}
    while (i <= ju) and (tr > minpos) do
      begin
        iminus1toi(i,tl,tr);
        {0.0 <= tl <= f(i) <= tr}
        sl := sl +< tl;
        sr := sr +> tr;
        {0.0 <= sl <= f(jl) + ··· + f(i) <= sr}
        i := i + 1
        {0.0 <= tl <= f(i − 1) <= tr, 0.0 <= sl <= f(jl) + ··· + f(i − 1)
         <= sr}
      end;
    if (i <= ju) then {tr <= minpos and therefore
                    0.0 < f(l) <= minpos, l = i − 1, ..., n}
      sr := sr +> ((ju − i + 1)* > minpos)
end{sumup};
{}
{}
{main program}
begin
    we := 1 ;          {we = 1 for 'go on', we = 0 for 'end'}
    while we <> 0 do
      begin
        {Heading}
        writeln('Enclosure of w = b(il,n,p) + ··· + b(i,n,p) + ··· + b(iu,n,p),');
        writeln('where');
        writeln('    b(i,n,p) = (n!/(i!∗(n − i)!)∗p∗∗i∗(1 − p)∗∗(n − i),');
        write('    il,i,iu,n : integer : 0 <= il <= i <= iu <= n,');
```

```
writeln('1 <= n <= maxint − 1,');
writeln('       p : real : 0.0 < p < 1.0 or');
writeln('       p = p1/p2, p1,p2 : integer : 1 <= p1 < p2.');
writeln ;
{}
{Input of n and computation of m := n + 1}
repeat
   write('n (: integer : 1 <= n <= maxint − 1) = ? ');
   read(n)
until (n >= 1) and (n <= maxint − 1);
writeln;
m := n + 1; { <= maxint}
{}
{Input of p or p1,p2 and computation of km, bu, bo, ub, ob:
 km = [m*p], 0.0 <= bu <= (1 − p)/p <= bo and
 0.0 <= ub <= p/(1 − p) <= ob}
repeat
writeln('Type 1 if you want to enter p as a real value,');
writeln('type 2 if you want to enter p as a ratio p1/p2');
write('(p1,p2 : integer : 1 <= p1 < p2)! ');
read(dec)
until (dec = 1) or (dec = 2);
writeln;
if dec = 1 {p real} then
   begin
      repeat
      write('p (: real : 0.0 < p < 1.0) = ? ');
      read(p)
      until (p > 0.0) and (p < 1.0);
      writeln;
      km := trunc(m* <p);
      writeln('[(',n,' + 1) * ',p,'] = ',km); writeln;
      cpl := 1 −< p ; cpr := 1 −> p ;
      bu := cpl/< p ; bo := cpr/> p ;
      {0.0 <= bu <= (1 − p)/p <= bo}
      ub := p/< cpr ; ob := p/> cpl
      {0.0 <= ub <= p/(1 − p) <= ob}
   end
else {dec = 2 i.e. p = p1/p2}
   begin
      repeat
      write('p1 p2 (: integer : 1 <= p1 < p2) = ? ');
      read(p1,p2)
      until (p1 >= 1) and (p1 < p2);
      writeln;
      km := etrunc(m,p1,p2);
```

```
    writeln('[(',n,' + 1) * (',p1,'/',p2,')] = ',km); writeln;
    del := p2 − p1; {del is a positive integer}
    bu := del /< p1 ; bo := del /> p1;
    {0.0 <= bu <= (1 − p)/p <= bo}
    ub := p1 /< del ; ob := p1 /> del
    {0.0 <= ub <= p/(1 − p) <= ob}
  end;
{}
{Input of il and iu}
repeat
write('il iu (: integer : 0 <= il <= iu <= ',n,') = ? ');
read(il,iu)
until (il >= 0) and (iu >= il) and (iu <= n);
writeln;
{}
{Computation of wl, wr : wl <= w <= wr}
{}
if iu < km then location := left
else {iu >= km}
  if il > km then location := right
  else {il <= km <= iu}
    location := middle;
{}
case location of
    left : {iu < km}
        begin
          tl := 1.0 ; tr := 1.0; {tl <= f(km) <= tr}
          sumdown(iu + 1,km,tl,tr,s1l,s1r);
          {0.0 <= tl <= f(iu + 1) <= tr,
            1.0 <= s1l <= f(iu + 1) + ··· + f(km) <= s1r}
          itoiminus1(iu + 1,tl,tr);
          {0.0 <= tl <= f(iu) <= tr}
          sumdown(il,iu,tl,tr,s2l,s2r);
          {0.0 <= tl <= f(il) <= tr,
            0.0 <= s2l <= f(il) + ··· + f(iu) <= s2r}
          if il > 0 then
            begin
              itoiminus1(il,tl,tr);
              {0.0 <= tl <= f(il − 1) <= tr}
              sumdown(0,il − 1,tl,tr,s3l,s3r);
              {0.0 <= s3l <= f(0) + ··· + f(il − 1) <= s3r}
            end
          else {il = 0}
            begin
              s3l := 0.0; s3r := 0.0
            end;
```

```
      if km < n then
        begin
          tl := 1.0 ; tr := 1.0 ; {tl <= f(km) <= tr}
          iminus1toi(km + 1,tl,tr);
          {0.0 <= tl <= f(km + 1) <= tr}
          sumup(km + 1,n,tl,tr,sl,sr);
          {0.0 <= sl <= f(km + 1) + ··· + f(n) <= sr}
        end
      else {km = n}
        begin
          sl := 0.0 ; sr := 0.0
        end;
      wl := s2l/<(s1r +> s2r +> s3r +> sr);
      wr := s2r/>(s1l +< s2l +< s3l +< sl)
      {0.0 <= wl <= w <= wr}
    end;
  right : {il > km}
    begin
      tl := 1.0 ; tr := 1.0 ; {tl <= f(km) <= tr}
      sumup(km,il − 1,tl,tr,s1l,s1r);
      {0.0 <= tl <= f(il − 1) <= tr,
       1.0 <= s1l <= f(km) + ··· + f(il − 1) <= s1r}
      iminus1toi(il,tl,tr);
      {0.0 <= tl <= f(il) <= tr}
      sumup(il,iu,tl,tr,s2l,s2r);
      {0.0 <= tl <= f(iu) <= tr,
       0.0 <= s2l <= f(il) + ··· + f(iu) <= s2r}
      if iu < n then
        begin
          iminus1toi(iu + 1,tl,tr);
          {0.0 <= tl <= f(iu + 1) <= tr}
          sumup(iu + 1,n,tl,tr,s3l,s3r);
          {0.0 <= s3l <= f(iu + 1) + ··· + f(n) <= s3r}
        end
      else {iu = n}
        begin
          s3l := 0.0 ; s3r := 0.0
        end;
      if km > 0 then
        begin
          tl := 1.0 ; tr := 1.0 ; {tl <= f(km) <= tr}
          itoiminus1(km,tl,tr);
          {0.0 <= tl <= f(km − 1) <= tr}
          sumdown(0,km − 1,tl,tr,sl,sr);
          {0.0 <= sl <= f(0) + ··· + f(km − 1) <= sr}
        end
```

```
        else {km = 0}
          begin
            sl := 0.0 ; sr := 0.0
          end;
        wl := s2l/ <(s1r +> s2r +> s3r +> sr);
        wr := s2r/ >(s1l +< s2l +< s3l +< sl)
        {0.0 <= wl <= w <= wr}
      end;
  middle : {il <= km <= iu}
    begin
      tl := 1.0 ; tr := 1.0 ; {tl <= f(km) <= tr}
      sumdown(il,km,tl,tr,s1l,s1r);
      {0.0 <= tl <= f(il) <= tr,
        1.0 <= s1l <= f(il) + ··· + f(km) <= s1r}
      if il > 0 then
        begin
          itoiminus1(il,tl,tr);
          {0.0 <= tl <= f(il − 1) <= tr}
          sumdown(0,il − 1,tl,tr,s2l,s2r);
          {0.0 <= s2l <= f(0) + ··· + f(il − 1) <= s2r}
        end
      else {il = 0}
        begin
          s2l := 0.0 ; s2r := 0.0
        end;
      if iu > km then
        begin
          tl := 1.0 ; tr := 1.0 ; {tl <= f(km) <= tr}
          iminus1toi(km + 1,tl,tr);
          {0.0 <= tl <= f(km + 1) <= tr}
          sumup(km + 1,iu,tl,tr,s3l,s3r);
          {0.0 <= tl <= f(iu) <= tr,
            0.0 <= s3l <= f(km + 1) + ··· + f(iu) <= s3r}
        end
      else
        begin
          s3l := 0.0 ; s3r := 0.0
        end;
      if iu < n then
        begin
          iminus1toi(iu + 1,tl,tr);
          {0.0 <= tl <= f(iu + 1) <= tr}
          sumup(iu + 1,n,tl,tr,s4l,s4r);
          {0.0 <= s4l <= f(iu + 1) + ··· + f(n) <= s4r}
        end
      else{iu = n}
```

```
                    begin
                        s4l := 0.0 ; s4r := 0.0
                    end;
                wl := (s1l +< s3l)/< (s1r +> s2r +> s3r +> s4r);
                wr := (s1r +> s3r)/> (s1l +< s2l +< s3l +< s4l)
                {0.0 <= wl <= w <= wr}
            end
    end; {case}
    {}
    if wr > 1.0 then wr := 1.0 ; {w <= 1.0}
    {}
    {}
    {Output}
    writeln(wl,'  <='); writeln ;
    if dec = 1 then
      begin
        writeln('b(',il,',',n,',',p,') + ··· + ');
        writeln;
        writeln('b(',iu,',',n,',',p,')   <=')
      end
    else
      begin
        writeln('b(',il,',',n,',',p1,'/',p2,') + ··· + ');
        writeln;
        writeln('b(',iu,',',n,',',p1,'/',p2,')   <=')
      end; writeln ;
    writeln(wr);
    writeln;
    {}
    {Continuation or Termination}
    write('Type 1 for go on, 0 for end ! ') ; read(we);
    writeln; writeln
  end {while}
end.
```

4. Examples

Originally the listed program was written in order to check the accuracy of the probabilities given in the tables IV/3, IV/4 and V/3 of the worth reading paper [7] of F. Ulmer. These probabilities were calculated by U. Tesche from approximations based on an extended Stirling formula.

Like in the following four typical examples it turned out that all entries of the tables are correct within the given decimal places.

Example 1
$n = 127200$, $p = 0.049$, $il = 6360$, $iu = n$
Tesches result: 0.050
Computed inclusion: 5.025567564200E-02 $\leq w \leq$ 5.025567566649E-02.

Example 2
$n = 1000$, $p = 0.055$, $il = 0$, $iu = 49$
Tesches result: 0.225
Computed inclusion: 2.253117770660E-01 $\leq w \leq$ 2.253117770756E-01.

Example 3
$n = 2000$, $p = 0.475$, $il = 1000$, $iu = n$
Tesches result: 0.0134
Computed inclusion: 1.336690330189E-02 $\leq w \leq$ 1.336690330517E-02.

Example 4
$n = 665000$, $p = 0.5$, $il = 0$, $iu = 332499$
Tesches result: 0.4995
Computed inclusion: 4.995107848902E-01 $\leq w \leq$ 4.995107866531E-01.

In all these examples n is much too large for deriving w from one of the existing statistical tables. In the Tables of the Binomial Probability Distribution mentioned earlier e.g., the scope of n is $\{2,\ldots,49\}$. On the other hand, deriving for large n approximations for w and numerically evaluable eror estimates from limit theorems (suited for some n, p, il, iu, and not suited for others) cannot serve as a basis for a transparent homogeneous algorithm with a wide range of application. The next examples will show that approximations of binomial probabilities based on the limit theorem of Moivre-Laplace can be poor even when the well known rules of thumb $n*p > 10$ and $n*p*(1-p) > 9$ are satisfied. But there can be observed also other anomalies in binomial probabilities given in wellknown textbooks.

Example 5 (from [1] p. 74/75)
$n = 10000$, $p = 0.005$, $il = iu = 40$
Moivre-Laplace approximation (given in [1]): 0.0206
"More precise approximation" (given in [1]): 0.0197
Computed inclusion: 2.143481170509E-02 $\leq w \leq$ 2.143481170561E-02.

Example 6 (from [4] p. 90 Tabelle 3a)
$n = 10000$, $p = 0.005$, $il = iu = 20$
Moivre-Laplace approximation (given in [4]): $0.66716 * 10^{-5}$
"Computed exact value" (given in [4]): $0.72444 * 10^{-6}$
Computed inclusion: 7.233764205778E-07 $\leq w \leq$ 7.233764206078E-07.

Example 7 (from [5] p. 59, Table 8)
$n = 100$, $p = 1/6$, $il = 14$, $iu = 20$
Moivre-Laplace approximation (given in [5]): 0.5
Computed inclusion: 6.481069042226E-01 $\leq w \leq$ 6.481069042308E-01.

Example 8 (from [6] p. 98)
$n = 8000$, $p = 0.024$, $il = 0$, $iu = 180$

Moivre-Laplace approximation (given in [6]): 0.2005
Computed inclusion: $2.013659474729\text{E-}01 \leq w \leq 2.013659475055\text{E-}01$.

Example 9 (due to U. Tesche)
$n = 1000$, $p = 0.1$, $il = 0$, $iu = 49$
Moivre-Laplace approximation: $> 5.09 * 10^{-8}$
Computed inclusion: $2.786710302119\text{E-}09 \leq w \leq 2.786710302593\text{E-}09$.

5. Concluding Remarks

Although a much more elegant version of the above program can be obtained using interval arithmetic and the powerful #expressions of PASCAL-XSC, [2], we preferred a version which can be easily checked also by a "normal programmer" and which can be translated without difficulties to any programming language having access to a floating point arithmetic supporting directed rounding for the algebraic operations $+$, $-$, $*$, $/$ and for the scalar product.

Finally it should be mentioned that cumulative hypergeometric probabilities can be included completely analogous to cumulative binomial probabilities. A suitable program is available. Theoretically it is even possible to include cumulative multinomial and poly-hypergeometric probabilities by an obvious iteration technique. In practice however a sufficient performance can be expected at most on parallel computers.

References

[1] Gnedenko, B. W.: Lehrbuch der Wahrscheinlichkeitsrechnung. Thun: Harri Deutsch 1980.
[2] Klatte, R., Kulisch, U., Neaga, M., Ratz, D. Ullrich, Ch.: PASCAL-XSC, Sprachbeschreibung mit Beispielen. Berlin: Springer 1991.
[3] Kulisch, U. (ed.): PASCAL-SC, a PASCAL extension for scientific computation, information manual and floppy disks, ATARI ST version. Stuttgart: Teubner 1987.
[4] Meschkowski, H.: Wahrscheinlichkeitsrechnung, Hochschultaschenbücher, Band 285/285a, Mannheim, Bibliographisches Institut, 1968.
[5] Savage, I. R.: Statistics: Uncertainty and behavior. Boston: Houghton Mifflin Company, 1968.
[6] Scheid, H.: Stochastik in der Kollegstufe, Lehrbücher und Monographien zur Didaktik der Mathematik, Band 6, Mannheim, Bibliographisches Institut 1986.
[7] Ulmer, F.: Der Lotteriecharakter des repräsentativen Querschnitts, Wahlprognosen und Meinungsumfragen und der Ablaßhandel mit Prozentzahlen. Zeitschrift für Markt-, Meinungs- und Zukunftsforschung, Heft 30/31, (1989).

Prof. Dr. G. Heindl
Fachbereich Mathematik
Universität Wuppertal
Gaußstrasse 20
D-W-5600 Wuppertal 1
Federal Republic of Germany

Computing, Suppl. 9, 93–100 (1993)

© Springer-Verlag 1993
Printed in Austria

Effective Evaluation of Hausdorff Distances for Finite Unions of Hyperrectangles*

K.-U. Jahn, Leipzig

Dedicated to Professor Dr. U. Kulisch on the occasion of his 60th birthday

Abstract — Zusammenfassung

Effective Evaluation of Hausdorff Distances for Finite Unions of Hyperrectangles. Based on the usual operations, relations and evaluations of function values for axis-parallel hyperrectangles, the corresponding objects for finite sets of hyperrectangles resp. their unions can be defined and implemented in a natural way. Algebraic properties remain valid. The aim of this paper is to give also methods for an effective evaluation of lower and upper distances as well as of Hausdorff distances for such sets. On this basis the distances and similarity and congruency grades for each pair of nonempty bounded subsets of $\mathbb{R}^n$ are approximately computable, and the convergence behaviour of general set-valued iteration processes can be investigated.

AMS Subject Classification: 65G10, 52A99

Key words: Hausdorff distances, similarity measures.

Effektive Berechnung von Hausdorff-Abständen für endliche Vereinigungen von Hyperrechtecken. Für endliche Mengen von achsenparallelen Hyperrechtecken bzw. deren Vereinigungen können Operationen, Relationen und Funktionswertberechnungen auf der Basis der für Hyperrechtecke erklärten eingeführt und implementiert werden. Algebraische Eigenschaften bleiben dabei weitgehend erhalten. In der vorliegenden Arbeit werden Methoden angegeben, auch untere und obere Abstände sowie insbesondere Hausdorff-Abstände für solche Mengen effektiv zu bestimmen. Dadurch können für beliebige Paare nichtleerer beschränkter Teilmengen von $\mathbb{R}^n$ deren Abstände sowie Ähnlichkeits- und Kongruenzgrade näherungsweise berechnet werden, und man kann Konvergenzuntersuchungen bei allgemeinen mengenwertigen Iterationsprozessen anstellen.

1. Introduction

Let $I(\mathbb{R}^n)$ be the set of all axis-parallel closed hyperrectangles in $\mathbb{R}^n$. Then, as shown in Jahn [5], for each metric $d\colon \mathbb{R}^n \times \mathbb{R}^n \to \mathbb{R}$ with property

$$\forall i(1 \leq i \leq n \to |x_i - y_i| \leq |u_i - v_i|) \to d(x, y) \leq d(u, v) \tag{1}$$

the Hausdorff distance

$$d^H(A, B) := \max \left\{ \sup_{a \in A} \inf_{b \in B} d(a, b), \sup_{b \in B} \inf_{a \in A} d(a, b) \right\} \tag{2}$$

---

* Received October 12, 1992; revised December 10, 1992.

of $A, B \in I(\mathbb{R}^n)$ can be computed in $O(n)$ time. The method is based on a partition of the set $\{1,\ldots,n\}$ into three parts in dependence of the relative positions of the boundary points $\underline{a}_i, \bar{a}_i, \underline{b}_i, \bar{b}_i$ of the i-th coordinates A_i and B_i of A and B, where $A = (A_1,\ldots,A_n)$, $A_i = [\underline{a}_i, \bar{a}_i]$.

Analogously, the upper distance

$$d_+(A, B) := \sup_{a \in A} \sup_{b \in B} d(a,b) \tag{3}$$

and the lower distance

$$d_{\#}(A, B) := \inf_{a \in A} \inf_{b \in B} d(a,b) \tag{4}$$

of A and B are computable with low costs.

For arbitrary nonempty bounded sets $A, B \subseteq \mathbb{R}^n$, it is not obvious how to compute $d^H(A, B)$, $d_+(A, B)$ and $d_{\#}(A, B)$, respectively. However, finite unions of intervals seem to be suitable sets for which the above-mentioned distances can be computed with tolerable costs. If suitable methods for computing their distances can be found, for more general bounded sets it would be possible to compute their distances approximately or to compute an enclosing interval for their distances by approximating the sets using finite unions of intervals.

2. Upper and Lower Distances of Finite Unions of Hyperrectangles

Let be

$$\mathbf{A} := \bigcup_{i=1}^{k} A^i, \qquad \mathbf{B} := \bigcup_{i=1}^{l} B^i,$$

where

$$A^i, B^j \in I(\mathbb{R}^n) \qquad \text{for } i \in \{1,\ldots,k\}, j \in \{1,\ldots,l\}.$$

Since

$$d_+(\mathbf{A}, \mathbf{B}) = \max_{1 \le i \le k} \sup_{a \in A^i} \max_{1 \le j \le l} \sup_{b \in B^j} d(a,b)$$

$$= \max_{i} \max_{j} \sup_{a \in A^i} \sup_{b \in B^j} d(a,b)$$

$$= \max_{i} \max_{j} d_+(A^i, B^j) \tag{5}$$

and

$$d_{\#}(\mathbf{A}, \mathbf{B}) = \min_{1 \le i \le k} \inf_{a \in A^i} \min_{1 \le j \le l} \inf_{b \in B^j} d(a,b)$$

$$= \min_{i} \min_{j} \inf_{a \in A^i} \inf_{b \in B^j} d(a,b)$$

$$= \min_{i} \min_{j} d_{\#}(A^i, B^j), \tag{6}$$

the computation of $d_+(\mathbf{A}, \mathbf{B})$ and $d_{\#}(\mathbf{A}, \mathbf{B})$ for metrices d with property (1) is easy.

So, if for arbitrary sets $X, Y \subseteq I(\mathbb{R}^n)$, finite unions $\mathbf{A}, \mathbf{B}$ as above and

$$\mathbf{V} := \bigcup_{i=1}^{p} V^i, \qquad \mathbf{W} := \bigcup_{i=1}^{q} W^i$$

are given with

$$\mathbf{A} \subseteq X \subseteq \mathbf{B}$$
$$\mathbf{V} \subseteq Y \subseteq \mathbf{W}, \tag{7}$$

we have

$$d_+(\mathbf{A}, \mathbf{V}) \leq d_+(X, Y) \leq d_+(\mathbf{B}, \mathbf{W})$$
$$d_\#(\mathbf{B}, \mathbf{W}) \leq d_\#(X, Y) \leq d_\#(\mathbf{A}, \mathbf{V}). \tag{8}$$

For sufficiently general sets X, Y, these bounds for $d_+(X, Y)$ resp. $d_\#(X, Y)$ can be made arbitrarily close:

Let be $d^H(\mathbf{A}, \mathbf{B}) = \alpha$; then for each $a \in \mathbf{A}$ there is a $b \in \mathbf{B}$ such that $d(a, b) \leq \alpha$ (since $\mathbf{A} \subseteq \mathbf{B}$, this is evident), and for each $b \in \mathbf{B}$ there is an $a \in \mathbf{A}$ with $d(a, b) \leq \alpha$. Assuming furthermore $d^H(\mathbf{V}, \mathbf{W}) = \beta$, there are $a \in \mathbf{A}$, $b \in \mathbf{B}$, $v \in \mathbf{V}$ and $w \in \mathbf{W}$ with

$$d_+(\mathbf{B}, \mathbf{W}) = d(b, w) \leq d(a, b) + d(a, w)$$
$$\leq \alpha + d(a, v) + d(v, w) \leq \alpha + \beta + d_+(\mathbf{A}, \mathbf{V}).$$

Therefore, in addition to (8) the inequality

$$0 \leq d_+(\mathbf{B}, \mathbf{W}) - d_+(\mathbf{A}, \mathbf{V}) \leq \alpha + \beta \tag{9}$$

holds. In a similar manner

$$0 \leq d_\#(\mathbf{A}, \mathbf{V}) - d_\#(\mathbf{B}, \mathbf{W}) \leq \alpha + \beta \tag{10}$$

can be proved.

How a nontrivial $A^i \in I(\mathbb{R}^n)$ with $A^i \subseteq X$ ($X \neq \varnothing$, X not a singleton) can be found?

Let be known a hyperrectangle Z with $X \subseteq Z$, and let be given mappings $f_1, \ldots,$ $f_r: Z \to \mathbb{R}$ and constants $\gamma_1, \ldots, \gamma_r$, such that

$$X := \{z \in Z \mid f_1(z) \leq \gamma_1 \wedge \cdots \wedge f_r(z) \leq \gamma_r\}. \tag{11}$$

Using a subdivision technique for Z, for each obtained interval $Z^j \subseteq Z$ the values of $F_1(Z^j), \ldots, F_r(Z^j)$ will be computed, where F_i is an interval extension of f_i (for terminology and realization cf. for instance Alefeld/Herzberger [1], Bauch et al. [2]). Then it holds:

(a) if $F_i(Z^j) \leq \gamma_i$ (i.e., the right-hand boundary point of $F_i(Z^j)$ is $\leq \gamma_i$) for all i, then $Z^j \subseteq X$
(b) if $F_i(Z^j) > \gamma_i$ (i.e., the left-hand boundary point of $F_i(Z^j)$ is $> \gamma_i$) for at least one i, then $Z^j \subseteq Z \setminus X$
(c) if none of the preconditions of (a) and (b) is fulfilled, then subdivide Z^j and consider its subintervals.

So, in dependence of the grade of subdivision, the set X ($X \neq \emptyset$) can be approximated by finite unions $\mathbf{A}, \mathbf{B}$ of intervals such that (7) holds and $d^H(\mathbf{A}, \mathbf{B}) \leq \alpha$. Thereby the difference $\mathbf{B} \setminus \mathbf{A}$ results from those Z^j which none of the preconditions of (a) and (b) fulfill.

Under weak assumptions which concern the mappings $f_1, \ldots, f_r$, the constant α can be chosen arbitrarily small.

3. Hausdorff Distances of Finite Unions of Hyperrectangles

The Hausdorff distance of $\mathbf{A}$ and $\mathbf{B}$ is not so easy computable as the upper resp. lower distances: Let us regard the part

$$\sup_{a \in A} \inf_{b \in B} d(a, b) \tag{12}$$

of (2). Then, without difficulties, it can be shown that

$$(12) = \max_{1 \leq i \leq k} \sup_{a \in A^i} \min_{1 \leq j \leq l} \inf_{b \in B^j} d(a, b)$$

$$\leq \max_{1 \leq i \leq k} \min_{1 \leq j \leq l} \sup_{a \in A^i} \inf_{b \in B^j} d(a, b). \tag{13}$$

The value of $\sup_{a \in A^i} \inf_{b \in B^j} d(a, b)$ is easily computable, as mentioned in section 1). So, if under (13) equality would hold, then the value of $d^H(\mathbf{A}, \mathbf{B})$ could be computed in $O(n*k*l)$ time. Unfortunately, this is not the case as shown by the following simple example:

Example 1
Let be $k = 1, l = 2, n = 2$ and
$A^1 := ([0, 1], [0, 7]), B^1 := ([1, 2], [0, 4]), B^2 := ([1, 2], [4, 7])$.
Then the value of (12) is 1 for $d = d_2$ and $d = d_\infty$, but

$$\sup_{a \in A^1} \inf_{b \in B^1} d_\infty(a, b) = 3, \qquad \sup_{a \in A^1} \inf_{b \in B^2} d_\infty(a, b) = 4,$$

so that 3 is the value of the right-hand term of (13). Taking d_2 instead of d_∞, it results

$$\sup_{a \in A^1} \inf_{b \in B^1} d_2(a, b) = \sqrt{3^2 + 1} = \sqrt{10},$$

$$\sup_{a \in A^1} \inf_{b \in B^2} d_2(a, b) = \sqrt{4^2 + 1} = \sqrt{17},$$

so that also in this case in (13) equality does not hold.

What can be done in order to get the correct value of (12)? It seems to be clear that further subdivisions of A could be helpful. Indeed, this is the case.

(a) Subdivision of $\mathbf{A}$ independently from $\mathbf{B}$
Let be

$$\gamma := \sup_{a \in A} \inf_{b \in B} d(a, b), \tag{14}$$

$$\delta_i := \sup_{a, b \in A^i} d(a, b) = d_+(A^i, A^i) \tag{15}$$

and i an arbitrary element of $\{1, \ldots, k\}$. Then there are $a^i \in A^i$ and $j(i) \in \{1, \ldots, l\}$ with $\inf_{b \in B^{j(i)}} d(a^i, b) \le \gamma$; for each $a \in A^i$ therefore the inequalities

$$\inf_{b \in B^{j(i)}} d(a, b) \le \inf_{b \in B^{j(i)}} (d(a, a^i) + d(a^i, b))$$

$$\le \inf_{b \in B^{j(i)}} (\delta_i + d(a^i, b)) \le \delta_i + \gamma \tag{16}$$

are valid; we have for each $i \in \{1, \ldots, k\}$

$$\sup_{a \in A^i} \inf_{b \in B^{j(i)}} d(a, b) \le \delta_i + \gamma \tag{17}$$

and

$$\min_j \sup_{a \in A^i} \inf_{b \in B^j} d(a, b) \le \delta_i + \gamma, \tag{18}$$

so that the condition

$$\max_i \min_j \sup_{a \in A^i} \inf_{b \in B^j} d(a, b) \le \max_i \delta_i + \gamma \tag{19}$$

holds. Putting

$$\tau := \max_i \min_j \sup_{a \in A^i} \inf_{b \in B^j} d(a, b) \tag{20}$$

and $\delta := \max_i \delta_i$, we have the following enclosure for γ:

$$\tau - \delta \le \gamma \le \tau. \tag{21}$$

It follows: The smaller the intervals A^i are, the better is the computed enclosure for γ. In order to be able to evaluate $\sup_{a \in A^i} \inf_{b \in B^j} d(a, b)$ with low costs, the metric d should have property (1).

(b) Subdivision of **A** in dependence of **B**
In example 1, if A^1 is subdivided into a new $A^1 := ([0, 1], [0, 4])$ and $A^2 := ([0, 1], [4, 7])$ (the new value of k is then 2), then under (13) equality holds for instance for $d = d_\infty$ and $d = d_2$. This can be validated by computation, but in case of d_∞ this also follows from the circumstance that for each $i \in \{1, \ldots, k\}$ there is a $j \in \{1, \ldots, l\}$, such that for each $a \in A^i$ there is a $b \in B^j$ with $|a_s - b_s| \le (12)$ for all coordinate directions s. A strategy for subdividing the A^i in order to get equality under (13) for $d = d_\infty$ can be derived: for each coordinate direction s and each A^i, if there are B^p and B^q with $\underline{a}_s^i < (\underline{b}_s^p + \overline{b}_s^q)/2 < \overline{a}_s^i$ or with $\underline{a}_s^i < (\underline{b}_s^q + \overline{b}_s^p)/2 < \overline{a}_s^i$, then subdivide A^i in

$$A^{i,1} := (A_1^i, \ldots, A_{s-1}^i, [\underline{a}_s^i, (\underline{b}_s^p + \overline{b}_s^q)/2], A_{s+1}^i, \ldots, A_n^i) \tag{22}$$

and in

$$A^{i,2} := (A_1^i, \ldots, A_{s-1}^i, [(\underline{b}_s^p + \overline{b}_s^q)/2, \overline{a}_s^i], A_{s+1}^i, \ldots, A_n^i) \tag{23}$$

and increment k; analogously in the second case. The process must be continued until no subdivisions in the above manner are possible. Then under (13) equality holds.

Example 2

Let be $n = 2$, $k = 1$, $l = 2$ and $A^1 := ([3,21],[2,10])$; $B^1 := ([0,7],[0,7])$, $B^2 := ([13,22],[3,8])$. Then

$$d_\infty^H(\mathbf{A}, \mathbf{B}) = \sup_{a \in \mathbf{A}} \inf_{b \in \mathbf{B}} d_\infty(a,b) = 3,$$

$$d_2^H(\mathbf{A}, \mathbf{B}) = \sup_{a \in \mathbf{A}} \inf_{b \in \mathbf{B}} d_2(a,b) = d_2((\xi, 10),(7,7)) = 3.958\ldots \text{ with } \xi := 115/12.$$

Subdividing A^1 in dependence of $\{B^1, B^2\}$, A^1 must be subdivided in x_1-direction at the values 10 and 11, and in x_2-direction at 4 and 5. On this way, A^1 will be replaced by 9 subintervals $A^1_{\text{new}}, A^2, \ldots, A^9$, and it is

$$\max_{1 \le i \le 9} \min_{1 \le j \le 2} \sup_{a \in A^i} \inf_{b \in B^j} d_\infty(a,b) = 3$$

(subdividing A^1 at $x_1 = 10$ and at $x_2 = 5$ would already be sufficient to get the same result).

However, with respect to the metric d_2, for the above $\mathbf{A}$, $\mathbf{B}$ it is

$$d_2^H(\mathbf{A}, \mathbf{B}) = \sup_{a \in \mathbf{A}} \inf_{b \in \mathbf{B}} d_2(a,b) = d_2((\xi, 10),(7,7)) = 3.958\ldots$$

($\xi = 115/12$), whereas

$$\max_{1 \le i \le 9} \min_{1 \le j \le 2} \sup_{a \in A^i} \inf_{b \in B^j} d_2(a,b) = \sqrt{18} = 4.242\ldots$$

The subdivision and point ξ are illustrated by the following picture:

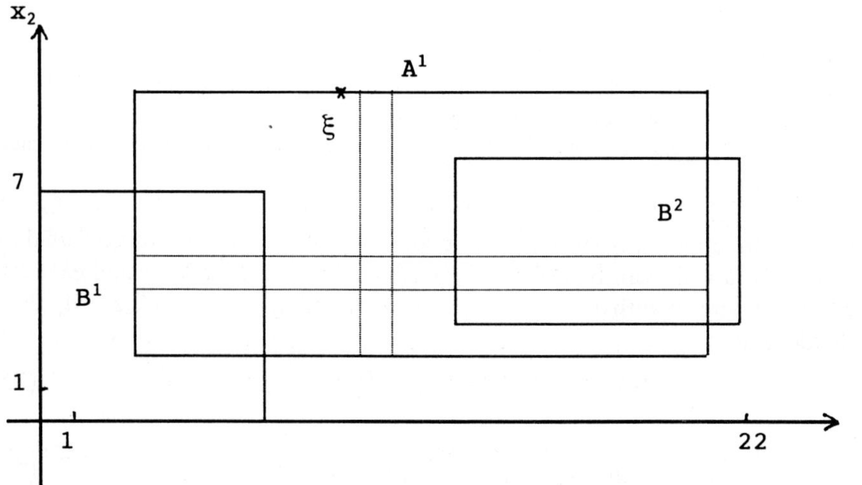

Therefore by the above-mentioned subdivision of **A** in dependence of **B** in general only the correct value of $\sup\limits_{a \in A} \inf\limits_{b \in B} d_\infty(a, b)$ can be computed. For other metrices than d_∞, it is not so easy to find a suitable subdivision.

4. Banach-like Fixed Point Theorems

Although d_+ is not a metric (however, symmetry and triangle inequality are valid) and the limit of a convergent sequence (with respect to d^H) of finite unions of hyperrectangles is in general not of the same kind, Banach-like fixed point theorems are available.

Let $\mathbf{M}^n$ be the set of all nonempty finite sets of hyperrectangles $X \in I(\mathbb{R}^n)$.

Theorem 1. *If $Q \subseteq \mathbb{R}^n$ is nonempty compact, $\mathbf{M}_Q := \{\mathbf{X} \in \mathbf{M}^n | \bigcup \mathbf{X} \subseteq Q\}$ and $f: \mathbf{M}_Q \to \mathbf{M}_Q$ is contractive with respect to d_+ and with contraction number q (where each $\mathbf{X} \in \mathbf{M}_Q$ can also be interpreted as $\bigcup \mathbf{X}$), then*

(a) *f has a uniquely determined fixed point $\mathbf{X}^* \in \mathbf{M}_Q$*
(b) *$\bigcup \mathbf{X}^*$ is a singleton*
(c) *for given $\mathbf{X}^{(0)} \in \mathbf{M}_Q$, if $\mathbf{X}^{(i+1)} := f(\mathbf{X}^{(i)})$,*

then
$$d_+(\mathbf{X}^{(i)}, \mathbf{X}^*) \le q^i/(1 - q)d_+(\mathbf{X}^{(0)}, \mathbf{X}^{(1)}). \qquad \blacksquare$$

This theorem is a generalization of an assertion in Jahn [3]. Now instead of intervals alone also finite sets of intervals resp. their unions are allowed in domain and range of f.

Therefore more general enclosing functions than interval functions are available for iteration processes, where the fixed points are real values or real-valued functions. For rational functions f, the computation of their values is described in Jahn [6].

Theorem 2. *Replace "d_+" in the preconditions of theorem 1 by "d^H". Then, for arbitrarily given $\mathbf{X}^{(0)} \in \mathbf{M}_Q$ and $\mathbf{X}^{(i+1)} := f(\mathbf{X}^{(i)})$, the following assertions hold:*

(a) *there is a uniquely determined nonempty compact set $X^* \subseteq Q$ such that the sequence $(\bigcup \mathbf{X}^{(i)})_{i < \infty}$ converges to X^**
(b) *$d^H(\bigcup \mathbf{X}^{(i)}, X^*) \le q^i/(1 - q)d^H(\bigcup \mathbf{X}^{(0)}, \bigcup \mathbf{X}^{(1)})$.* $\qquad \blacksquare$

In practical computation, the set X^* can be arbitrarily close approximated by the finite unions $\bigcup \mathbf{X}^{(i)}$ of intervals. In this manner, also compact-valued functions as fixed points of contractive mappings are computable. Such functions are regarded in Jahn [4].

If, additionally, $\bigcup f(\mathbf{X}^{(0)}) \subseteq \text{int} \bigcup \mathbf{X}^{(0)}$, then for inclusion monotone mappings f (which in practice often occur) a monotone decreasing sequence $(\mathbf{X}^{(i)})_{i < \infty}$ will be generated.

Unfortunately, unions of finite sets of intervals are in general not convex. Therefore Brouwer- and Schauder-like fixed point theorems as often used in interval mathematics (cf. for instance Kaucher/Miranker [7], Kulisch [8], Kulisch/Miranker [9]) are not available.

5. Similarity Measures

Hausdorff distances can be used for evaluating congruency and similarity measures.

Let X and Y be two bounded nonempty subsets of $\mathbb{R}^n$. Assume that their interval hulls $\square X$ and $\square Y$ are computable (for instance, by a subdividing technique as described in section 2). Translating these hulls together with their contents in such a way that the lower left-most corners are placed in the origin and then scaling the hulls (again with contents) with greatest factor without changing the interval hull of $\square X \cup \square Y$, the transformed X resp. Y will be renamed as X_{new} resp. Y_{new}. As a possible similarity measure $sim(X, Y)$ for X and Y then the number

$$1/(1 + d^H(X_{new}, Y_{new})) \tag{24}$$

can be used.

The proposed method works very well for sets X and Y representable as finite unions of (axis-parallel) hyperrectangles. If X and Y are themselves hyperrectangles, then $sim(X, Y) = 1$ iff the ratios of corresponding edge-lengths of X and Y all have the same amount.

References

[1] Alefeld, G., Herzberger J.: Introduction to interval computations. New York: Academic Press 1983.
[2] Bauch, H., Jahn, K.-U., Oelschlägel, D., Süsse, H., Wiebigke, V.: Intervallmathematik. Leipzig: Teubner 1987.
[3] Jahn, K.-U.: Punktkonvergenz in der Intervallrechnung. ZAMM 55, 606–608 (1975).
[4] Jahn, K.-U.: Enclosure functions for the solutions of initial value problems. Proceedings of the Second Polish Symposium on Interval & Fuzzy Mathematics, 71–75 (1987).
[5] Jahn, K.-U.: Evaluation of Hausdorff distances in interval mathematics. Computing 45, 69–77 (1990).
[6] Jahn, K.-U.: Finite sets of hyperrectangles and applications. to appear in Proceedings of SCAN'91.
[7] Kaucher, E., Miranker, W. L.: Self-validating numerics for function space problems. New York: Academic Press 1984.
[8] Kulisch, U. (ed.): Wissenschaftliches Rechnen mit Ergebnisverifikation. Berlin: Akademie-Verlag 1989.
[9] Kulisch, U., Miranker, W.L.: A new approach to scientific computation. New York: Academic Press 1983.

Prof. Dr. K.-U. Jahn
FB Mathematik-Informatik der TH Leipzig
Karl Liebknechtstrasse 132
D-O-7030 Leipzig
Federal Republic of Germany

Computing, Suppl. 9, 101–115 (1993)

© Springer-Verlag 1993
Printed in Austria

A Verified Computation of Fourier-Representations of Solutions for Functional Equations*

E. Kaucher and **C. Baumhof**, Karlsruhe

Dedicated to Professor U. Kulisch on the occasion of his 60th birthday

Abstract — Zusammenfassung

A Verified Computation of Fourier-Representations of Solutions for Functional Equations. A generalisation of the functoid concept for the representation of functions in $L^2(-\pi, \pi)$, in particular for periodic functions, is introduced. In such so-called *hyper functoids* the spectrum of a function can be included with arbitrary accuracy. This calculus can be used for example to solve periodical differential and integral equations. An example shows the computed inclusion of the spectrum of a solution.

AMS Subject Classification: 42A10, 42O4, 65G, 65L, 65R

Key words: Fourier-expansion, verified computation, series, functoids, functional equations.

Verifizierte Berechnung der Fourier-Entwicklung der Rösung von Funktionalgleichungen. Eine Verallgemeinerung des Funktoidkonzeptes für die Darstellung von $L^2(-\pi, \pi)$-Funktionen, insbesondere von periodischen Funktionen wird als sogenanntes *Hyperfunktoid* eingeführt, in welchem im Gegensatz zum üblichen Funktoid das Spektrum einer Funktion durch enge Fehlerschranken eingegrenzt werden kann. Dieses Kalkül wird eingesetzt zur Lösung z.B. von periodischen Differential-und Integralgleichungen und es wird an einem einfachen Beispiel die Spektraleinschließung berechnet.

1. Introduction

In applied mathematics there is often the need to solve functional equations like integral or differential equations. The discretisation of the operators of such problems leads to approximations for the solution function at discrete points, and there are mostly no informations about the errors and the behaviour of the function between these points.

One approach for the automatic control of such discretisation errors is the idea of self validating methods in context with *ultra-arithmetic*, which bases on series expansion techniques as an arithmetic methodology. Thus, computation in a function space can be done with the original problem. For this we need an appropriate basis for the function space, e.g. the monom basis ($t^k, k \in \mathbb{N}$), the Chebyshev basis ($T_k(t), k \in \mathbb{N}$), or the Fourier basis ($\sin kt, \cos kt, k \in \mathbb{N}$).

---

* Received October 15, 1992; revised December 21.

If the series are made finite, the resulting structure is called *functoid*. In an *interval functoid* the objects are sets of functions whose graph lies between two boundary functions almost everywhere (a.e.).

In some cases it is unsatisfactory to work with finite series. If infinite series are involved, the resulting structure is called *hyper functoid*. Since infinitely many coefficients cannot be stored individually, the coefficients have to be represented as a function of the coefficient index in a finite-dimensional way.

A motivation for using infinite series is the following example: given a Fourier functoid with base $\Phi_N := \{1, \sin t, \cos t, \ldots, \sin Nt, \cos Nt\}$, $N \in \mathbb{N}$ fixed, $t \in [-\pi, \pi]$. Then Φ_N is not closed with regard to integration, because

$$\int 1 \, dt = t \sim 2 \sum_{k=1}^{\infty} \frac{(-1)^{k+1}}{k} \sin kt$$

In the simplest case, where $\dfrac{(-1)^k}{k}$ is representable in the basis of the coefficient space, $\int 1 \, dt$ is exactly representable in the hyper functoid. The deeper problem is that integration in Fourier bases leads to saw tooth-like functions. In the example above there is a discontinuous function with jump 2π. Therefore, in the Fourier functoid, $\int 1 \, dt$ can only be enclosed by $[-\pi, \pi]$, while in the Fourier hyper functoid, $\int 1 \, dt$ can be exactly represented or enclosed with arbitrary quality.

As a second example, consider an interval hyper functoid with the monom basis. If for $f(x) = \sum_{i=0}^{\infty} a_i x^i$ can be verified $a_i \in A_i$ for all $i \geq 0$, then the enclosed function f is verified to be analytic if the A_i deliver a lower bound for the radius of convergence. This sort of verification is impossible in an interval functoid.

Here, a Fourier hyper functoid for the function space $L^2(-\pi, \pi)$ is discussed. A class of representations for the Fourier coefficients is considered, and formulas for the functional operations $+$, $-$, $\cdot$ and $\int$ are given. Finally the calculus is applied for solving an initial value problem.

2. Arithmetic Concepts

In this section the concept of *hyper functoids* is introduced. A description of the underlying computer arithmetic with a floating point number system S can be found in [6], a description of the *functoid* concept in [3] and [4].

Let $\mathcal{M}$ be a separable and arithmetical Hilbert space with the basis $B = \{\varphi_1, \varphi_2, \ldots\}$ and the operations $\Omega := \{+, -, \cdot, /, \int\}$. A function $f \in \mathcal{M}$ can then be represented as

$$f = \sum_{i=1}^{\infty} a_i \varphi_i \tag{2.1}$$

In functoids the series $\{a_i\}$ representing f is cut off at an $i = N$. Now we consider a different approach.

3.3 Two Ansatzes for the Coefficient Space

According to (2.3) we have for a Fourier hyper functoid representation

$$S\mathcal{M} = \left\{ f(t) = \sum_{k=-\infty}^{\infty} c(k)e^{ikt} : c(k) \in S_N R\mathcal{M} \right\} \tag{3.6}$$

A possible idea for representing the infinitely many coefficients $c(k)$ is as follows: We assume that the coefficients $c(k)$ are a bounded function on $\mathbb{Z}$. Since necessarily $|c(k)| \to 0$ for $|k| \to \infty$ we take as an obvious choice (and for reasons of easy description) the monomials $\psi_p(k) = \dfrac{(-1)^k}{k^p}$. This leads to the first ansatz

$$c(k) = \begin{cases} c_0 & : k = 0 \\ \\ (-1)^k \displaystyle\sum_{p=1}^{M} a_p \frac{1}{k^p} & : k \neq 0 \end{cases} \tag{3.7}$$

where $M \geq 1$ fixed, and $c_0 \in \mathbb{R}$. From $c(-k) = \overline{c(k)}$ follows the condition that a_p is real for even p and imaginary for odd p. All operations from $\Omega = \{+, -, \cdot, \int\}$ are finitely representable using this ansatz. This ansatz is further dealt with in [1].

The example of the Fourier coefficients for the exponential function which can be written as

$$c(k) = \frac{1}{\pi}\sinh \pi(-1)^k\left(\frac{i}{k} + \frac{1}{k^2} - \frac{i}{k^3} - \frac{1}{k^4} + \frac{i}{k^5} + \frac{1}{k^6} - \cdots \right) \quad \text{for } |k| \neq 0, 1$$

shows that there exist a finite number of exception values. Thus, an improved ansatz has to take into account a fixed number $R > 0$ of exception values for the $c(k)$:

$$c(k) = \begin{cases} c_k & : 0 \leq k \leq R \\ \\ \overline{c_{-k}} & : -R \leq k \leq -1 \\ \\ (-1)^k \displaystyle\sum_{p=1}^{M} a_p \frac{1}{k^p} & : |k| > R \end{cases} \tag{3.8}$$

with $c_0 \in \mathbb{R}$, $c_k \in \mathbb{C}$ $(k \geq 1)$, $a_p \in \mathbb{R}$ for $2 \mid p$, $ia_p \in \mathbb{R}$ for $2 \nmid p$.

For $R = 0$ this is the same as (3.7). Unfortunately, for $R > 0$ the multiplication is no longer finitely representable.

4. Evaluation of the Series $\Phi_{p,q,R}(k)$

For the representation of the multiplication with ansatz (3.8) the series

$$\Phi_{p,q,R}(k) := \sum_{\substack{j=-\infty \\ |j| > R \\ |k-j| > R}}^{\infty} \frac{1}{(k-j)^p}\frac{1}{j^q}, \qquad p, q, R \in \mathbb{N}, p, q \geq 1, R \geq 0, k \in \mathbb{Z} \tag{4.1}$$

must be expanded in powers of $\dfrac{1}{k}$, which is done in two steps in sections 4.1 and 4.2. For this the following property is used (cf. [1]):

$$\sum_{\substack{j=-\infty \\ |j|>R \\ |k-j|>R}}^{\infty} \frac{1}{(k-j)^m} \overset{k-j\to j}{=} \sum_{\substack{j=-\infty \\ |j|>R \\ |k-j|>R}}^{\infty} \frac{1}{j^m} = \sum_{\substack{j=-\infty \\ |j|>R}}^{\infty} \frac{1}{j^m} - \sum_{j=\max(R+1,k-R)}^{k+R} \frac{1}{j^m}$$

$$= \left\{\begin{array}{ll} 0 & : 2\nmid m \\[2mm] 2b_1\left(\dfrac{m}{2},R\right) & : 2\mid m \end{array}\right\} - \sum_{j=\max(R+1,k-R)}^{k+R} \frac{1}{j^m} \qquad (4.2)$$

where (cf. [5, pp. 245, 247])

$$b_1(s,R) := \sum_{j=R+1}^{\infty} \frac{1}{j^{2s}} = \frac{(-1)^{s-1}(2\pi)^{2s}B_{2s}}{2(2s)!} - \sum_{j=1}^{R} \frac{1}{j^{2s}} \qquad (4.3)$$

and for later use

$$b_2(s,R) := \sum_{j=R+1}^{\infty} \frac{(-1)^j}{j^{2s}} = \frac{(-1)^s(2^{2s-1}-1)\pi^{2s}B_{2s}}{(2s)!} - \sum_{j=1}^{R} \frac{(-1)^j}{j^{2s}} \qquad (4.4)$$

and B_k are the *Bernoulli numbers*, defined by $\dfrac{x}{e^x-1} = \sum_{k=0}^{\infty} \dfrac{B_k}{k!} x^k$

4.1　Partial Fraction Decomposition of $\dfrac{1}{(k-j)^p}\dfrac{1}{j^q}$

In the first step a partial fraction decomposition of $\dfrac{1}{(k-j)^p}\dfrac{1}{j^q}$ is needed for the expansion of $\Phi_{p,q,R}(k)$. This decomposition results as follows:

$$\frac{1}{(k-j)^p}\frac{1}{j^q} = \sum_{r=1}^{p} \frac{\alpha_{p-r}}{k^{p+q-r}}\frac{1}{(k-j)^r} + \sum_{s=1}^{q} \frac{\beta_{q-s}}{k^{p+q-s}}\frac{1}{j^s} \qquad (4.5)$$

leads to a linear system of equations for $\alpha_0, \ldots, \alpha_{p-1}, \beta_0, \ldots, \beta_{q-1}$:

$$\beta_0 = 1 \qquad \text{(equation 0)}$$

$$(-1)^t\binom{r}{t}\alpha_r = 0 \qquad (0 \le r \le p-1, 0 \le t \le r)$$

$$(-1)^u\binom{p}{u}\beta_s = 0 \qquad (0 \le s \le q-1, 0 \le u \le p, (u,s) \ne (0,0))$$

where the mth equation $(1 \le m \le p+q-1)$ is the summation of all terms with $m = t+q = \text{const}$ and $m = u+s = \text{const}$.

For $p \le q$ the solution is given by

$$\beta_0 = 1$$

$$\beta_i = \sum_{k=\max(i-p,0)}^{i-1} (-1)^{i+k+1} \binom{p}{i-k} \beta_k, \qquad i = 1 \ldots q-1$$

$$\alpha_{p-1} = \beta_{q-1} \tag{4.6}$$

$$\alpha_{p-i-1} = \sum_{k=0}^{i} (-1)^k \binom{p}{p-i+k} \beta_{q-k-1}$$

$$- \sum_{k=0}^{i-1} \binom{p-k-1}{p-i-1} \alpha_{p-k-1}, \qquad i = 1 \ldots p-1$$

Note that the solution is integer and therefore can principally be calculated exactly. As the system of equations does not depend on k, the roles of $k-j$ and j in (4.5) can be interchanged in the case $p > q$, thus the same solution results as for $p \le q$ with α and β interchanged.

$$4.2 \ \textit{Expansion of } \Phi_{p,q,R}(k)$$

Now in the second step $\Phi_{p,q,R}(k)$ can be expanded in powers of $\dfrac{1}{k}$. The resulting expansion will only be valid for $|k| \ge 2R + 1$, for $|k| \le 2R$ the series itself must be evaluated. It is sufficient to consider positive k, as (3.8) already guarantees the correct values for negative k.

$k = 0$:

$$\Phi_{p,q,R}(0) = \sum_{\substack{j=-\infty \\ |j|>R}}^{\infty} \frac{1}{(-j)^p} \frac{1}{j^q} = (-1)^p \sum_{\substack{j=-\infty \\ |j|>R}}^{\infty} \frac{1}{j^{p+q}}$$

$$= \begin{cases} 0 & : p+q \text{ odd} \\ (-1)^p b_1 \left(\dfrac{p+q}{2}, R \right) & : p+q \text{ even} \end{cases} \tag{4.7}$$

$k \ge 1$:

$$\Phi_{p,q,R}(k) = \sum_{\substack{j=-\infty \\ |j|>R \\ |k-j|>R}}^{\infty} \frac{1}{(k-j)^p} \frac{1}{j^q}$$

$$\overset{\substack{(4.5) \\ (4.2)}}{=} \sum_{r=1}^{p} \frac{\alpha_{p-r}}{k^{p+q-r}} \left[- \sum_{j=\max(R+1,k-R)}^{k+R} \frac{1}{j^r} + \begin{cases} 0 & : 2 \nmid r \\ 2b_1\left(\dfrac{r}{2}, R\right) & : 2 | r \end{cases} \right]$$

$$+ \sum_{s=1}^{q} \frac{\beta_{q-s}}{k^{p+q-s}} \left[- \sum_{j=\max(R+1,k-R)}^{k+R} \frac{1}{j^s} + \begin{cases} 0 & : 2 \nmid s \\ 2b_1\left(\dfrac{s}{2}, R\right) & : 2 | s \end{cases} \right]$$

$$=: f_1^{(R)}(p, q, k) \tag{4.8}$$

The $f_1^{(R)}(p, q, k)$ are needed for $1 \le p, q \le M, 1 \le k \le 2R$. After choosing M and R they are once computed and then stored.

$k \ge 2R + 1$: here an expansion in powers of $\dfrac{1}{k}$ can be given (derived in [1]):

$$
\Phi_{p,q,R}(k) = 2 \sum_{\substack{r=1 \\ 2|r}}^{p} \frac{\alpha_{p-r}}{k^{p+q-r}} b_1\left(\frac{r}{2}, R\right) + 2 \sum_{\substack{s=1 \\ 2|s}}^{q} \frac{\beta_{q-s}}{k^{p+q-s}} b_1\left(\frac{s}{2}, R\right)
$$

$$
- (2R + 1)\left(\sum_{r=1}^{p} \alpha_{p-r} + \sum_{s=1}^{q} \beta_{q-s} \right)\frac{1}{k^{p+q}}
$$

$$
- 2 \sum_{l=1}^{\infty} \left(\left(\sum_{r=1}^{p} \alpha_{p-r} \binom{r-1+2l}{r-1} \right) + \sum_{s=1}^{q} \beta_{q-s} \binom{s-1+2l}{s-1} \right)
$$

$$
\times \left(\sum_{t=1}^{R} t^{2l} \right) \frac{1}{k^{p+q+2l}} \tag{4.9}
$$

From this representation the coefficients $f_2^{(R)}(p, q, r)$ of $\dfrac{1}{k^r}$ can be obtained and stored. Remark: An estimation for the infinite part of the series is still missing, but is principally possible. It requires some more intensive considerations which will be published in a separate paper.

5. Formulation of a Hyper Functoid

We proceed to describe the Fourier hyper functoid resulting from ansatz (3.8) with a rounding specified below.

5.1 The Hyper Functoid Operations

The operations $\Omega := \{+, -, \cdot, \int\}$ are expressed according to (3.8) using only the ansatz coefficients. Let $f, g, h \in S\mathcal{M}$ be represented by their Fourier coefficients $c(k)$, $d(k)$, $b(k)$ with the ansatz coefficients a_p^f, a_p^g, a_p^h and c_k^f, c_k^g, c_k^h, resp.

<u>Addition:</u> using (3.2) follows

$$
b(k) = c(k) \pm d(k)
$$

$$
\Rightarrow \quad c_k^h = c_k^f \pm c_k^g, \qquad 0 \le k \le R
$$

$$
a_p^h = a_p^f \pm a_p^g, \qquad 1 \le p \le M
$$

<u>Integration:</u> using (3.4) follows

$$
b(k) = -i\frac{c(k)}{k} + (-1)^k \frac{ic(0)}{k}
$$

$$
\Rightarrow \quad c_k^h = -i\frac{c_k^f}{k} + (-1)^k \frac{ic_0^f}{k}, \qquad 1 \le k \le R
$$

$$a_p^h = -ia_{p-1}^f, \qquad 2 \leq p \leq M + 1$$

$$a_1^h = ic_0^f$$

$$b(0) = c_0^h = -2 \sum_{k=1}^{\infty} \frac{\operatorname{Im} c(k)}{k} = -2 \sum_{k=1}^{R} \frac{\operatorname{Im} c_k^f}{k} - 2 \sum_{\substack{k=R+1}}^{\infty} (-1)^k \frac{1}{k} \sum_{\substack{p=1 \\ 2\mid p}}^{M} \operatorname{Im} a_p^f \frac{1}{k^p}$$

$$\overset{(4.4)}{=} -2 \left(\sum_{k=1}^{R} \frac{\operatorname{Im} c_k^f}{k} + \sum_{\substack{p=1 \\ p+1=2s \,\text{even}}}^{M} \operatorname{Im} a_p^f b_2(s, R) \right)$$

Here a_{M+1}^h occurs, so a rounding is necessary.

<u>Multiplication</u>: from (3.3) follows $b(k) = \sum_{j=-\infty}^{\infty} c(k-j)d(j)$. Four cases must be distinguished depending on how far the intervals $|k - j| \leq R$ and $|j| \leq R$ intersect. A detailed derivation of the following formulas is given in [1].

1. $k = 0$

$$c_0^h = \sum_{j=-\infty}^{\infty} c(-j)d(j)$$

$$= 2 \sum_{p=1}^{M} \sum_{\substack{q=1 \\ 2\mid p+q}}^{M} (-1)^p a_p^f a_q^g b_1 \left(\frac{p+q}{2}, R \right) + 2 \sum_{j=1}^{R} \operatorname{Re}(\overline{c_j^f} c_j^g) + c_0^f c_0^g$$

2. $1 \leq k \leq R$

$$c_k^h = \sum_{j=-\infty}^{\infty} c(k-j)d(j)$$

$$= \sum_{p=1}^{M} \sum_{q=1}^{M} (-1)^k a_p^f a_q^g f_1^{(R)}(p, q, k)$$

$$+ \sum_{p=1}^{M} a_p^f \sum_{j=R+1}^{k+R} \frac{(-1)^j}{j^p} \overline{c_{j-k}^g} + \sum_{q=1}^{M} a_q^g \sum_{j=R+1}^{k+R} \frac{(-1)^j}{j^q} \overline{c_{j-k}^f}$$

$$+ \sum_{j=k-R}^{-1} c_{k-j}^f \overline{c_{-j}^g} + \sum_{j=0}^{k} c_{k-j}^f c_j^g + \sum_{j=k+1}^{R} \overline{c_{j-k}^f} c_j^g$$

3. $R + 1 \leq k \leq 2R$

As the expansion of $\Phi_{p,q,R}(k)$ from section 4.2 is valid only for $|k| \geq 2R + 1$, the c_k^h must be computed for $R + 1 \leq k \leq 2R$ as well.

$$c_k^h = \sum_{j=-\infty}^{\infty} c(k-j)d(j)$$

$$= \sum_{p=1}^{M} \sum_{q=1}^{M} (-1)^k a_p^f a_q^g f_1^{(R)}(p, q, k) + \sum_{j=k-R}^{R} c_{k-j}^f c_j^g$$

$$+ (-1)^k \sum_{p=1}^{M} a_p^f \left(\sum_{j=1}^{R} (-1)^j \overline{c_j^g} \frac{1}{(k+j)^p} + \sum_{j=0}^{k-R-1} (-1)^j c_j^g \frac{1}{(k-j)^p} \right)$$

$$+ (-1)^k \sum_{q=1}^{M} a_q^g \left(\sum_{j=1}^{R} (-1)^j \overline{c_j^f} \frac{1}{(k+j)^q} + \sum_{j=0}^{k-R-1} (-1)^j c_j^f \frac{1}{(k-j)^q} \right)$$

4. $2R + 1 \leq k$

Here the expansion of $\Phi_{p,q,R}(k)$ in powers of $\dfrac{1}{k}$ from (4.9) is valid.

$$
\sum_{j=-\infty}^{\infty} c(k-j)d(j) = \sum_{r=1}^{\infty} (-1)^k \left(\sum_{p=1}^{M} \sum_{q=1}^{M} a_p^f a_q^g f_2^{(R)}(p,q,r) \right) \frac{1}{k^r}
$$

$$
+ \sum_{p=1}^{M} (-1)^k \left(a_p^f \left(c_0^g + 2 \sum_{j=1}^{R} (-1)^j \operatorname{Re} c_j^g \right) \right) \frac{1}{k^p}
$$

$$
+ \sum_{q=1}^{M} (-1)^k \left(a_q^g \left(c_0^f + 2 \sum_{j=1}^{R} (-1)^j \operatorname{Re} c_j^f \right) \right) \frac{1}{k^q}
$$

$$
+ \sum_{t=2}^{\infty} (-1)^k \left(\sum_{p=1}^{\min(t-1,M)} a_p^f \binom{t-1}{p-1} \sum_{j=1}^{R} (-1)^j \right.
$$

$$
\times \, (c_j^g + (-1)^{t-p}\overline{c_j^g}) j^{t-p} + \sum_{q=1}^{\min(t-1,M)} a_q^g \binom{t-1}{q-1}
$$

$$
\times \, \sum_{j=1}^{R} (-1)^j (c_j^f + (-1)^{t-q}\overline{c_j^f}) j^{t-q} \bigg) \frac{1}{k^t}
\tag{5.1}
$$

The result is an (infinite) expansion in powers of $\dfrac{1}{k}$ where the a_r^h can be obtained by adding the coefficients of $\dfrac{1}{k^r}$ from the four sums.

So $h = f \cdot g$ is represented by

$$
b(k) = \begin{cases}
c_k^h & : 0 \leq k \leq 2R \\[2mm]
\overline{c_{-k}^h} & : -2R \leq k \leq -1 \\[2mm]
\displaystyle\sum_{p=1}^{\infty} a_p^h \frac{1}{k^p} & : |k| > 2R
\end{cases}
$$

Not only the sum over $a_p^h \dfrac{1}{k^p}$ must be shortened to length M, but also the shortened sum must be valid for $|k| > R$. These problems are discussed in section 5.3.

Division:
The division is performed according to section 3.2. The stopping criterion for the iteration is $\|1 - u_i g\| < \varepsilon$ with

$$
\|f\| := \max \left\{ |c_0|, |c_1|, \ldots, |c_R|, \frac{|a_1|}{R}, \frac{|a_2|}{R^2}, \ldots, \frac{|a_M|}{R^M} \right\}
\tag{5.2}
$$

5.2 Function Evaluation

The function evaluation is done with the following formula derived in [1]:

$$
f(t) = \sum_{k=-\infty}^{\infty} c(k)e^{ikt} = c_0 + \sum_{p=1}^{M} b_p B_p \left(\frac{t+\pi}{2\pi} \right)
$$

$$+ 2 \sum_{k=1}^{R} \left(\operatorname{Re} c_k - (-1)^k \sum_{\substack{p=1 \\ 2|p}} \operatorname{Re} a_p \frac{1}{k^p} \right) \cos kt$$

$$- 2 \sum_{k=1}^{R} \left(\operatorname{Im} c_k - (-1)^k \sum_{\substack{p=1 \\ 2\nmid p}} \operatorname{Im} a_p \frac{1}{k^p} \right) \sin kt \qquad (5.3)$$

where the $B_p(x)$ are the *Bernoulli polynomials defined by* $\dfrac{te^{xt}}{e^t - 1} = \sum_{n=0}^{\infty} B_n(x) \dfrac{t^n}{n!}$, and

the b_p can be computed from the a_p.

5.3 Rounding

The result of an integration or a multiplication is in general a function which is not in $S\mathcal{M}$. The rounding must map this result back to $S\mathcal{M}$. Thus the following problem must be solved:

Let be given a function f with $N \geq M$ and $S \geq R$ and

$$c^f(k) = \begin{cases} c_k^f & : 0 \leq k \leq S \\[2ex] (-1)^k \sum_{p=1}^{N} a_p^f \frac{1}{k^p} & : |k| > S \end{cases}$$

Then we look for a function g with

$$c^g(k) = \begin{cases} c_k^g & : 0 \leq k \leq R \\[2ex] (-1)^k \sum_{p=1}^{M} a_p^g \frac{1}{k^p} & : |k| > R \end{cases}$$

so that g approximates f as good as possible.

Integration delivers an f with $S = R$. Here the rounding cuts off the sum, that is $c_k^g := c_k^f$ for $0 \leq k \leq R$, $a_p^g := a_p^f$ for $1 \leq p \leq M$. The error is

$$\left| \sum_{p=M+1}^{N} a_p^f \frac{1}{k^p} \right| \leq \sum_{p=M+1}^{N} |a_p^f| \frac{1}{R^p} \qquad (|k| > R)$$

$$\leq \frac{1}{R^{M+1}} \max_{p=M+1}^{N} |a_p^f| \sum_{p=0}^{N-M-1} \frac{1}{R^p}$$

$$\leq \frac{1}{R^{M+1}} \max_{p=M+1}^{N} |a_p^f| \frac{1}{1 - \dfrac{1}{R}} \qquad (\text{if } R > 1)$$

$$= \frac{1}{R^M (R - 1)} \max_{p=M+1}^{N} |a_p^f|$$

which decreases when R and M are increasing.

Multiplication gives an f with $S = 2R$. Here the sum in the ansatz of g must approximate the sum in the ansatz of f for $|k| > S$ and the c_k^f for $R + 1 \le k \le S$. So a_p^g should be chosen so that

$$(-1)^k \sum_{p=1}^{M} a_p^g \frac{1}{k^p} \approx (-1)^k \sum_{p=1}^{N} a_p^f \frac{1}{k^p} \qquad \text{for } S + 1 \le k$$

$$(-1)^k \sum_{p=1}^{M} a_p^g \frac{1}{k^p} \approx c_k^f \qquad \text{for } R + 1 \le k \le S$$

holds. Additionally, $a_p^g \in \mathbb{R}$ for even p and $ia_p^g \in \mathbb{R}$ for odd p must be fulfilled. The following method has been chosen:

Determine $a_p'^g \in \mathbb{R}$, $1 \le p \le M$ to be the least-squares-solution of the following two overdetermined linear systems for the even and the odd $a_p'^g$

$$(-1)^k \sum_{\substack{p=1 \\ 2|p}}^{M} a_p'^g \frac{1}{k^p} = \operatorname{Re} \sum_{p=1}^{M} a_p^f \frac{(-1)^k}{k^p} \qquad \text{for } S + 1 \le k \le S + M$$

$$(-1)^k \sum_{\substack{p=1 \\ 2|p}}^{M} a_p'^g \frac{1}{k^p} = \operatorname{Re} c_k^f \qquad \text{for } R + 1 \le k \le S$$

$$(-1)^k \sum_{\substack{p=1 \\ 2 \nmid p}}^{M} a_p'^g \frac{1}{k^p} = \operatorname{Im} \sum_{p=1}^{M} a_p^f \frac{(-1)^k}{k^p} \qquad \text{for } S + 1 \le k \le S + M$$

$$(-1)^k \sum_{\substack{p=1 \\ 2 \nmid p}}^{M} a_p'^g \frac{1}{k^p} = \operatorname{Im} c_k^f \qquad \text{for } R + 1 \le k \le S$$

and set

$$c_k^g := c_k^f \qquad \text{for } 1 \le k \le R$$

$$a_p^g := \begin{cases} i: 2 \nmid p \\ 1: 2 \nmid p \end{cases} \cdot a_p'^g \qquad \text{for } 1 \le p \le M$$

5.4 Formulations for the Interval Hyper Functoid

For the interval hyper functoid all operations occuring in the previous formulas are to be replaced by rounded interval operations. A more detailed analysis is necessary in the formulation of the directed roundings.

In case of integration the error is as follows:

$$|(f - g)(t)| = \left| \sum_{k=-\infty}^{\infty} c^f(k)e^{ikt} - c^g(k)e^{ikt} \right| \le 2 \sum_{p=M+1}^{N} |a_p^f| \sum_{k=R+1}^{\infty} \frac{1}{k^p}$$

$\displaystyle\sum_{k=R+1}^{\infty} \frac{1}{k^p}$ is known for even p, and with $k^{2n+1} \ge k^{2n}$ ($k \ge R + 1$) is

$$\sum_{k=R+1}^{\infty} \frac{1}{k^{2n+1}} \le \sum_{k=R+1}^{\infty} \frac{1}{k^{2n}} = b_1(n, R)$$

So

$$|(f - g)(t)| \in \left(2 \sum_{p=M+1}^{N} |a_p^f| b_1\left(\left\lfloor \frac{p}{2} \right\rfloor, R\right)\right) \cdot [-1, 1]$$

This term is added to the constant term and inclusion for the integration is obtained.

In case of multiplication things get more difficult. If in the rounding $a_1^g := a_1^f$ is set, the error is

$$|(f - g)(t)| = \left| 2 \operatorname{Re} \sum_{k=R+1}^{\infty} (c^f(k) - c^g(k))e^{ikt} \right|$$

$$\le 2 \sum_{k=R+1}^{S} \left| c_k^f - (-1)^k \sum_{p=1}^{M} a_p^g \frac{1}{k^p} \right| + 2 \sum_{p=2}^{M} |a_p^f - a_p^g| b_1\left(\left\lfloor \frac{p}{2} \right\rfloor, R\right)$$

$$+ 2 \sum_{p=M+1}^{L} |a_p^f| b_1\left(\left\lfloor \frac{p}{2} \right\rfloor, R\right) + 2 \sum_{k=S+1}^{\infty} \left| \sum_{r=L+1}^{\infty} a_r^f \frac{1}{k^r} \right| \tag{5.4}$$

with a $L \ge 2R$.

For $r \ge L + 1 \ge 2R + 1$ follows for $h = f \cdot g$ using (4.9) and (5.1) [$r = p + q + 2l$ in (4.9) and $r = t$ in (5.1)]:

$$a_r^h = -2 \sum_{\substack{p=1 \\ r-p-q \text{ even}}}^{M} \sum_{q=1}^{M} a_p^f a_q^g \left(\sum_{s=1}^{p} \alpha_{p-s} \binom{s-1+r-p-q}{s-1} \right.$$

$$\left. + \sum_{s=1}^{q} \beta_{q-s} \binom{s-1+r-p-q}{s-1} \right) \sum_{t=1}^{R} t^{r-p-q}$$

$$+ 2 \sum_{\substack{p=1 \\ 2|r-p}}^{M} a_p^f \binom{r-1}{p-1} \sum_{j=1}^{R} (-1)^j \operatorname{Re} c_j^g j^{r-p}$$

$$+ 2 \sum_{\substack{p=1 \\ 2\nmid r-p}}^{M} a_p^f \binom{r-1}{p-1} \sum_{j=1}^{R} (-1)^j \operatorname{Im} c_j^g j^{r-p}$$

$$+ 2 \sum_{\substack{q=1 \\ 2|r-q}}^{M} a_q^g \binom{r-1}{q-1} \sum_{j=1}^{R} (-1)^j \operatorname{Re} c_j^f j^{r-q}$$

$$+ 2 \sum_{\substack{q=1 \\ 2\nmid r-q}}^{M} a_q^g \binom{r-1}{q-1} \sum_{j=1}^{R} (-1)^j \operatorname{Im} c_j^f j^{r-q}$$

This has to be substituted for a_r^f in (5.4). However, due to the remark to (4.9) an upper bound for the infinite series is still missing. A verified multiplication is principally possible, but it requires some more intensive considerations which are to be published in a separate paper.

6. Application

A Fourier hyper functoid can be used to compute the periodical solution of functional, differential and integral equations. Furthermore, it can be used to compute the periodical continuation of a function with arbitrary period T which may be defined by functional, differential or integral equations in a period interval T.

As an example we compute the inclusion of a periodically continued solution with period 2π of the initial value problem

$$u''(t) = \frac{1}{\pi^2}(u(t) + \cosh 1), \qquad u(0) = 1 - \cosh 1,$$

$$u'(0) = 0, \qquad -\pi \leq t \leq \pi$$

The equivalent integral equation is

$$u(t) = \int_0^t \int_0^t \frac{1}{\pi^2}(u(t) + \cosh 1)\, dt\, dt + 1 - \cosh 1 \tag{6.1}$$

with the solution

$$u(t) = \cosh\left(\frac{t}{\pi}\right) - \cosh 1, \qquad -\pi \leq t \leq \pi$$

An approximation is calculated by direct iteration of (6.1) with the starting value $u_0 := 0$ and the stopping criterion $\|u_{i+1} - u_i\| < 10^{-15}$, $\|f\|$ according to (5.2). Then we take the starting function $U_0 := u_{i+1} \cdot [1 - \varepsilon, 1 + \varepsilon]$, $\varepsilon = 10^{-8}$ expecting that the exact solution is contained in the resulting function set. Then the interval iteration

$$U_{i+1} := \int_0^t \int_0^t \frac{1}{[\pi]^2}(U_i + [\cosh 1])\, dt\, dt + 1 - [\cosh 1]$$

is performed with enclosing intervals $[\pi]$ for π and $[\cosh 1]$ for $\cosh 1$ until $U_{i+1} \subseteq U_i$ holds. (For L^2-problems one has to take weaker conditions such as $\int U_{i+1}\phi\, dt \subseteq \int U_i\phi\, dt$ for all test functions ϕ of a test space H). The Schauder fixed-point theorem then guarantees the existence of the exact solution of (6.1) in U_{i+1}.

The computation has been done with $M = R = 10$. After 10 iteration steps the approximation has ~ 13 correct digits. The inclusion criterion is fulfilled after 5 interval steps, and after 15 more interval steps the improved inclusion delivers the following results (real intervals are characterized by small digits in the last positions where the lower digits are for the left bound and the upper digits for the right bound):

$$c_0 = -0.367879441171144^{173}_{276}D + 00$$

$$c_1 = (-0.108118120060194^{73}_{95}D + 00, 0.0000000000D + 00)$$

$$\vdots$$

$$c_{10} = (0.118952252128052^{91}_{68}D - 02, 0.0000000000D + 00)$$

$$a_2 = 0.119072775957878^{26}_{06}D + 00$$

$$\vdots$$

$$a_{10} = 0.125491264934469^{525}_{439}D - 04$$

($a_p = 0$ for odd p). The spectral values $c(k)$ can then be calculated from these numbers using (3.8).

A function evaluation at $t = 0$ gives the exact solution $u = -0.54308063\ldots$ and the computed inclusion $f = -0.5430806348152^{3968}_{4781}$. A more accurate inclusion is, even with higher M or R, unlikely to be obtained because of the given mantissa length of the arithmetic.

In [2] the same problem has been formulated as a boundary value problem and the solution has been enclosed in the Fourier interval functoid

$$\mathrm{IS}_N \mathscr{M} = \left\{ A_0 + \sum_{k=1}^{N} A_k \cos kt + \sum_{k=1}^{N} B_k \sin kt : A_0, \ldots, A_N, B_1, \ldots, B_N \in \mathrm{IS} \right\}$$

with $N = 100$. All B_k are 0, and e.g. $A_0 = -3.^{651738760376}_{705850070386} \cdot 10^{-1}$ (compare to c_0). These inclusions are substantially coarser than the inclusions obtained in the hyper functoid. A function evaluation is not given in [2].

References

[1] Baumhof, C.: Untersuchungen zur Implementierung eines Fourier-Hyperfunktoids, Diplomarbeit, Institut für Angewandte Mathematik, Karlsruhe 1990.

[2] Heiser, W.: Lösung von Funktionalproblemen in einem Fourierfunktoid mit Verifikation, Diplomarbeit, Institut für Angewandte Mathematik, Karlsruhe 1987.

[3] Kaucher, E. W., Miranker, W. L.: Self-validating numerics for function space problems. New York London: Academic Press 1984.

[4] Kaucher, E. W.: Self-validating computations of ordinary and partial differential equations. In: Kaucher, E., Kulisch, U., Ullrich Ch. (eds.) Computer arithmetic. Stuttgart: BG Teubner 1987.

[5] Knopp K.: Theorie und Anwendung der unendlichen Reihen, 5. Aufl. Berlin, Heidelberg: Springer 1964.

[6] Kulisch, U. W., Miranker, W. L.: Computer arithmetic in theory and practice. New York London: Academic Press 1981.

E. Kaucher
Ch. Baumhof
Institut für Angewandte Mathematik
Universität Karlsruhe
Englerstrasse 2
D-W-7500 Karlsruhe
Federal Republic of Germany

Computing, Suppl. 9, 117–127 (1993)

Computing
© Springer-Verlag 1993
Printed in Austria

The Cluster Problem in Global Optimization: the Univariate Case[1,*]

B. Kearfott and **K. Du**, Lafayette

Dedicated to Professor U. Kulisch on the occasion of his 60th birthday

Abstract — Zusammenfassung

The Cluster Problem in Global Optimization the Univariate Case. We consider a branch and bound method for enclosing all global minimizers of a nonlinear C^2 or C^1 objective function. In particular, we consider bounds obtained with interval arithmetic, along with the "midpoint test," but no acceleration procedures. Unless the lower bound is exact, the algorithm without acceleration procedures in general gives an undesirable cluster of intervals around each minimizer. In this article, we analyze this problem in the one dimensional case. Theoretical results are given which show that the problem is highly related to the behavior of the objective function near the global minimizers and to the order of the corresponding interval extension.

AMS Subject Classification: 65K10, 65H20, 90C26.

Key words: Branch and bound principle, inclusion function, interval extensions, midpoint test, global optimization, order of an interval extension.

Das Cluster Problem bei globalen Optimierungsaufgaben. Wir betrachten ein Bisektionsverfahren zur Einschließung aller globalen Minimalpunkte für stetig differenzierbare bzw. zweimal stetig differenzierbare Zielfunktionen. Das Verfahren basiert auf den Grundlagen der Intervallarithmetik und benutzt den Mittelpunkttest ohne zusätzliche Beschleunigungsverfahren. Es ist bekannt, daß bei Benutzung von Bisektionsverfahren Cluster von Intervallen in der Nähe globaler Minimalpunkte auftreten. Dieses Problem wird hier im eindimensionalen Fall näher untersucht. Insbesondere sind theoretische Resultate angegeben, die zeigen, wie die Anzahl der Intervalle in diesen Clustern vom Verhalten der Zielfunktion in der Nähe der globalen Minimalpunkte sowie von der Güte der benutzten Einschließungsfunktion abhängt.

1. Introduction and Basic Concepts

Our underlying problem is:

$$\text{find all global minimizers to } f(x)$$
$$\text{subject to} \quad x \in \mathbf{X}, \tag{1}$$

where $\mathbf{X} \subset \mathbb{R}$ is an interval. We denote the global minimum as f^* and the set of global minimizers as $\mathscr{X}^*$.

---

[1] This material is based on upon work supported by the National Science Foundation under Grand No. CCR-9203730.

* Received June 1, 1992; revised December 18, 1992.

In this paper, we study the branch and bound principle to enclose the solution set $\mathscr{X}^*$ of (1). Our analysis deals with algorithms similar to Algorithm 3, p. 111 of [9]. Also, as in [9], we will use interval arithmetic to obtain the bounds.

This paper deals with the phenomenon of clusters of small intervals around global minimizers which such algorithms produce, but which the algorithm cannot eliminate. We refer to this phenomenon as the *cluster problem*. Discussion of clustering in a branch and bound method in a particular context in the multidimensional case appears in [6]; however, the cause of the phenomenon studied there is different from the cause in the one-dimensional case studied here. In this paper, we consider the phenomenon, using interval extensions, in the one dimensional case.

Throughout, we denote non-interval quantities by lower case letters and interval quantities by upper case boldface letters. We will occasionally use a lower case bold letter to denote a non-interval function value which has been bounded using interval arithmetic. For introductions to interval arithmetic, see e.g. [8], [5], or [1].

Inclusion functions. Using interval arithmetic, we may extend an objective function f to interval values such that $f(\mathbf{X}^{(1)})$ contains the entire range of f over interval $\mathbf{X}^{(1)}$. Let $\mathbf{F}(\mathbf{X}^{(1)})$ denote the interval extension of f evaluated over the interval $\mathbf{X}^{(1)}$. If there is a constant K such that

$$w(\mathbf{F}(\mathbf{X}^{(1)})) - w(f(\mathbf{X}^{(1)})) \leq Kw(\mathbf{X}^{(1)})^{\alpha},$$

where $f(\mathbf{X}^{(1)})$ denotes the exact range of $f(x)$ over $\mathbf{X}^{(1)}$ and $w(\mathbf{X}^{(1)})$ denotes the width of $\mathbf{X}^{(1)}$, then we say that $\mathbf{F}(\mathbf{X}^{(1)})$ is an order α inclusion function for $f(x)$. When α is 1 or 2, we call the inclusion first order or second order, respectively.

The basic algorithm. The following algorithm is similar to Algorithm 3, p. 111 of [9]. Some of the notations in the algorithm are

$$\text{mid}(\mathbf{X}): \text{ the midpoint of } \mathbf{X};$$

$$\text{ub}(\mathbf{X}): \text{ the upper bound of } \mathbf{X};$$

$$\text{lb}(\mathbf{X}): \text{ the lower bound of } \mathbf{X}.$$

Algorithm 1.
0. *Input the original interval $\mathbf{X}$, the inclusion function $\mathbf{F}$ of f, and additional parameters used in the termination criteria.*
1. *Set $\mathbf{Y} := \mathbf{X}$.*
2. *Calculate $\mathbf{F}(\mathbf{Y})$ and $\tilde{f} := ub\mathbf{F}(c)$ where $c := mid\,\mathbf{Y}$.*
3. *Set $y := lb\mathbf{F}(\mathbf{Y})$.*
4. *Initialize the list $L := \{(\mathbf{Y}, y)\}$.*
5. *Bisect $\mathbf{Y}$ to obtain intervals $\mathbf{V}_1$ and $\mathbf{V}_2$ such that $\mathbf{Y} = \mathbf{V}_1 \cup \mathbf{V}_2$, then delete $\mathbf{Y}$ from the list.*
6. *Calculate $\mathbf{F}(\mathbf{V}_1)$ and $\mathbf{F}(\mathbf{V}_2)$.*
7. *Set $v_i := lb\mathbf{F}(\mathbf{V}_i)$ for $i = 1, 2$.*
8. *Enter the pairs $(\mathbf{V}_1, v_1)$ and $(\mathbf{V}_2, v_2)$ at the end of the list.*
9. *Choose a pair $(\tilde{\mathbf{Y}}, \tilde{y})$ of the list which satisfies $\tilde{y} \leq z$ for all pairs $(\mathbf{Z}, z)$ of the list.*
10. *Discard all pairs $\{\mathbf{Z}, z\}$ from the list that satisfy $z > \tilde{f}$ (midpoint test).*

11. *If the termination criteria hold go to 15.*
12. *Denote the first pair of the list by $(\mathbf{Y}, y)$. Then set $c := mid\,\mathbf{Y}$ and $\tilde{f} :=$ $\min(\tilde{f}, ub\,\mathbf{F}(c))$.*
13. *Go to 5.*
14. *End.*

Step 10 is crucial for our purposes, since we are studying the power of the midpoint test to discard intervals which do not contain global minimizers.

2. The Cluster Problem Around Global Minimizers

Because inclusion function $\mathbf{F}$ in general overestimate the range of $f(x)$, the midpoint test may not be able to reject all the intervals which do not contain global minimizers. This can be illustrated with the following example.

$$f(x) = (x - 1)^2 = x^2 - 2x + 1 \qquad \text{for } x \in [0, 2]$$

and

$$\mathbf{F}(\mathbf{X}) = \mathbf{X}^2 - 2\mathbf{X} + 1.$$

We know

$$\mathscr{X}^* = \{1\} \quad \text{and} \quad f^* = 0.$$

The unique minimizer occurs at the endpoint of the adjacent intervals produced during the bisection process. Suppose at some stage, the length of each interval in the list L in Algorithm 1 is ε, so that the intervals immediately adjacent to the minimizer 1 are

$$[1 - 2\varepsilon, 1 - \varepsilon], \quad [1 - \varepsilon, 1], \quad [1, 1 + \varepsilon], \quad \text{and} \quad [1 + \varepsilon, 1 + 2\varepsilon].$$

We will now investigate the interval $[1 + \varepsilon, 1 + 2\varepsilon]$. (The same reasoning can be applied to the interval $[1 - 2\varepsilon, 1 - \varepsilon]$.)

On $[1 + \varepsilon, 1 + 2\varepsilon]$,

$$\begin{aligned}
\mathbf{F}([1 + \varepsilon, 1 + 2\varepsilon]) &= [1 + \varepsilon, 1 + 2\varepsilon]^2 - 2[1 + \varepsilon, 1 + 2\varepsilon] + 1 \\
&= [1 + 2\varepsilon + \varepsilon^2, 1 + 4\varepsilon + 4\varepsilon^2] - [2 + 2\varepsilon, 2 + 4\varepsilon] + 1 \\
&= [-2\varepsilon + \varepsilon^2, 2\varepsilon + \varepsilon^2],
\end{aligned}$$

so $f^* \in \mathbf{F}([1 + \varepsilon, 1 + 2\varepsilon])$ for *all* sufficiently small ε; thus, $[1 + \varepsilon, 1 + 2\varepsilon]$ can never be rejected. Hence, no matter how small ε is, the number of intervals in the list will always be greater than 2.

3. Our Analysis of the Cluster Problem

Here, we will consider the cluster problem for the one-dimensional case.

General sufficient condition for an interval to be rejected. Suppose that $x^* \in \mathscr{X}^*$ is a particular minimizer, and suppose that $\mathbf{X}^{(1)} = (x_1, x_2)$ is an interval to the right of the minimizer x^* in a region sufficiently close to x^* to guarantee that f is monotonic between x^* and x_2. (See Fig. 1.) Then a sufficient condition for $\mathbf{X}^{(1)}$ to be rejected by the midpoint test is

$$w(\mathbf{F}(\mathbf{X}^{(1)})) - w(f(\mathbf{X}^{(1)})) < f(x_1) - f(x^+). \tag{2}$$

where x^+ is the current midpoint which is used in the midpoint test. This formula follows directly from the above assumptions and the geometry. (See Fig. 1 and its caption.)

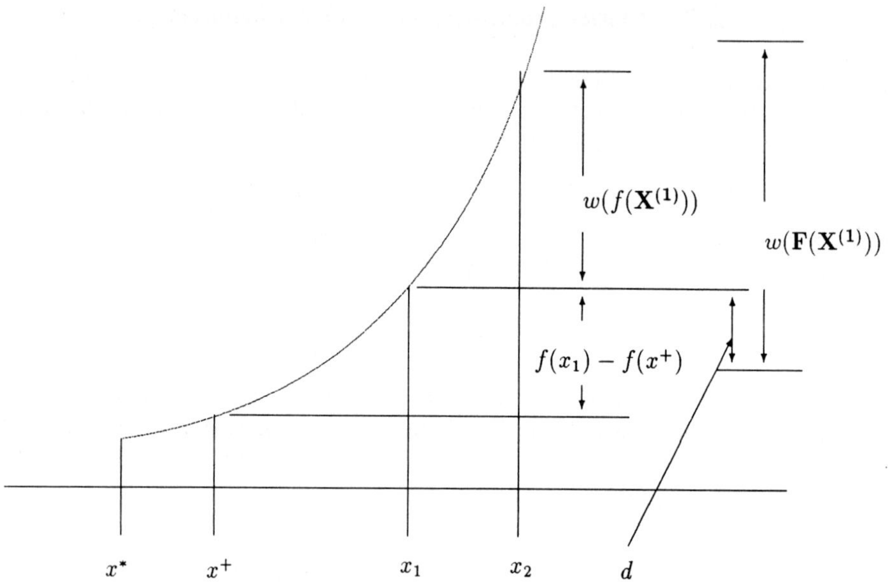

Figure 1. $d \leq w(\mathbf{F}(\mathbf{X}^{(1)})) - w(f(\mathbf{X}^{(1)}))$

Since $f(x^*) \leq f(x^+)$, we may suppose that x^+ is a point near x^*. (Again, see Fig. 1.) If x^+ happens to equal x^*, we have:

$$w(\mathbf{F}(\mathbf{X}^{(1)})) - w(f(\mathbf{X}^{(1)})) < f(x_1) - f(x^*). \tag{2'}$$

Major theorem and some corollaries. Suppose at some stage (when ε is small enough), the intervals corresponding to the list L in Algorithm 1 are

$$[x^* - (m+1)\varepsilon, x^* - m\varepsilon], \ldots, [x^* - \varepsilon, x^*],$$

$$[x^*, x^* + \varepsilon], \ldots, [x^* + n\varepsilon, x^* + (n+1)\varepsilon],$$

Here, for simplicity, we assume that each interval in the list has the same width ε. This assumption will not change the essence of our discussion. Also, in the numerical experiments in the next section, we found that, for sufficiently small ε, the algorithm actually produced a list with only one or two box sizes, for boxes near a particular global minimizer. It is possible to explain the mechanism which makes this so.

Theorem 1. *Suppose that the objective function $f(x)$: $\mathbf{X} \subset \mathbb{R} \to \mathbb{R}$ has two continuous derivatives and a unique global minimizer x^* in the interior of $\mathbf{X}$, where x^* is on the boundary of two adjacent intervals produced by the algorithm. (i.e., suppose x^* is at a common endpoint of two adjacent intervals). Let*

$$\mu_1 = \min_{x \in [x^*, x^* + n\varepsilon]} f''(x) > 0,$$

where n is the index of the rightmost interval in the final list, and let $\mathbf{F}$ denote an order α interval extension of f. Also, assume that the best upper bound $f(x^+)$ for the global minimum happens to equal the global minimum itself $f(x^*)$. Assume that each interval in the list has width ε. Then the maximum number of intervals left in the list and to the right of x^* is equal to

$$RN = \left\lfloor \sqrt{\frac{2K}{\mu_1}} \cdot \sqrt{\varepsilon^{\alpha-2}} \right\rfloor + 1 \tag{3}$$

where $\lfloor r \rfloor$ stands for the integer part of the positive real number r.

Proof. Since $\mu_1 > 0$, the pattern is as Fig. 1. Thus, we may base the proof on (2′). From the assumptions, f has a Taylor expansion about x^* of the form

$$f(x) = f(x^*) + f'(x^*)(x - x^*) + \frac{f''(\xi)}{2}(x - x^*)^2,$$

for some $\xi \in (x^*, x)$. Since the global minimizer x^* is at an interior point, $f'(x^*) = 0$ and $f''(x^*) > 0$. Thus, when $x = x^* + n\varepsilon$, the above Taylor expansion becomes

$$f(x^* + n\varepsilon) - f(x^*) = \frac{f''(\xi)}{2}(n\varepsilon)^2.$$

Also, let $\mathbf{X}^{(1)}$ be as in (2) or (2′) with $w(\mathbf{X}^{(1)}) = \varepsilon$. Then, since the interval extension $\mathbf{F}$ of f is of order α,

$$w(\mathbf{F}(\mathbf{X}^{(1)})) - w(f(\mathbf{X}^{(1)})) < Kw(\mathbf{X}^{(1)})^\alpha.$$

In the worst case, when $f''(\xi) = \mu_1 > 0$,

$$Kw(\mathbf{X}^{(1)})^\alpha < \frac{\mu_1}{2}(n\varepsilon)^2 \le \frac{f''(\xi)}{2}(n\varepsilon)^2$$

is equivalent to the sufficient condition (2′). Solving the above inequality for n, we get

$$n > \sqrt{\frac{2K}{\mu_1}} \cdot \sqrt{\varepsilon^{\alpha-2}} \ge \sqrt{\frac{2K}{f''(\xi)}} \cdot \sqrt{\varepsilon^{\alpha-2}}.$$

The conclusion then follows. $\square$

Theorem 1b. *Under assumptions analogous to Theorem 1, the maximum number of intervals in the final list to the left of x^* is*

$$LN = \left\lfloor \sqrt{\frac{2K}{\mu_2}} \cdot \sqrt{\varepsilon^{\alpha-2}} \right\rfloor + 1 \tag{3′}$$

where

$$\mu_2 = \min_{x \in [x^*-m\varepsilon, x^*]} f''(x),$$

where m is index of the leftmost interval in the final list.

Theorem 1b is proven entirely analogously to Theorem 1, with the mirror image of Fig. 1, and with corresponding analogues to (2) and (2′).

Corollary 1. *The maximum number of intervals left in the final list is*

$$N = 2 \max\{RN, LN\}.$$

Corollary 2. *If*

(1) $\alpha < 2$, *then there may exist a severe cluster, i.e. the number of intervals in the list associated with x^* may increase without bound as ε becomes small;*
(2) $\alpha = 2$, *then the cluster is not serious but there may always be a constant number $N > 1$ of intervals in the list associated with x^*, no matter how small ε is;*
(3) $\alpha > 2$, *then there is no cluster, i.e. for sufficiently small ε there is only one box in the portion of the list associated with the global minimizer x^* on each side of x^*.*

Remarks.

Remark 1. In the above corollary we say "there may ..." rather than "there must" because the theorem gives an upper bound, and not a precise value, for the number of intervals. In problems in which the lower bound given by the interval extension **F** is exact, there will be no cluster even though **F** is not of high order. Note, however, that in the experiments in the next section, the bounds correspond very closely with the actual results.

Remark 2. If, in contrast to the above, we suppose $f''(x^*) = 0$ (i.e. $\mu_1 = \mu_2 = 0$) and $f'''(x^*) > 0$ in a small neighborhood of x^*, but the other conditions remain the same, then a similar analysis can be carried out. In this case, however, there may exist a severe cluster when $\alpha < 3$. It is not difficult to obtain one-dimensional interval extensions up to order 7, and it is inexpensive to obtain such extensions up to order 5; see [3]. These extensions may be used in this case.

Remark 3. We may use a similar analysis to find a bound on the number of boxes N in the cluster when x^* is an endpoint of **X** instead of an interior point. We obtain the following conclusions:

(i) If $f'(x^*) = 0$, then the results will be the same as in corollary 2.
(ii) Suppose $f'(x^*) \neq 0$ and suppose that x^* occurs at an endpoint of the initial interval. Without loss of generality, suppose x^* is the left endpoint. Then, if f has a continuous first derivative, we may apply the general sufficient condition (2) and a first order Taylor expansion to obtain

$$N > \frac{K}{f'(\xi)} \cdot \varepsilon^{(\alpha-1)}, \tag{4}$$

where ξ is a point between the left endpoint of $\mathbf{X}^{(1)}$ and x^+ of (2). Since $f'(x^*) \neq 0$ and $f \in C^1$, $1/f'(\xi)$ will be bounded. Thus,

(a) when $\alpha < 1$, there may exist a severe cluster;
(b) when $\alpha = 1$, the maximum number of intervals in the final list tends to a constant;
(c) when $\alpha > 1$, no cluster exists.

Remark 4. When x^* is neither an endpoint of the initial interval nor an endpoint of one of the intervals produced from bisection of the initial interval, the analysis is slightly more complicated to write down. Formula (4) in Remark 3 is still valid, but $1/f'(\xi)$ is no longer bounded. Instead, since $x^+ \to x^*$ as $\varepsilon \to 0$, $f'(\xi)$ will approach $f'(x^*) = 0$. When ε is sufficiently large, we may use (4), but when ε is sufficiently small, the machine can no longer distinguish the difference between x^* and x^+, and (3) may be applied. When ε is small enough, we obtain the same conclusion as in corollary 2. Thus, the cluster problem is of the same magnitude as in the case stated in (3) and (3').

Remark 5. The only requirement for the above conclusions is that the objective function be twice continuously differentiable or continuously differentiable in some neighborhood of the global minimizer. Thus the conclusions apply to a large range of functions.

3. Numerical Results

We implemented Algorithm 1 in FORTRAN-SC (see [2]) on an IBM 3090. We terminated the procedure when all of the intervals in the list had the same width ε, which is less than the specified tolerance ε'. Below, we give numbers of intervals in the final lists together with the corresponding stopping tolerance ε' (the first lines in the tables) and the given initial intervals (the first columns in the tables).

Example 1. $f(x) = (x - 1)^2$, with $x^* = 1$ and $f^* = 0$.

(A) We used the interval extension

$$\mathbf{F(X) = X(X - 2) + 1}$$

with convergence order 1. Results are shown in Table 1.

Table 1

initial	$\varepsilon' = 10^{-2}$	$\varepsilon' = 10^{-3}$	$\varepsilon' = 10^{-4}$	$\varepsilon' = 10^{-5}$	$\varepsilon' = 10^{-6}$	$\varepsilon' = 10^{-7}$
[0, 2]	22	64	256	724	2048	8192
[1, 3]	11	32	128	362	1024	4096
[1.01, 5]	10	24	45	50	51	51
[0.101, 5]	20	82	231	655	2617	7402

When [0, 2] is the initial interval, x^* will be a midpoint of one of the intervals and thus a common point of subsequently generated neighboring subintervals. Our results show that the number of intervals in the final list increases with no sign of boundedness.

When $[1, 3]$ is the initial interval, x^* is actually its left endpoint, and $f'(x^*) = 0$. The results show that the number of intervals in the final list increases with no sign of boundedness.

When $[1.01, 5]$ is the initial interval, x^* is at its left endpoint, but $f'(x^*) \neq 0$. The results show that the number of intervals in the final list is almost constant as a function of ε' when ε' is small enough.

When $[0.101, 5]$ is the initial interval, x^* is an interior point of it. The results show that the number of intervals in the final list increases and shows no sign of boundedness.

To see how well the estimate (3) corresponds quantitatively to the algorithm's actual behavior, let us compare ratios of different N's corresponding to different ε's. Theoretically, from formula (3) or (3′), we have

$$N_1/N_2 \approx \sqrt{\varepsilon_1}/\sqrt{\varepsilon_2}$$

To make the results easy to interpret we use a different set of ε: if $\varepsilon_1/\varepsilon_2 = 4$ then $\sqrt{\varepsilon_1}/\sqrt{\varepsilon_2} = 2$. Results are shown in Table 1′. It is easy to see that the results fit our analysia almost exactly.

Table 1′

initial	$\varepsilon_1' = 10^{-3}$	$\varepsilon_2' = 1/4 \cdot 10^{-3}$	$\varepsilon_3' = 1/16 \cdot 10^{-3}$	$\varepsilon_4' = 1/64 \cdot 10^{-3}$
$[0, 2]$	64	128	256	512
$[0.101, 5]$	82	164	327	655

(B) We used the interval extension

$$\mathbf{F}(\mathbf{X}) = f(c) + f'(c)(\mathbf{X} - c) + \frac{f''(c)}{2}(\mathbf{X} - c)^2,$$

where c is the midpoint of $\mathbf{X}$. This extension is of order 2. Results are shown in Table 2.

Table 2

initial	$\varepsilon' = 10^{-2}$	$\varepsilon' = 10^{-3}$	$\varepsilon' = 10^{-4}$	$\varepsilon' = 10^{-5}$	$\varepsilon' = 10^{-6}$	$\varepsilon' = 10^{-7}$
$[0, 2]$	2	2	2	2	2	2
$[1, 3]$	2	2	2	2	2	2
$[1.01, 5]$	2	2	2	2	2	2
$[0.101, 5]$	2	3	2	2	3	3

Table 2 shows that no serious cluster occurred. In fact, since the global optimum was on a boundary of two sub-intervals for the first two initial intervals, no cluster at all occurred in those cases. Empirically, this would be evidence that the interval extension could *possibly* be of order higher than 2 when applied to this particular function (depending on K and μ_1 in (3)).

We also tried the mean value extension:

$$\mathbf{F(X)} = f(c) + \mathbf{F'(X)(X} - c),$$

which is also of order 2. The results occur in Table 2'.

Table 2'

initial	$\varepsilon' = 10^{-2}$	$\varepsilon' = 10^{-3}$	$\varepsilon' = 10^{-4}$	$\varepsilon' = 10^{-5}$	$\varepsilon' = 10^{-6}$	$\varepsilon' = 10^{-7}$
[0, 2]	6	6	6	6	6	6
[1, 3]	4	4	4	4	4	4
[1.01, 5]	4	4	2	2	2	2
[0.101, 5]	6	5	6	5	5	5

The behavior shown in Table 2' is typical of an interval extension of order 2: the numbers of intervals in the final lists are all constants greater than 2 when ε is small enough.

Example 2. $f(x) = x^2 - \sin(x)$, with $x^* = 0.45018361129$.

(A) We used the interval extension

$$\mathbf{F(X) = X^2 - \sin(X)},$$

which is of order 1. Results are in Table 3.

Table 3

initial	$\varepsilon' = 10^{-2}$	$\varepsilon' = 10^{-3}$	$\varepsilon' = 10^{-4}$	$\varepsilon' = 10^{-5}$	$\varepsilon' = 10^{-6}$	$\varepsilon' = 10^{-7}$
[0.45, 5]	7	27	76	228	1165	4671
[0.46, 5]	6	15	19	20	20	20

Since the minimizer is no longer a rational number, we are only able to consider the case when the initial interval contains the minimizer and the case when the initial interval does not contain the minimizer. The results in the above table clearly indicate that, for the first initial interval, the number of intervals in the final list shows no sign of convergence as ε becomes smaller and smaller, but for the second initial interval, the number of intervals in the final list is constant for small enough ε.

To see the accuracy of our formulas (3) and (3'), let us compare ratios of different N's corresponding to different ε's, as we did in part (A) of example 1. Results are shown in Table 3'. The results in this table also fit our analysis very well.

Table 3′

initial	$\varepsilon_1' = 10^{-3}$	$\varepsilon_2' = 1/4 \cdot 10^{-3}$	$\varepsilon_3' = 1/16 \cdot 10^{-3}$	$\varepsilon_4' = 1/64 \cdot 10^{-3}$
[0.45, 5]	27	53	109	228

(B) We used the order-2 interval extension

$$\mathbf{F}(\mathbf{X}) = f(c) + \mathbf{F}'(\mathbf{X}) \cdot (\mathbf{X} - c),$$

where c is the midpoint of $\mathbf{X}$.

Table 4

initial	$\varepsilon' = 10^{-2}$	$\varepsilon' = 10^{-3}$	$\varepsilon' = 10^{-4}$	$\varepsilon' = 10^{-5}$	$\varepsilon' = 10^{-6}$	$\varepsilon' = 10^{-7}$
[0.45, 5]	4	4	5	5	5	6
[0.46, 5]	4	4	2	2	2	2

The above results show that, when the interval extension is of order 2, the number of intervals in the final list will be almost constant with respect to ε, provided the initial interval contains the minimizer. They also show that no cluster exists when the minimizer is one of the endpoints of the initial interval and is not a critical point. These are precisely the conclusions of our theoretical analysis.

4. Conclusions and Future Work

Practical algorithms use more than bisection to decrease the size of intervals, and use more than the midpoint test to eliminate boxes from the list. Nonetheless, the above theoretical development can serve as a guide for the construction of efficient practical algorithms.

The above results, both theoretical and numerical, show that interval extensions of order at least 2 should be used if only the midpoint test is used to discard intervals. This is especially true when the initial interval contains the global minimizer. Additionally, we have the following possibilities for improvement.

1. Use acceleration devices such as an interval Newton method. (Note that this is standard practice in interval arithmetic-based branch and bound algorithms; see, for example [4] for one of the earlier explanations of this process.)
2. Use higher order interval extensions. When $f''(x^*) \neq 0$, an order 3 or higher extension should result in *no* cluster, even without acceleration procedures.
3. When $f'(x^*) = f''(x^*) = 0$, use an interval extension of order at least 3. In this case, this may be important, since acceleration devices may not function efficiently.

A similar behavior can be shown both theoretically and in practice when $f: \mathbf{X} \subset \mathbb{R}^n \to \mathbb{R}^n$ for $n > 1$. However, it may be harder to obtain higher order interval extensions in $n > 1$ dimensions; see [3], or the improved version in [7], §2.4. Also, other phenomena can cause clusters when $n > 1$; see [6]. These topics are the subject of work in progress.

References

[1] Alefeld, G., Herzberger, J.: Introduction to interval computations. New York: Academic Press 1983.

[2] Bleher, J. H., Rump, S. M., Kulisch, U., Metzger, M., Ullrich, C., Walter, W.: Fortran-SC—A study of a Fortran extension for engineering scientific computation with access to ACRITH. Computing 39, 93–110 (1987).

[3] Cornelius, H., Lohner, R.: Computing the range of values of real functions with accuracy higher than second order. Computing 33, 331–347 (1984).

[4] Hansen, E. R.: Global optimization using interval analysis—the multidimensional case. Numer. Math. 34, 247–270 (1980).

[5] Moore, Ramon E.: Methods and applications of interval analysis. Philadelphia: SIAM 1979.

[6] Morgan, A. P., Shapiro, V.: Box bisection for solving second-degree systems and the problem of clustering. ACM Trans. Math. Software 13, 152–167 (1987).

[7] Neumaier, A.: Interval methods for systems of equations. Cambridge, England: Cambridge University Press 1990.

[8] Ratschek, H., Rokne, J.: Computer methods for the range of functions. Chichester, England: Horwood 1984.

[9] Ratschek, H., Rokne, J.: New computer methods of global optimization. New York: Wiley 1988.

B. Kearfott
K. Du
Department of Mathematics, U.S.L., Box 4-1010,
Lafayette, LA 70504-1010, U.S.A.

Computing, Suppl. 9, 129–146 (1993)

Computing
© Springer-Verlag 1993
Printed in Austria

Developing Expert Systems for Validating Numerics*

S. König and **C. P. Ullrich**, Basel

Dedicated to Professor U. Kulisch on the occasion of his 60th birthday

Abstract — Zusammenfassung

Developing Expert Systems for Validating Numerics. The selection of an appropriate method out of a software library and its application to a given problem require a lot of expert knowledge. Usually there are more than one mathematically equivalent self-validating methods with completely different behavior on a computer. Therefore only a quantitative valuation of the available software modules rather than qualitative can lead to a satisfactory result. First the design of a shell suitable for the derivation of expert systems dealing with different domains of numerics is discussed. The main idea is the decoupling of the logic and the domain-specific part allowing an easy adaptation to a new field or extension with additional methods and selection criteria. Then a functional model for knowledge acquisition and representation under quantitative aspects is elaborated and enhanced by considering execution time problems arising from portations. The procedure is demonstrated by a comprehensive example dealing with two methods for systems of linear equations.

AMS Subject Classification: 65G10, 65G099, 65B99, 68G99

Key words: Conpled systems, expert systems, automation of software selection, self-validating methods.

Die Auswahl einer geeigneten Methode aus einer Software-Bibliothek und deren Anwendung auf ein gegebenes Problem erfordern umfangreiches Expertenwissen. Oftmals stehen mehrere mathematisch äquivalente selbstverifizierende Methoden zur Verfügung, welche sich jedoch auf einem Computer völlig unterschiedlich verhalten können. Daher kann nur eine quantitative und nicht die qualitative Bewertung verfügbarer Software-Module zu einem befriedigenden Selektionsergebnis führen. Zunächst wird der Entwurf einer Shell diskutiert, welche die Ableitung von Expertensystemen für verschiedene Gebiete der Numerik erlaubt. Die Hauptidee ist die Entkopplung des logischen vom domänenspezifischen Teil, um die Anpassung an ein neues Gebiet oder die Erweiterung mit zusätzlichen Methoden und Auswahlkriterien so einfach wie möglich zu gestalten. Anschliessend wird ein funktionales Modell für Wissenserwerb und -darstellung unter quantitativen Gesichtspunkten vorgestellt und durch die Berücksichtigung von Änderungen der Ausführungszeiten, welche bei Portierungen der Software entstehen, verbessert. Die Auswirkungen und Vorzüge dieses Modells können leicht verständlich am Beispiel von zwei Methoden zur Lösung linearer Gleichungssysteme demonstriert werden.

1. Support for Algorithm Selection and Application

The urgent need for numerical expert systems arises out of the considerable increase of newly developed numerical methods and consequently out of the yearly growing numerical software libraries. So the selection of a suitable method and its application to a given problem requires more and more expert knowledge. Up to now there

---

* Received September 21, 1992; revised November 27, 1992.

Table 1. Algorithm $A1$—a self-validating linear system solver

1. Compute an approximation of the inverse $R \approx A^{-1}$
2. Compute an enclosure $\mathbf{B}$ of the residuum $I - R * A$
3. $x_0 := R * b$;
   Repeat

$$x_{k+1} := x_k + R * (b - A * x_k)$$

   until corrections-are-too-small or potential-divergence; $x := x_{k+1}$
4. Compute an enclosure $\mathbf{y}_0$ of the defect $R * (b - A * x)$
5. Repeat with diameter vector $d(\mathbf{y}_k)$ of $\mathbf{y}_k$;

$$\mathbf{y}_k := \mathbf{y}_k + [-\text{eps}, +\text{eps}] * d(\mathbf{y}_k)$$

$$\mathbf{y}_{k+1} := \mathbf{y}_0 + \mathbf{B} * \mathbf{y}_k$$

   until $\mathbf{y}_{k+1}$ is-in-the-inner-of $\mathbf{y}_k$ or too-many-loops
6. If inclusion-is-achieved then display the solution $x + \mathbf{y}_{k+1}$ else display an appropriate message

are different attempts to solve this task by knowledge-based systems. One approach, for example, is NAXPERT, which selects under considerations of qualitative aspects ([8]). NAXPERT may be consulted to find a subroutine in a library providing a number of methods applicable to a variety of numerical problems as zeroes of polynomials, quadrature, linear algebra problems, optimization, differential equations, etc. The selection is based on the matching of sets of keywords. During a dialogue with the user all essential keywords of the current numerical problem are checked if they apply to an available routine. Further, additional compatible keywords are considered to roughly rank the possible pieces of software, the so-called consistency count.

Validating numerical methods make the situation even more confusing since quantitative aspects have to be considered additionally. Mathematically equivalent methods can give verified results with a certain accuracy consuming quite different CPU-times on a computer or one method is able to compute a result while the other method cannot give one at all. Therefore a reasonable application of those methods should not consist in a final selection of an algorithm but rather have the possibility of controlling the complete solving process, influencing or even changing the method. Apart from deciding between different algorithms, the control within a single method should depend on the required accuracy to provide for maximum performance. There are some actions (stopping criteria of iterations, computation with multiple mantissas, inflation strategy, etc.) to improve scalability of high precision and fast runtime. For this reason, only an integrated system can achieve optimum performance.

For illustration we choose the solution of systems of linear equations $Ax = b$. Let us consider the decision between two algorithms as an example for detailed explanations. In Table 1, we present algorithm $A1$ [7] which iteratively computes an enclosure of the solution from an initial approximation. The verification of the solution enclosure is derived from a fixed point theorem. Algorithm $A2$ is an interval

Table 2. Algorithm $A2$—interval version of Gaussian elimination

> 1. Compute interval matrices **L**, **U**
>   with $\mathbf{l}_{ij}, \mathbf{u}_{ji} = 0$ for $i < j$ and $A \subset \mathbf{L} * \mathbf{U}$ (with pivoting)
> 2. Solve $\mathbf{L} * \mathbf{y} = b$, i.e. compute an interval vector $\mathbf{y} \supseteq \{y | \mathbf{L} * y = b\}$
> 3. Solve $\mathbf{U} * \mathbf{x} = \mathbf{y}$, i.e. compute an interval vector $\mathbf{x} \supseteq \{x | \mathbf{U} * x = \mathbf{y}\}$
> 4. Repeat
>
> $$\tilde{x} := \mathrm{mid}(\mathbf{x});$$
>
> $$\text{Solve } \mathbf{L} * \mathbf{y} = [b - A * \tilde{x}];$$
>
> $$\text{Solve } \mathbf{U} * \mathbf{x} = \mathbf{y}$$
>
> until accuracy-of-**x**-sufficient or too-many-loops
> 5. If inclusion-is-achieved then display the solution **x**
>   else display an appropriate message

version of the usual Gaussian elimination including column pivoting and defect iteration (see Table 2).

Both $A1$ and $A2$ are equivalent for a system like NAXPERT, but their concrete implementations $S1$ and $S2$, resp., show different pros and cons. For $A \in \mathbb{R}^{n \times n}$, $b \in \mathbb{R}^n$ software module $S1$ requires $2n^3 + O(n^2)$ operation[1], while S_2 needs only $\frac{2}{3}n^3 + O(n^2)$. In practice the difference in execution times is machine-dependent (see Section 4). Further, the results of $S1$ are much more precise than those of $S2$. In contrast to $S1$, the accuracy of $S2$ decreases with the problem size n. Additionally, $S2$ is more sensitive to the numerical condition which is often unknown. In [3] an easy to compute indicator for the resulting accuracy is given which is based on the absolute values of the components of matrix A. Consider the variance of these numbers:

$$delta := \frac{\max_{i,j} |a_{ij}|}{\frac{1}{n^2} \sum_{i,j} |a_{ij}|}$$

For eight types of random test matrices we display their *delta* and the achieved accuracies of $S2$ in Fig. 1: With growing *delta* the average accuracy first increases but then decreases for higher values (*delta* > 60). $S1$ achieves maximum accuracy in all cases.

There exist many variants of both algorithms that provide a trade-off between runtime and precision. We will not discuss them in this paper but briefly mention some aspects here:

- total pivoting at $S2$: it increases the runtime about 20 percent and improves the achieved accuracy by some digits (depending on *delta*)
- If the coefficients of matrix A are intervals, the costs of algorithm $S2$ do not change, but they are doubled for $S1$.

---

[1] 1 operation = 1 multiplication + 1 addition.

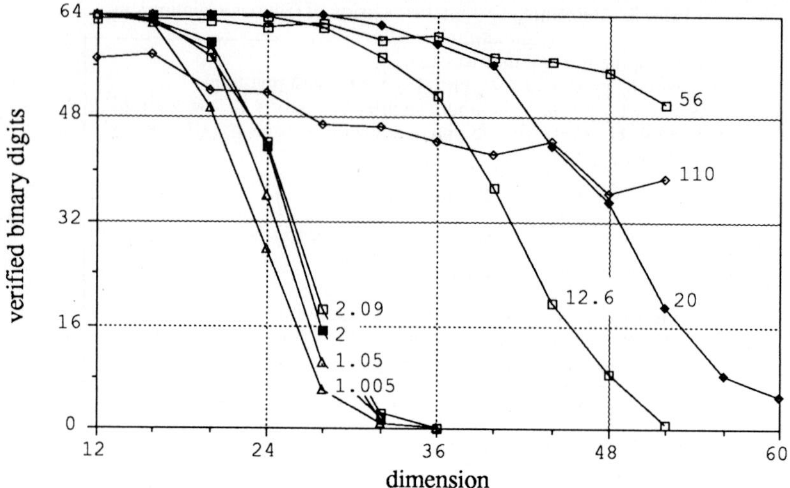

Figure 1. Achieved accuracy of $S2$ for matrices with different *delta*

- The sequence of operations may be changed for algorithm $S2$. If we apply Crout's strategy, we can use the scalar product of maximum accuracy (which is slower, but more precise).
- Defect iteration (step 3) and inflation strategy (step 5) of algorithm $S1$ may be modified with respect to the required accuracy to decrease runtime [2, 3].

2. Design of an Expert System Shell

In this section we will elaborate the structure of an expert system shell with respect to the field of (validating) numerics in general. The idea is to separate features that are common to all numerical problems from those that are specific to certain subdomains. A shell should include the first while real expert systems are derived by adding the latter. Design guidelines are modularity and transparency which aim at maintenance, extendability and portability.

For illustration we shortly discuss input/output routines when dealing with linear systems. The numerical input data consist of the matrix of coefficients and the right-hand side vector. Usually, such data are stored in files as result of previous processing steps. However there are many common ways to interpret a data stream like a file for initialization of a matrix. Depending on the source, an expert system must be able to accept sparse matrices of different compressed storage schemes, single components of differential sizes and formats (for example floating-point numbers of different precision or as ASCII strings), etc. At least the expert system must be easily extendable to allow adaptations to new situations by a developer. It is necessary to allow domain-specific procedural attachments to a general interface for providing full flexibility at numerical input and output. Then, it is the expert system developer's task to provide a routine that maps the expected input from file

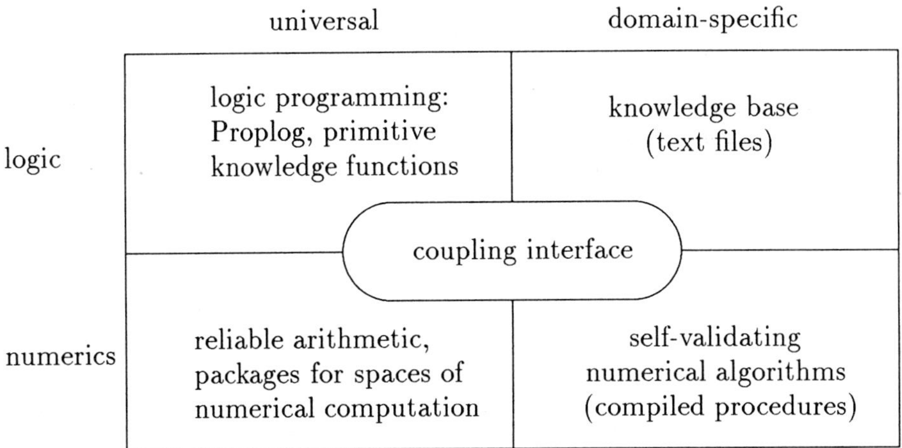

Figure 2. Modularity of the source code

or keyboard, etc. onto an internal numerical data structure which is used by the available solving routines. During a consultation the user is asked to choose one of the implemented input routines which may ask him for a filename, for example.

We will see that the same mechanism is used for the integration of further validating algorithms and since it is embodied in the expert system shell it is transferred to all its derivations. The main thought is that on the one hand knowledge-based and numerical methods must be brought together to solve the software selection problem and on the other hand they must be separated as much as possible to provide complete flexibility.

Figure 2 gives a classification of all involved system parts. Besides the distinction *universal—domain-specific*, the partition *logical—numerical methods* serves to characterize the contents of the modules. The left half of Fig. 2 represents the expert system shell. By adding the modules of the right half an expert system for a certain field of numerical problems is derived. The whole problem solving environment can be used in three modes:

1. In *consultation sessions* a user presents numerical problems to be solved. Choice and application of an appropriate routine are done by the system based on some analysis of the problem. A flexible dialogue is provided that is complemented by procedural attachments and by explanations and justifications of questions posed by the system.
2. The *extension* of an existing expert system by a domain expert is strongly supported.
3. The most fundamental usage is the *development of a new expert system* as derivation from the expert system shell.

During a consultation each derivation of the proposed shell applies a three stage process: In Phase 1 a qualitative calculus reduces the set of all integrated problem solving routines to the subset of applicable ones. Propositional logic combined with user interaction is used for this task. Hereby the degree of presence of characteristic features of the numeric subdomain in the current problem is determined (either via user input or via attached procedures). In Phase 2 the suitability of the methods that passed Phase 1 is quantified using primitive knowledge functions for knowledge modelling. We will discuss an example in Section 3. Phase 3 is the numerical execution of the self-validating procedural algorithm top-ranked by Phase 2.[2]

Tests showed that the costs of Phases 1 and 2 for selecting a numerical routine are at least one order of magnitude below the execution time of the most expensive method that solves the original problem.

Implementation of the logic parts may be organized domain-independent due to the complete separation of code and data which is typical for knowledge-based systems. The static knowledge of a domain may be stored in editable text files in a transparent and portable way, the inference strategies providing dynamic knowledge deduction though can be hard-coded by procedural programming.—The numerical part (lower half of Fig. 2) consists of self-validating methods for a special domain (right box) that should be based on general packages for the spaces of numerical computation (left box), i.e. libraries for interval, complex, complex interval arithmetic and for vectors and matrices of these scalar types which are programmed and compiled applying the procedural or object-oriented software construction paradigm. In this part existing wide-spread packages like LINPACK may be integrated for computation of approximations because validating methods often include this task as a subtask.

Understanding the coupling interface is most important for manipulation of this heterogeneous programming system. Its role is well characterized by Fig. 2: it connects the qualitative logical with the quantitative numerical world. This is achieved by definition of a central class of objects[3], called *ProblemObj*. Its instances hold the numerical data of the problem to be solved as well as intermediate results of Phase 1 and additional user specifications (see Fig. 3).

In code belonging to the logical phases instances of *ProblemObj* are to be used as atomic symbols. Access to components of this structured type is not allowed. Hence, no domain dependencies are carried into the modules of the upper left box of Fig. 2.—Numerical routines (lower right box) though regard these variables as a mere collection of domain-dependent mathematical data (lower left box) which are of interest for all further internal operations and subroutine calls. The transition from one usage to the other is controlled by some procedure types that also belong to the coupling interface module and have a formal parameter of type *ProblemObj*.

---

[2] Detailed descriptions and the mathematical background of methods that yield a precise enclosure of the true solution and prove its existence and uniqueness (if possible) at the same time can be found in other contributions of this volume and their references.

[3] In this section the term *class* is used in the sense of object-oriented programming, embracing structured data types and methods providing operations on instances of these data types.

ProblemObj

<table>
<tr><td>

numerical data

matrix a; vector b; vector x;

environmental data

required accuracy; input accuracy; symmetry;

</td></tr>
</table>

Figure 3. Data of a problem object (here for linear systems of equations)

These are routines for reading, solving and writing a problem as well as small procedural attachments for checking the presence of a characteristic feature.

The advantage of this design becomes obvious when developing an expert system for another domain. This only entails the redefinition of class *ProblemObj* located in the coupling interface and requires no changes to the code of the upper left box. Just a recompilation makes the complete logic part work on another knowledge domain. Then, instances of the procedure types mentioned above are to be programmed with respect to the new problem domain.

During each consultation one attached procedure for definition of the numerical problem and one for its solution is chosen and executed. The selection of the initialization routine is embedded in the dialogue of Phase 1, the automation of the selection of the solving routine is the essential contribution of the expert system's symbolic part.

The symbol of a routine is related to the corresponding executable code by owning a variable of a specific procedure type. Insertion of an interpretation routine for another data format or the integration of an additional solving routine just requires—besides the real programming of the new algorithm with conformal interface—the creation of another symbol. Since each extension of the expert system's numerical abilities requires this very easy but necessary action, clarity and openness of the corresponding part of code is most important. The presented coupling interface obviously achieves the maximum separation of logic and numerics with respect to practical programming while allowing far-going mathematical interlacing. The multiple transition between logic and numerics is technically possible, but it should be applied with caution for the reasons explained in [4].

Let us finally remember the introductory input/output example. We can state that our approach is powerful enough to deal with all fields of numerical computation conveniently. Suppose an expert system for validated integration of a real valued function of one independent variable over a finite interval. Here, the input of the function's formula via keyboard is realistic. An arbitrary extendable set of problem reading routines may be supplied with such an expert system, each of which implements one rule how to map a character string onto one of the data structures

internally used by the considered integration routines, i.e. from infix/postfix notation a binary tree is built. Direct reading of a binary file does not result in exceptions
to our concept, either.

3. Knowledge Representation and Acquisition

In Section 1 the difference between qualitative and quantitative valuation of numerical software modules has been discussed. In [5] it is shown that the latter needs a
functional mechanism for knowledge representation; a shell for knowledge-based
systems is presented which generates an ordered list of applicable library routines
for the solution of a mathematical problem posed by the user. The order is determined by a quantitative estimation of the suitability of the software. A functional
knowledge representation mechanism is developed which is able to express the
suitability of a piece of code with respect to a set of characteristic features of a
problem, their grade of presence and the relations between them. As an application
a *Selection Advisor for Initial Value Software* (SAIVS) is presented. An interesting
aspect of SAIVS is the step into the direction of processing practical facts instead
of formal algorithms based on mathematical theorems and models. Here we find
programs executable on existing machines which are weighed one against the other.
Our approach goes even further because the discrepancy between formal algorithms
and the corresponding implemented routines is much deeper in validating numerics
than in classical scientific computation as discussed in Section 1. Therefore, only by
great flexibility of the expert system can the practical relevance of the stored
knowledge (on logical as well as numerical level) be conserved under portations or
enhancements of the hardware situation by technical progress.

We start a detailed discussion with the citation of the *Software Selection Problem*
([5] in [1], p. 425f):

A problem domain P with characteristic features $F = \{f_1, \ldots, f_n\}$;
A set if software modules (codes) $S = \{s_1, \ldots, s_k\}$;
A set of performance objectives $O = \{o_1, \ldots, o_m\}$;
A suitability function $\Omega: \mathbb{R}^m \to \mathbb{R}$ reflecting the subjective preferences of a domain
expert for the relative importance of the performance objectives O;
Find performance functions $H_{i_j}: P \to \mathbb{R}$, $i = 1, 2, \ldots, k$, $j = 1, 2, \ldots, m$ such that
for any $p \in P$, $H_{i_j}(p)$ computes the degree to which the objective o_j is satisfied by
the application of s_i to p. The overall suitability of s_i for p is computed by applying
Ω to H_{i_j}, $j = 1, 2, \ldots, m$.

In the following, human knowledge mostly existing as a compendium rather than in
single facts has to be split into explicit knowledge units which can be coded, manipulated by a computer and understood by non-experts. In [5] the work concen-
trates upon acquisition of expert knowledge and therefore $m = 1$ is used without
loss of generality. For determining the functions $H_1, \ldots, H_k$ are partitioned corresponding to the characteristic features $f_j \in F$, $j = 1, \ldots, n$. Each H_i is represented as
the composition of four types of *primitive knowledge functions* (PK functions):

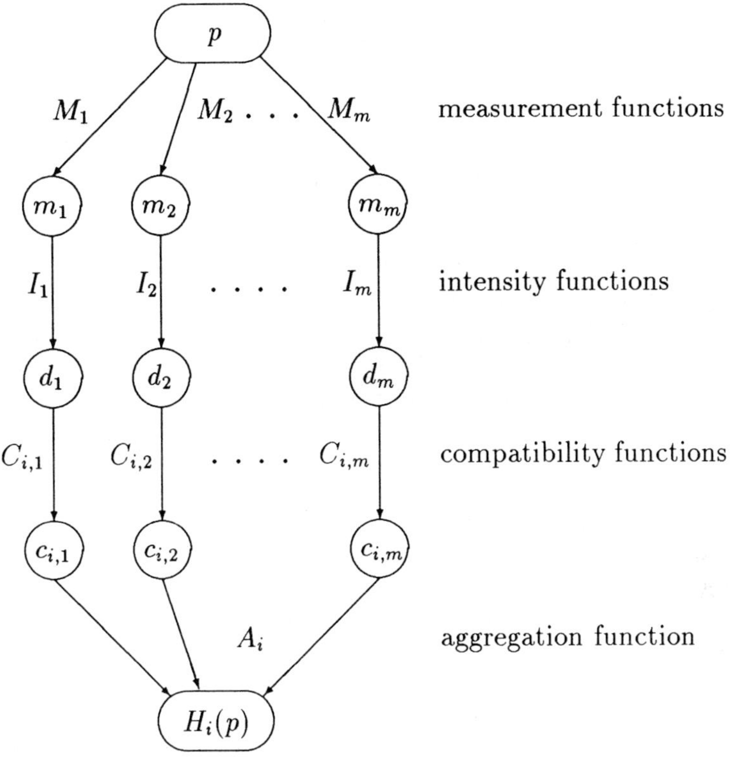

Figure 4. Software evaluation network (SEN)

- *measurement functions* M_j quantifying the degree of presence of characteristic feature f_j
- *intensity functions* I_j mapping the values m_j of the measurement functions to a standard scale (e.g. $L = [0, 1]$)
- *compatibility functions* $C_{i,j}$ describing the relationships between the behaviour of code s_i and the degree of presence d_j of characteristic feature f_j
- *aggregation functions* A_i which represent the overall behaviour of software s_i — based on the aggregate effect of the occuring features f_j

The relation between these functions is the usual composition. The functions $C_{i,j}$ and A_i both map the standard interval L onto itself. The final aggregation function A_i yields increasing values as suitability of method s_i increases. This model can be visualized as *software evaluation network* (Fig. 4) and the knowledge base can be understood as a set of such graphs—one for each software module s_i.

The functions M_j receive their values by a question to the user, an internal computation, a call of a complex external procedure, etc. These values are objective data.

The functions I_j make the knowledge base clearer and more modular since all software feature relations are mapped to a standard scale and furthermore changes of the M_j no longer effect the $C_{i,j}$.

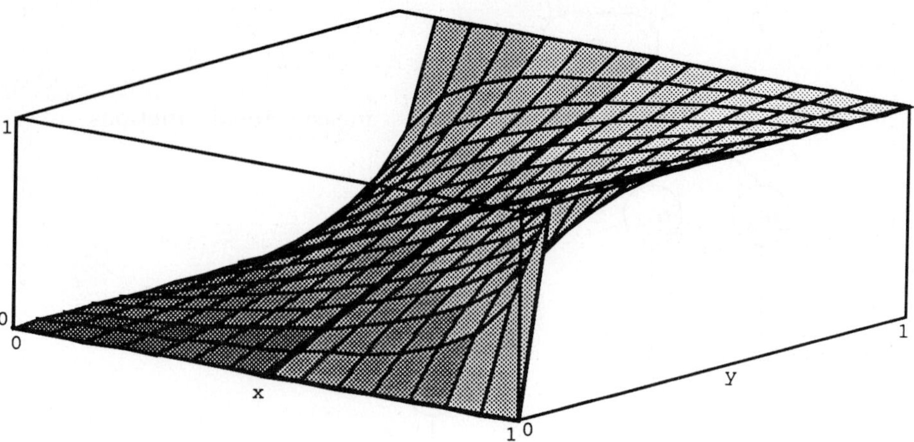

Figure 5. Aggregation of compensatory influence $\alpha(x, y) = \dfrac{xy}{xy + (1 - x)(1 - y)}$

Small results of $C_{i,j}$ show that the performance of code s_i behaves contrary to the intensity of the presence of feature f_j and vice versa. If the result is the midpoint L_M of L, neutral intensity with respect to f_j is stated, i.e. the performance of code s_i is not influenced. During knowledge acquisition this fact is very useful for the isolation of the effect of compensating characteristic features.

The construction of the A_i is done by a further decomposition to so-called *primitive aggregation operators* which assign a specific value combining the effects of a pair of features. Domination of the smaller or bigger value is modelled by $\min(x, y)$ or $\max(x, y)$. If a compensatory effect exists the result is defined by $\alpha(x, y)$. The choice of $\alpha(x, y)$ is neither obvious nor unique. In [5] formal considerations led to a mapping α (see Fig. 5) such that (L, α) is an Abelian group which is partially ordered and fulfills further desirable relations, for example:

$$(x < L_M \wedge y > L_M) \quad \Rightarrow \quad x < \alpha(x, y) < y$$

$$\alpha(x, 0) = 0, \qquad \alpha(x, L_M) = x, \qquad \alpha(x, 1) = 1$$

Repeated applications of the primitive aggregation operators to the compatibility values and aggregation results finally generate a value of L which is the expert's estimation of suitability of code s_i for solving problem p.

In SAIVS continuous functions are approximated by piecewise linear functions. Then continuous and discrete functions are implemented by a set of ordered pairs of values which makes knowledge acquisition and representation simple, clear and reliable.

Altogether the above approach to solving the software selection problem impresses by its constructive procedure. As a module of an intelligent software system it contributes much to achieve the aim formulated in Section 1, since particularly the possibility to express nonlinear relations in the numerical context is very valuable.

a) characteristic feature	`Prop = matrixStructure`
	`Type = enumeration`
	`Attr = dense`
	`Attr = triangular`
	`Attr = banded`
	`Attr = sparse`
b) Proplog rule	`Fact = Interval Gaussian Algorithm`
	`Cond = dense`
	`Cond = (dimension < 100)`
c) intensity function	`Prop = loss of accuracy`
	`x,fx =    3.0   0.0`
	`x,fx =   10.0   0.35`
	`x,fx =   18.0   1.0`

Figure 6. Examples from the knowledge base

For the discussion of details we now look at a symbolic knowledge base for systems of linear equations which can be split into the five text files *Property*, *Rule*, *Meta-Rule*, *Intensity* and *AlgoPKF*. Although the prototype currently includes more solving methods, we only consider knowledge related to the codes $S1$ and $S2$ of Section 1.

The first file is relevant in Phase 1 and 2, the second and third file only relate to Phase 1, the last two to Phase 2. Furthermore, the symbolic knowledge base includes the symbols for all solving methods and read and write routines defined in the coupling module; however these have to be defined before compilation since they include references to procedures and therefore build the bridge from the knowledge-based to the numerical part. In the coupling module an instance is created for each algorithm of the system consisting of a reference to the corresponding code and the name of the algorithm. From a knowledge-based data processing point of view the mentioned objects of the coupling module are symbols which are used in rules in the same way as the characteristic features and their attribute values defined in the Property.

The contents of file *Property* determines the segmentation of problems of a numerical domain into problem classes basing on characteristic features. In our prototype these are only attributes of the coefficient matrix (see Fig. 6). Recorded are: dimension, required accuracy, loss of accuracy, structure, type of the coefficients, symmetry, diagonally dominance, *delta*, bandwidth, sparsity. Other reasonable characteristic features are: spectral radius, arithmetic sign, reducibility, number of right-hand sides, number of matrices with same non-zero pattern, etc. The last two features refer to series of linear systems to be solved. In most cases of characteristic features with enumerable domain mathematical theory uniquely determines which attributes can occur.

The file *Rule* consists of Horn clauses modelling mathematical expert knowledge. Their number is still small and the structure of the implemented knowledge is more "wide" than "deep". This means that there are no long conclusion chains but mainly dependencies of type "Algorithm X is applicable, if condition1, condition2, …" are valid. Mostly the conditions are verified by queries to the user or procedural attachments rather than by deductions.

The text file *MetaRule* stores rules which represent information for managing the dialogue with the user: For controlling the resolution state and deciding which property to resolve next further symbols are automatically generated based on the rules of this file. These special symbols serve to define the sequence of questions. Features that are resolved or irrelevant at a certain point of inference due to previous answers are suppressed for queries. For instance interval valued components of a matrix imply that the matrix is not symmetric, hence this property is resolved, too. Further, meta rules can be used to allow direct execution of an available problem solving routine as soon as the numerical data are input.—This way the possibility for very flexible navigation facilities can be provided.

In the files *Intensity* and *AlgoPKF* mainly PK functions are defined in textual form (see Fig. 6c). The processing of these functions is demonstrated by our concrete example.

In the beginning the intention was to reproduce the objectives elaborated in [3] for the decision between software modules $S1$ and $S2$:

> **if** $dim \leq 6$
> **or** $dim \leq 9$ **and** $delta > 1.05$ **and** $loss > 4$
> **or** $dim \leq 12$ **and** $10 < delta < 50$ **and** $loss > 7$
> **then** invoke software module $S2$

So we have the characteristic features:

- system size *dim*
- variance of absolute values *delta*
- difference between input[4] and output accuracy *loss* (as number of decimal digits)

Already the attempt to define an intensity function for *dim* shows that the statement "large system" is relative. A dimension which is "large" for dense systems is "medium" or even "small" for sparse systems. As future extensions the introduction of additional characteristic features (for example dimFull and dimSparse) could be helpful: the user can input dim, the system assigns this value internally to the new characteristic features and all primitive knowledge functions use either the one or the other. Then sparsity could be determined first and then an appropriate intensity function for system size could be applied.—The internal assignment can be generalized to an automated determination of a characteristic feature. For example the

---

[4] Actually we consider the interval versions of both algorithms.

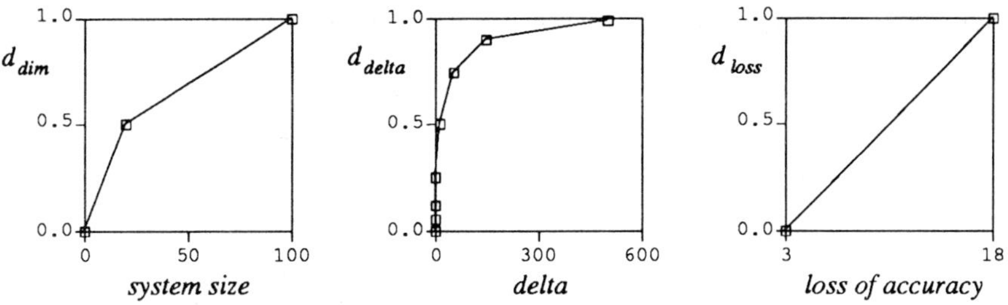

Figure 7. Intensity functions of case study for software module $S2$

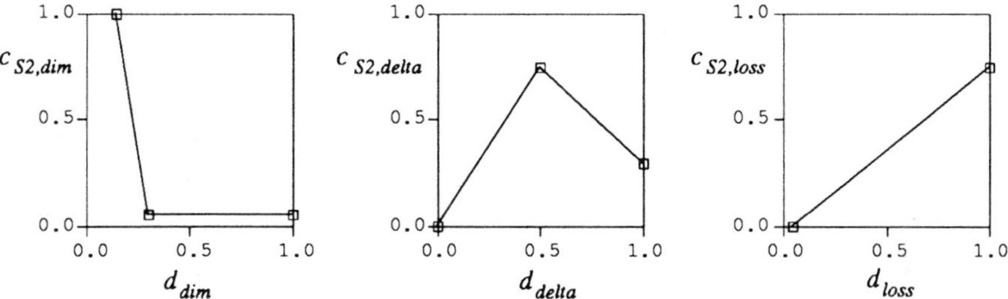

Figure 8. Compatibility functions of case study for software module $S2$

feature *loss* could be immediately computed by one subtraction if the precision of the input data and the user required accuracy are known.

So we consider at first only dense matrices and define I_{dim}, I_{delta} and I_{loss} as in Fig. 7.

Next we must choose the compatibility functions for $S2$. Familiar with the above objective the functions $C_{S2,dim}$, $C_{S2,delta}$ and $C_{S2,loss}$ seem to be reasonable (Fig. 8).

With the aggregation function

$$A_{S2}(c_{S2,dim}, c_{S2,delta}, c_{S2,loss}) := \max(c_{S2,dim}, \alpha(c_{s2,delta}, c_{s2,loss}))$$

we finally get a real-valued quantification of the applicability of software module $S2$ for given m_{dim}, m_{delta} and m_{loss}.

Up to now the actual runtime of a software module is not yet introduced into the considerations. Routine $S1$ is slower than $S2$, while earning value 1 at all three characteristic features. We assume that a feature *runtime* could be embedded such that $S1$ always receives 0.6 as global valuation. Then we can test our approach by comparing the above boolean decision to apply $S2$ with the valuation above or below this value. Independently varying the dimension from 1 to 25, the accuracy requirement from 2 to 14 and *delta* over 16 values between 1.03 and 300 we produce 5200 test examples. It turns out that both approaches come to the same result in 93 percent of the test cases (1716 pros and 3139 cons for $S2$).

The assumption in the example that a value can be kept constant during aggregation or multiplied in an appropriate manner is not correct in general. If two methods show the same behaviour in most cases but one method halves its effort for symmetric matrices and the second does not, then one intuitively intends to say that the first one is better suited for the symmetric case with factor 2 than the other one. This way of thinking can not be represented directly using *min, max, α*.

In order to implement heuristics in the PK function calculus correctly it is essential to get a "feeling" for the compensation operator $\alpha(x, y)$. Figure 5 shows a 3-dimensional image. The appropriate definition of the points of neutral or extreme intensity and compatibility is particularly important. Then we obtain a reasonable global behaviour as shown by the prototype.

4. Extendability and Portability

To support the types 2 and 3 of possible applications (see Section 2) the highest priority in the design of the problem solving environment is assigned to the extendability of the system. A developer, usually a domain expert, can extend the collection of algorithms, insert additional noteworthy characteristic features of the problem space and add rules or primitive knowledge functions which express relations between these objects. Extensions of the knowledge base will be automatically involved in all following consultations.—For achieving this goal the minimization of the interface between modules of different system parts endeavored to reduce modifications of existing code and to avoid recompilations as far as possible.

Extensions and changes to the knowledge base of the logic part (Phase 1 and 2) do not require any compilation or binding. Without any programming an expert system engineer is able to adapt the characteristic features and the logical rules in an executable system according to mathematical theory—just by editing the text files *Property*, *Rule* and *MetaRule*.

Extensions of Phase 2 knowledge results in additions of primitive knowledge functions according to desired heuristics—just by editing the text files *Property*, *Intensity* and *AlgoPKF*. Interpreting the knowledge base as a set of software evaluation networks, the insertion of a new code s_{k+1} means establishing a further software evaluation network, i.e. n new compatibility functions $C_{k+1,1}, C_{k+1,2}, \ldots,$ $C_{k+1,n}$ and a new aggregation function A_{k+1} must be defined. The insertion of an additional characteristic feature f_{n+1} requires a new measurement function M_{n+1}, a new intensity function I_{n+1}, k new compatibility functions $C_{1,n+1}, C_{2,n+1}, \ldots, C_{k,n+1}$ and the adaptation of $A_1, A_2, \ldots, A_k$. In particular the definition of the $C_{i,j}$ is difficult since all the other $C_{i,j}$ must be considered to assure that all relations are correct. Special scenario construction algorithms are discussed in [5] to support the knowledge engineer who is not used to isolating knowledge components in terms of PK functions.

Since the stored knowledge for Phase 3 consists of programs, each change means "programming, compiling and binding". The execution of numerical routines is invoked by logical calculations. Procedures of Phase 3 must be linked to symbols which are manipulated in Phases 1 and 2. These special symbols are localized in the coupling module which must be modified.

A highly developed expert system considers the runtime of the numerical software modules as an essential characteristic feature as discussed in Section 3. This requires an a priori runtime estimation for each algorithm that passed Phase 1. The conventional approach in numerics is to calculate an estimation by counting all involved floating-point operations. For example, the factorization of a matrix by Gauss elimination has $dim * (dim - 1) * (2 * dim - 1)/6$ additions and multiplications and $dim * (dim - 1)/2$ divisions. In theory the necessary total runtime can be easily calculated by inserting the actual runtimes of the single operations on the used computer. In practice large deviations of this theoretical value may occur during execution of a corresponding software module since the estimation is only valid under the assumption that all the other time contributions during program execution (as there are subroutine calls, loops, jumps, comparisons, restoring, etc.; further on parallel machines: communication, synchronisation, etc.) are negligible compared to the time consumed by floating-point operations. However this is not true in many cases since mathematical coprocessors, fast caches, powerful vector units, etc. essentially increase the Mflops rate.

Validating numerical computations introduce an additional parameter: The calculation of precise enclosures often requires the optimal scalar product instead of the common floating-point operations. In contrast to these the scalar product is mostly provided in software yielding a runtime disadvantage of factor 2 to 10. This fact is not registered by operation count. Furthermore, if the floating-point arithmetic does not fulfill the requirements of validating numerics, a proper arithmetic completely implemented in software must be used changing the relations between the runtimes of single operations and consequently of all software modules applying those operations.

An experiment confirms these considerations. Five different hardware and software configurations were used to factorize a random interval matrix of dimension 25 ($\frac{2}{3} dim^3$ operations), to calculate an approximate inverse (dim^3 operations) and to enclose the resulting residual ($2\,dim^3$ operations). Table 3 displays the results. All values are standardized with respect to the runtime for the inverse. One would expect that also the values in the first line and in the third line are equal but the actual deviation is up to 400 percent. The result is even worse if machines other than personal computers are compared.

Table 3. Configuration dependent runtimes of validating numerical routines

	Mac II fx with coprocessor with assembler	Mac II fx with coprocessor no assembler	Mac II fx no coprocessor no assembler	Sam (68020)	Atari (68000)
factorization	2.7	4.1	2.1	0.9	1.1
inversion	1	1	1	1	1
residue	3.9	7.5	3.2	1.9	3.0

Therefore, an adequate automatic adaptation of the symbolic knowledge base after a portation to another computer environment maintains its relevance. For that purpose the expert system itself must observe the behavior of the computer configuration, store experiences in a suitable manner and consider them later on. Since these are purely quantitative reflections, only Phase 2 is involved.

Between a portation and the first consultation a system-independent "installation program" should be executed once to automate the acquisition of information on actual runtimes—this may be regarded as the last step of any portation. Then all information necessary for this extraordinary run is read from the symbolic knowledge base and some numerical problems are solved to measure execution times which are added to the knowledge base by the system. Since complete software modules are tested, the data are obtained within a relevant context (compare previous text and Table 3). In this way a further relation of PK function type for each routine is defined which may be embedded into the quantitative evaluation scheme presented in Section 3 and allows estimations of runtimes during a consultation.

The more experience is collected on runtimes during the installation process, the more time is spent. In real life a reasonable limit will exist (i.e. some minutes or hours) so that only a small number of constellations from the multi-dimensional space of runtime-relevant characteristic features can be tested. Further, increasing the number of problem solving routines additionally decreases the time available for each method.

Assuming the runtime depends on just one characteristic feature (in our exemplary software modules S1 and S2 this is the dimension of the matrix), numerical problems of different values for this quantity are solved and the gained relation between this property and the runtime with discrete values is used to define a continuous function that approximates the true dependency for arbitrary values of this characteristic feature. For obvious reasons we choose the class of piecewise linear functions for modelling runtime estimations.[5]

In practice we can define the values to be tested similar to the independent parameters of PK functions in file *AlgoPKF*: one value per line. During execution of the installation program the system puts the measured runtimes behind the test value into the same line without assistance. Here is an extract of the prototype's *AlgoPKF*.

$$\text{Time} = \text{dimension}$$

$$
\begin{array}{lll}
p, t = & 8 & 133 \\
p, t = & 18 & 650 \\
p, t = & 51 & 8917 \\
p, t = & 95 & 50400
\end{array}
$$

---

[5] For advantages compared with other functions, e.g. polynomials, see [4].

For the software selection a runtime estimation of each applicable routine is needed. The above data are interpreted and evaluated as **PK** functions[6] for a certain routine.

The estimated value replaces the measurement result $m_{runtime}$ within the SEN scheme (see Fig. 4). The intensity function $I_{runtime}$ must map the interval from the smallest to the largest estimated runtime t_{max} of all applicable routines onto $L = [0, 1]$. For the prototype we chose the very simple, but sound definition

$$d_{runtime} = I_{runtime}(t) = \frac{t}{t_{max}}.$$

Next compatibility functions $C_{s_i, runtime}$ must be defined. How different runtime intensities are judged depends on the user, for example:

- runtimes may be indifferent to him since he is only interested in verification of results
- runtimes may be indifferent to him only if they are shorter than a certain bound and not acceptable otherwise
- he judges the algorithm the less suitable, the longer it takes to solve the problem

As a first approach it is appropriate to use the same mapping $C_{runtime}$ for all routines since all users demand short execution times. We defined $C_{runtime}$ such that the most probable situation is modelled. From experiments with the prototype a formula arised which is defined by two pairs of values: $(0, 0.75)$, $(1, 0.25)$ or

$$C_{runtime}(d_{runtime}) = 0.75 - \frac{d_{runtime}}{2}.$$

Using $C_{runtime}$ yields plausible results if compensatory interaction of runtime compatibility and overall suitability without respect to runtime is assumed.

In the prototype the runtime relevant feature need not be the dimension. It may be arbitrarily chosen from the set of characteristic features F for each integrated routine and must be coordinated with the test data generating routine. For instance, when estimating the runtime of sparse linear system solving routines the sparsity or the envelope may be much more appropriate.

The method described in this section is almost embedded into the SEN scheme and works fine as long as the number of operations is roughly fixed a-priori. This is not fulfilled when solving linear systems with iterative algorithms. Even good understanding of these methods can not prevent large variation due to data dependencies which are not captured by the approach presented above. Methods with two runtime relevant parameters were developed and successfully applied in the prototype ([4]).

---

[6] The function is continuously extended with the gradient of the first (last) interval below (above) the first (last) tested value.

References

[1] Houstis, E. N., Rice, J. R., Vichnevetsky, R. (eds.): Expert systems for scientific computing. Proceedings of the 2nd IMACS International Conference on Expert Systems for Numerical Computing (Purdue University, USA, 1990). Amsterdam: North-Holland 1992.

[2] König, S., Ullrich, C. P.: Towards an expert system for solving systems of linear equations. Proceedings of Beijing International Conference on System Simulation and Scientific Computing *I*, 1–5 (1989).

[3] König, S., Ullrich, C. P.: An expert system for the economical application of self-validating methods for linear equations. In [1], pp. 195–220.

[4] König, S.: Über Expertensysteme für wissenschaftliches Rechnen mit Ergebnisverifikation. Dissertation, Univesität Basel, 1992.

[5] Lucks, M., Gladwell, I.: A functional representation for software selection expertise. In [1], pp. 421–459.

[6] Maier, D., Warren, D. S.: Computing with logic. Menlo Park: Benjamin/Cummings 1988.

[7] Rump, S. M.: Solving algebraic problems with high accuracy. In: Kulisch, U. Miranker W. L. (eds.) A new approach to scientific computation, pp. 51–120. New York: Academic Press 1983.

[8] Schulze, K. Cryer, C. W.: NAXPERT: A prototype expert system for numerical software. SIAM J. Sci. and Stat. Comp. *9*, 503–515 (1988).

S. König
C. P. Ullrich
Institut für Informatik
Universität Basel
Mittlere Strasse 142
CH-4056 Basel
Switzerland

Computing, Suppl. 9, 147–159 (1993)

Computing
© Springer-Verlag 1993
Printed in Austria

Computation of Interval Bounds for Weierstrass' Elliptic Function $\wp(z)$*

W. Krämer and **B. Barth**, Karlsruhe

Dedicated to Professor U. Kulisch on the occasion of his 60th birthday

Abstract — Zusammenfassung

Computation of Interval Bounds for Weierstrass' Elliptic Function $\wp(z)$. A method to enclose the values of Weierstrass' elliptic function $\wp(z) = \wp(z|g_2, g_3)$ for arbitrary complex invariants g_2, g_3 or arbitrary given zeros of the characteristic polynomial (arbitrary period lattices) is presented. The function is approximated by its truncated Laurent series at zero. An error bound is derived for the remainder term. If necessary, the periodicity of $\wp$, the homogeneity relations and the addition formulas are used to perform the reduction of the argument and the corresponding adaptation of the result.

AMS Subject Classification: 33E05, 65D20, 65G10

Key words: Elliptic functions, enclosure methods, XSC-languages.

Berechnung von Intervalleinschließungen für Werte der Weierstraßschen elliptischen Funktion $\wp(z)$. Eine Methode zur Intervalleinschließung von Werten der Weierstraßschen elliptischen Funktion $\wp(z) = \wp(z|g_2, g_3)$ für beliebige komplexe Invarianten g_2, g_3 bzw. beliebig vorgegebene Nullstellen des charakteristischen Polynoms (d. h. beliebige Periodengitter) wird vorgestellt. Die Funktion wird durch ihre abgebrochene Laurent-Reihe um den Nullpunkt approximiert. Für den Fehlerterm wird eine Fehlerabschätzung hergeleitet. Eine eventuell notwendige Argumentreduktion und die zugehörige Ergebnisanpassung werden mit Hilfe der Periodizität von $\wp$, der Additionstheoreme und der Homogenitätsbeziehungen durchgeführt.

1. Introduction

Elliptic functions play an important role in many technical applications. These functions arise in two dimensional problems in hydrodynamic, in the field of Elasticity Theory and in the field of Electrical Engineering treating conformal mappings of polygonal regions onto the unit circle. For example, they are directly needed in calculating electrical filters [19].

Weierstrass' elliptic functions are also relevant in the computation of the potentials of crystal grids [11], in the motion of a rigid body with one fixed point [5] and in applications of the theory of type-II superconductors and the computation of vortices that arise by rotation of superfluid helium II [8].

---

* Received November 16, 1992; revised December 21, 1992.

Frequently, numerical values of Weierstrass' $\wp$-function are computed using tables for the computation of elliptic functions and elliptic integrals ([7], [17], [20], [21], [23] ...) or using nomograms [18]. For the direct computation on a machine some polynomial approximations can be found in literature [1]. The few material found in literature describing numerical methods for the computation of the $\wp$-function, is restricted to real invariants and gives no error estimation. Only one exception is known by the authors: Eckhardt ([9], [10]) describes a numerical method for the computation of the equianharmonic and the lemniscatic case giving reliable error estimations depending on the data format used. However, he uses a very special method which can not be extended to the more general cases of general real or general complex invariants.

In the following we will describe a method for the computation of interval bounds (save bounds) for the values of the $\wp$-function for arbitrary real and arbitrary complex invariants. For the implementation of our proposed method we take advantage of programming languages especially well suited for the needs of scientific computations and computations with (automatic) result verification. These so called XSC-languages (PASCAL-XSC, ACRITH-XSC, C-XSC) offer among other things a complex interval arithmetic to the user ([2], [14], [16]).

2. General Remarks on Elliptic Functions

Let us cite from [1]:

An elliptic function is a single-valued doubly periodic function of a single complex variable which is analytic except at poles and whose only singularities in the finite plane are poles. If ω and ω' are a pair of (primitive) half-periods of such a function $f(z)$, then $f(z + 2m\omega + 2n\omega') = f(z)$, m and n being integers. Such the study of any such function can be reduced to consideration of its behavior in a fundamental period parallelogram (**FPP**). An elliptic function has a finite number of poles (and the same number of zeros) in a FPP; the number of such poles (zeros) (an irreducible set) is called the **order** of the function (poles and zeros are counted according to their multiplicity). All other poles (zeros) are called congruent to the irreducible set. The simplest (nontrivial) elliptic functions are of order two. One may choose as the standard function of order two either a function with two simple poles (Jacobi's choice) or one double pole (Weierstrass' choice) in a FPP.

3. Weierstrass' Elliptic Function $\wp(z)$

Let 2ω, $2\omega'$ be a fixed pair of complex numbers (primitive periods) with $\Im(\omega/\omega') > 0$ and

$$\omega_{mn} = 2m\omega + 2n\omega', \qquad m, n \text{ integers}.$$

Weierstrass' function $\wp(z) = \wp(z|2\omega, 2\omega')$ is an elliptic function of periods 2ω, $2\omega'$ which is of order two, has a double pole at $z = 0$, the principal part of the function at this pole being z^{-2}, and for which $\wp(z) - z^{-2}$ is analytic in a neighborhood of, and vanishes at $z = 0$. An explicit formula is [25]

$$\wp(z) = \wp(z|2\omega, 2\omega') = \frac{1}{z^2} + {\sum_{m,n}}' \left[\frac{1}{(z - 2\omega_{mn})^2} - \frac{1}{(2\omega_{mn})^2} \right]. \tag{1}$$

The notation $\sum'$ means that the summation has to be taken over all integers m, n with the exception $m = n = 0$. With the abbreviations

$$g_2 = g_2(2\omega, 2\omega') := 60 {\sum}' \frac{1}{(2\omega_{mn})^4}, \tag{2}$$

$$g_3 = g_3(2\omega, 2\omega') := 140 {\sum}' \frac{1}{(2\omega_{mn})^6}, \tag{3}$$

the Laurent series expansion of $\wp(z)$ in a neighborhood of the origin is [1]

$$\wp(z) = \frac{1}{z^2} + \sum_{k=2}^{\infty} c_k z^{2k-2}. \tag{4}$$

The values g_2 and g_3 are called **invariants**. All coefficients c_k in (4) are expressible using these invariants:

$$c_2 = \frac{g_2}{20}, \qquad c_3 = \frac{g_3}{28}, \qquad c_k = \frac{3}{(k-3)(2k+1)} \sum_{i=2}^{k-2} c_i c_{k-i} \qquad k \geq 4. \tag{5}$$

This motivates the notation $\wp(z) = \wp(z|2\omega, 2\omega') = \wp(z|g_2, g_3)$.

The first derivative $\wp'(z)$ is given by the following Laurent-series:

$$\wp'(z) = \frac{-2}{z^3} + \sum_{k=2}^{\infty} (2k - 2)c_k z^{2k-3}. \tag{6}$$

This function is an elliptic function of order three. The half-periods ω, ω', $\omega'' := \omega + \omega'$ are the three simple zeros

$$\wp'(\omega) = 0, \qquad \wp'(\omega') = 0, \qquad \wp'(\omega'') = 0 \tag{7}$$

in the nondegenerated FPP. The function values of $\wp(z)$ at the half-periods are denoted by

$$e_1 := \wp(\omega), \qquad e_2 := \wp(\omega''), \qquad e_3 := \wp(\omega'). \tag{8}$$

A differential equation for $\wp(z)$ is as follows:

$$(\wp'(z))^2 = 4\wp^3(z) - g_2 \wp(z) - g_3. \tag{9}$$

With the notation $s := \wp(z)$ the **characteristic polynomial** $p(s) := 4s^3 - g_2 s - g_3$ has the three distinct zeros e_1, e_2, e_3 (these values are given by (8)) and it follows by (9):

$$e_1 + e_2 + e_3 = 0$$

$$e_1 e_2 + e_2 e_3 + e_3 e_1 = -g_2/4 \tag{10}$$

$$e_1 e_2 e_3 = g_3/4.$$

The quantity

$$\Delta := g_2^3 - 27g_3^2 = 16(e_2 - e_3)^2(e_3 - e_1)^2(e_1 - e_2)^2 \tag{11}$$

is called the **discriminant**.

Later on we also need the **homogeneity relations**

$$\wp(tz|t2\omega, t2\omega') = \frac{1}{t^2}\,\wp(z|2\omega, 2\omega'), \tag{12}$$

$$\wp'(tz|t2\omega, t2\omega') = \frac{1}{t^3}\,\wp'(z|2\omega, 2\omega') \tag{13}$$

or equivalently

$$\wp(z|2\omega, 2\omega') = t^2\,\wp(tz|t2\omega, t2\omega'), \tag{14}$$

$$\wp'(z|2\omega, 2\omega') = t^3\,\wp'(tz|t2\omega, t2\omega') \tag{15}$$

as well as some **addition formulas**

$$\wp(u + v) = \frac{1}{4}\left[\frac{\wp'(u) - \wp'(v)}{\wp(u) - \wp(v)}\right]^2 - \wp(u) - \wp(v) \tag{16}$$

$$\wp(2u) = \frac{(\wp^2(u) + \tfrac{1}{4}g_2)^2 + 2g_3\,\wp(u)}{4\wp^3(u) - g_2\,\wp(u) - g_3} \tag{17}$$

$$\wp'(2u) = \frac{-4(\wp'(u))^4 + 12\wp(u)(\wp'(u))^2\wp''(u) - (\wp''(u))^3}{4(\wp'(u))^3}. \tag{18}$$

4. Numerical Computation of $\wp(z)$

In the neighborhood of the origin the Laurent series expansion (4), (5) for the $\wp$-function is used. The power series in (4) converges for values of z lying in a circle of radius $|2\omega^*|$ and the origin as midpoint. The quantity $2\omega^*$ denotes the period with smallest modulus.

Expansion (4) is split up into one part with the first $N - 1$ summands ($N \geq 2$) and the remainder term R_N:

$$\wp(z) = \frac{1}{z^2} + \sum_{k=2}^{N} c_k z^{2k-2} + R_N. \tag{19}$$

The function $\wp(z)$ will be approximated by $\wp_N(z)$, with

$$\wp_N(z) := \frac{1}{z^2} + \sum_{k=2}^{N} c_k z^{2k-2}, \tag{20}$$

leading to the approximation error

$$R_N := \sum_{k=N+1}^{\infty} c_k z^{2k-2} = \sum_{k=N+1}^{\infty} c_k(z^2)^{k-1} = z^{2N}\sum_{k=0}^{\infty} c_{N+1+k}(z^2)^k. \tag{21}$$

An upper bound for the approximation error will now be derived. For this purpose we define the series (γ_k) with $\gamma_2 = \gamma_3 := 1$ and

$$\gamma_k := \frac{3}{(k-3)(2k+1)} \sum_{i=2}^{k-2} \gamma_i \gamma_{k-i}, \qquad k \geq 4. \tag{22}$$

If $|c_2| \leq \gamma_2$ and $|c_3| \leq \gamma_3$ holds, the same relation is true for all corresponding elements, i.e.

$$0 \leq |c_k| \leq \gamma_k \qquad \text{for } k \geq 2. \tag{23}$$

Additionally, the series (γ_k) is monotone decreasing, i.e.

$$\gamma_i \leq \gamma_j \qquad \text{for } i > j \geq 2 \tag{24}$$

with $\lim_{k \to \infty} \gamma_k = 0$. By induction one can show (see below)

$$\gamma_{k+1} \leq \frac{2k+1}{2k+3} \gamma_k \leq \gamma_k, \qquad k \geq 3, \tag{25}$$

which proofs these results.

Proof: For $k = 3$ the relation

$$\gamma_{k+1} = \gamma_4 = \tfrac{1}{3} \leq \gamma_3 = \gamma_k$$

holds. For $k > 3$ we find:

$$\gamma_{k+1} = \frac{3}{(k-2)(2k+3)} \sum_{i=2}^{k-1} \gamma_i \gamma_{k+1-i}$$

$$= \frac{3}{(k-2)(2k+3)} (\gamma_2 \gamma_{k-1} + \gamma_3 \gamma_{k-2} + \cdots + \gamma_{k-2} \gamma_3 + \gamma_{k-1} \gamma_2)$$

Let M be the index with the property

$$\gamma_M \gamma_{k-M} = \min_{2 \leq i \leq k-2} \{\gamma_i \gamma_{k-i}\}.$$

This gives

$$\gamma_{k+1} = \frac{3}{(k-2)(2k+3)} (\gamma_2 \gamma_{k-1} + \gamma_3 \gamma_{k-2} + \cdots + \gamma_M \gamma_{k-M+1}$$

$$+ \gamma_{M+1} \gamma_{k-M} + \cdots + \gamma_{k-2} \gamma_3 + \gamma_{k-1} \gamma_2).$$

Using the relation $\gamma_{i+1} \leq \gamma_i$ for $2 \leq i \leq k - 2$ yields

$$\gamma_{k+1} \leq \frac{3}{(k-2)(2k+3)} (\gamma_2 \gamma_{k-2} + \gamma_3 \gamma_{k-3} + \cdots + \gamma_M \gamma_{k-M}$$

$$+ \gamma_M \gamma_{k-M} + \cdots + \gamma_{k-3} \gamma_3 + \gamma_{k-2} \gamma_2)$$

which is equivalent to

$$\gamma_{k+1} \leq \frac{3}{(k-2)(2k+3)} \left[\gamma_M \gamma_{k-M} + \sum_{i=2}^{k-2} \gamma_i \gamma_{k-i} \right].$$

Since $\gamma_M \gamma_{k-M}$ is equal to the least of the $k - 3$ summands it follows

$$\gamma_M \gamma_{k-M} \le \frac{1}{k-3} \sum_{i=2}^{k-2} \gamma_i \gamma_{k-i}$$

and therefore

$$\gamma_M \gamma_{k-M} + \sum_{i=2}^{k-2} \gamma_i \gamma_{k-i} \le \frac{k-2}{k-3} \sum_{i=2}^{k-2} \gamma_i \gamma_{k-i} \, .$$

Using the definition of γ_k yields

$$\frac{k-2}{k-3} \sum_{i=2}^{k-2} \gamma_i \gamma_{k-i} = \frac{k-2}{k-3} \left[\frac{(k-3)(2k+1)}{3} \gamma_k \right] = \frac{(k-2)(2k+1)}{3} \gamma_k \, .$$

This finally shows that inequality (25).

The proof may be extended to starting values $\gamma_2 = \gamma_3 := S$, with $\gamma_4 = \gamma_2^2/3 \le \gamma_3$, i.e. to starting values S with $0 \le S \le 3$.

Using (25) repeatedly we find

$$\Gamma_{N+1} := \max_{k > N} \{|c_k|\} \le \gamma_{N+1} \le \prod_{k=4}^{N} \gamma_4 \frac{2k+1}{2(k+1)+1} = \frac{3}{2N+3} \gamma_2^2 \qquad (26)$$

and for z with $|z| < 1$:

$$|R_N| \le |z^2|^N \sum_{k=0}^{\infty} |c_{N+1+k}||z^2|^k \le \frac{|z^2|^N}{1 - |z^2|} \max_{k > N} \{|c_k|\} = \frac{\Gamma_{N+1} |z^2|^N}{1 - |z^2|} \, . \qquad (27)$$

Remark: In the case of real invariants, the error estimation can be sharpened ([3]).

The estimation (25) is only valid for $|c_2| \le 3$ and $|c_3| \le 3$. For the invariants g_2, g_3 this means that $|g_2| \le 60$ and $|g_3| \le 84$ has to be satisfied.

This being not the case, the FPP given by 2ω, $2\omega'$ can be altered using a suitable factor t in the homogeneity relation (12).

We choose the factor t in the following way:

$$\frac{|g_2|}{t^4} \le 60 \qquad \text{and} \qquad \frac{|g_3|}{t^6} \le 84$$

or equivalently

$$t \ge \sqrt[4]{\frac{|g_2|}{60}} \qquad \text{and} \qquad t \ge \sqrt[6]{\frac{|g_3|}{84}} \, . \qquad (28)$$

For the concrete implementation $t := 2^n$ (exact representable in IEEE data format) is used with

$$n \ge \max \left\{ \frac{1}{4} \log_2 \left(\frac{|g_2|}{60} \right), \frac{1}{6} \log_2 \left(\frac{|g_3|}{84} \right) \right\} \, . \qquad (29)$$

For the new invariants $\dot{g}_2 := g_2/t^4$ and $\dot{g}_3 := g_3/t^6$ we can dominate the moduli of the coefficients (c_k) of the Laurent series by the corresponding coefficients (γ_k).

Now, for arguments z with $|z| < 1$ we can estimate the maximum truncation error by (27).

The modification of the invariants leads to the new fundamental periods $t2\omega$, $t2\omega'$, the absolute values of which are greater than 1. So, there are always arguments z in the new FPP with $|z| > 1$. For those arguments we can not apply the estimation (27) directly. In this situation we first compute an enclosure for $\wp(z/2^n)$ with n being chosen in such a way that $|z/2^n| < 1$ (notice, that in the actual implementation $z/2^n$ is computed by an exponent shift of z without any rounding error). Finally we apply (17) n-times to get a verified enclosure of $\wp(z)$.

5. Argument Reduction for z Outside the FPP

For the computation of enclosures of $\wp(z)$ for arguments z which are not elements of the fundamental period parallelogram, we first have to perform an argument reduction. For this reduction the fundamental periods 2ω, $2\omega'$ must be known (we assume, that the actual function $\wp$ is given by the numerical values of its invariants g_2, g_3. This seems to cover most applications since in general $\wp$ is given by its differential Eq. (9) i.e. by g_2 and g_3). The computation of enclosures of the fundamental periods is done in two steps:

Step 1: compute the zeros $\{e_1, e_2, e_3\}$ of the characteristic polynomial $p(s) = 4s^3 - g_2 s - g_3 = 0$.
Step 2: compute 2ω, $2\omega'$ via complete elliptic integrals of the first kind using the zeros found in step 1.

The solutions of the equation $4s^3 - g_2 s - g_3 = 0$ $(g_2^3 - 27g_3^2 \neq 0)$ are denoted by e_1, e_2, e_3. The numbering is done in such a manner that the quantities

$$k^2 := \frac{e_2 - e_3}{e_1 - e_3}, \qquad k'^2 := \frac{e_1 - e_2}{e_1 - e_3}, \qquad (30)$$

are of modulus < 1.

The values ω and ω' with

$$\omega := \frac{K}{\sqrt{e_1 - e_3}}, \qquad \omega' := \frac{iK'}{\sqrt{e_1 - e_3}} \qquad (31)$$

determine a fundamental period parallelogram associated with the invariants g_2, g_3. The complex square root always denotes the principal value, i.e. $\Re(\sqrt{z}) \geq 0$. See [24], p. 30 ff.

For the computation of the complete elliptic integrals of the first kind

$$K = K(k) := \int_0^{\pi/2} \frac{d\phi}{\sqrt{1 - k^2 \sin^2 \phi}} \qquad \text{and} \qquad K' = K(k') := K(\sqrt{1 - k^2}) \qquad (32)$$

we use the series expansion[1] ([24], p. 50)

$$K(k) = \frac{\pi}{2} \sum_{n=0}^{\infty} \frac{[(\frac{1}{2})_n]^2}{(n!)^2} k^{2n} = \frac{\pi}{2} \sum_{n=0}^{\infty} \left(\prod_{m=1}^{n} \frac{2m-1}{2m} \right)^2 k^{2n}$$

$$= \frac{\pi}{2} \left[1 + \left(\frac{1}{2} \right)^2 k^2 + \left(\frac{1 \cdot 3}{2 \cdot 4} \right)^2 k^4 + \left(\frac{1 \cdot 3 \cdot 5}{2 \cdot 4 \cdot 6} \right)^2 k^6 + \cdots \right], \tag{33}$$

which is convergent for arguments k lying in the unit circle.

The numbering of the zeros of $p(s)$ can be done in such a way, that $|k|$, $|k'| < 1$ and $0 \leq \Re(k)$, $\Re(k') < 1$ (k, k' are defined by (30)).

For values $|k|$ or $|k'|$ very near to 1 the rate of convergence in (33) is very poor. In such a case, the so called Landen transformation (see [1])

$$\dot{k} = \frac{1-k'}{1+k'} = \frac{(1-k')(1+k')}{(1+k')^2} = \frac{1-k'^2}{(1+k')^2} = \frac{k^2}{(1+k')^2} \tag{34}$$

$$\dot{k}' = \sqrt{1 - \left(\frac{1-k'}{1+k'} \right)^2} = \sqrt{\frac{(1+k')^2 - (1-k')^2}{(1+k')^2}} = \frac{2\sqrt{k'}}{1+k'} \tag{35}$$

can be applied. If the real parts of k and k' are positive and smaller than 1 it holds

$$\left| \frac{k^2}{(1+k')^2} \right| < |k^2| < |k|. \tag{36}$$

Applying the transformation repeatedly the value of k rapidly tends to 0, $k' = \sqrt{1 - k^2}$ to 1 and $K = K(k)$ to $K(0) = \frac{\pi}{2}$.

With the notation

$$k_0' := k', \qquad k_{n+1}' := \frac{2\sqrt{k_n'}}{1+k_n'} \tag{37}$$

and an iterative application of

$$K(\dot{k}) = K\left(\frac{1-k'}{1+k'} \right) = \frac{1+k'}{2} K(k) \tag{38}$$

we get (see [12]):

$$K(k) = \frac{\pi}{2} \prod_{n=0}^{\infty} \frac{2}{1+k_n'}. \tag{39}$$

Now we can use the computed enclosures of ω and ω' to get an enclosure of the reduced argument. We choose integers m and m' in such a manner, that the argument $\dot{z} = z/t$ is a point lying in that period parallelogram determined by the corners

---

[1] The notation $(a)_n$ means the product $a(a+1)(a+2)\ldots(a+n-1)$.

$$m2\omega + m'2\omega',$$
$$m2\omega + (m' + 1)2\omega',$$
$$(m + 1)2\omega + (m' + 1)2\omega' \qquad \text{and} \tag{40}$$
$$(m + 1)2\omega + m'2\omega'.$$

The four points

$$\dot{z} - (m2\omega + m'2\omega'),$$
$$\dot{z} - (m2\omega + (m' + 1)2\omega'),$$
$$\dot{z} - ((m + 1)2\omega + (m' + 1)2\omega'), \tag{41}$$
$$\dot{z} - ((m + 1)2\omega + m'2\omega')$$

are all congruent to $\dot{z}$. As the reduced argument $\ddot{z}$ we chose the point with smallest modulus.

6. Algorithm

The numerical computation of $\wp(z)$ is decomposed into two main steps:

- Computation and initialization of all values which do not depend on the argument of the considered function $\wp(z)$.
- Computation of the function value for a special value of the complex argument z.

In our description we assume that the special function is given by numerical values of its invariants g_2 and g_3. (Equivalently, we can fix the function by numerical values of the zeros e_1, e_2, e_3 of the characteristic polynomial.)

Initialization

- **Given:** the invariants g_2, g_3 (see (2) and (9)).
- Computation of the zeros e_1, e_2, e_3 of the characteristic polynomial
- Computation of a fundamental period parallelogram $2\omega, 2\omega'$
- Handling of special cases of the $\wp$-function using the FPP and the invariants
  - lemniscatic case
  - pseudo-lemniscatic case
  - equianharmonic case
  - real invariants
  - general case (only this case is considered in the text above; the other cases are distinguished by the value of the discriminant (11), see [1])
- Computation of the factor t using (28), (29).
- Computation of the new invariants $\dot{g}_2 = g_2/t^4$ and $\dot{g}_3 = g_3/t^6$.
- Computation of the new fundamental periods $2\dot{\omega} = t2\omega$ and $2\dot{\omega}' = t2\omega'$
- Computation of the coefficients c_k of the Laurent-series expansion and of the estimation $\Gamma_{N+1} = \max_{k > N} \{|c_k|\}$ (see (26)) depending on $\dot{g}_2$ and $\dot{g}_3$.

Computation of function values

- **Given:** z, ω, ω' **wanted:** $\wp(z) = \wp(z|2\omega, 2\omega')$
- Computation of the altered argument $\dot{z} = tz$. The function value
$$\wp(\dot{z}|2\dot{\omega}, 2\dot{\omega}') = \frac{1}{t^2}\,\wp(z|2\omega, 2\omega') \text{ will be computed.}$$

- Argument reduction (if necessary) using the double periodicity with periods $2\dot{\omega}, 2\dot{\omega}'$:
$$\ddot{z} = \dot{z} + m2\dot{\omega} + m'2\dot{\omega}', \quad m, m' \in Z \text{ are determined using (41).}$$

- Application of the addition theorem (if necessary) to decrease the modulus of the argument until it is smaller than 1. The number of bisections is denoted by k.
- Computation of an enclosure of $\wp(\ddot{z}/2^k)$ using the truncated Laurent-series in combination with the error estimation.
- Repeated application of the addition theorem (k-times) leads to an enclosure of $\wp(\ddot{z})$.
- Adaptation of the result to the original argument z and the original fundamental periods $2\omega, 2\omega'$:
$$\wp(z) = \wp(z|2\omega, 2\omega') = t^2\wp(\dot{z}|2\dot{\omega}, 2\dot{\omega}') = t^2\wp(\ddot{z}|2\dot{\omega}, 2\dot{\omega}').$$

Remarks:

Enclosures of the zeros of the characteristic polynomial are computed in the following way: An approximation for one of the zeros is determined using Newton's method. As a starting point we use Cardan's solution of the cubic polynomial. The second and third zero are approximated as zeros of the quadratic remainder polynomial. All approximation values are refined by Newton's method.

To get enclosures the (point-) approximations are used as the center of corresponding (inflated) intervals. These intervals are used as starting intervals for the Interval-Newton method which gives verified enclosures of the zeros.

If $|\ddot{z}| < 1$ and $\max\{|c_2|, |c_3|\} < 3$ the truncation error can be computed using (27) by

$$|R_N| \leq \frac{\gamma_{N+1}|z^2|^N}{1 - |z^2|}.$$

Using the addition formulas leads to a modified argument of the power series which is smaller than $S < 1$ in modulus.

The function value $\wp(\ddot{z})$ is computed as the sum of the value of the power series and the value $1/\ddot{z}^2$. Due to $|1/\ddot{z}| > 1/S > 1$ the rounding error of this operation is always greater than $eps := b^{-m}$, with b being the base and m being the number of mantissa digits of the internal representation of floating-point numbers. Therefore, the truncation error should have the same magnitude.

For the IEEE-format we have base $b = 2$ and $m = 53$ mantissa bits. The error bound should be approximately $2^{-53} \approx 1.1 \cdot 10^{-16}$. For the implementation we use $S := 0.87$ and $N := 64$. This leads to an appropriate upper bound of the truncation error.

The addition theorem (16) can be used (instead of (17)) to compute enclosures of $\wp(z)$ in different ways. The intersection of all of these enclosures may result in a sharper enclosure of the function value.

The implementation of the routine for the computation of an enclosure of $\wp(z)$ has been done in PASCAL-XSC [14]. PASCAL-XSC is a PASCAL extension for scientific computation. Especially the module concept, the concept of functions with arbitrary result type and the availability of a complex interval arithmetic have been very helpful. The relations given by (7), (8) and (10) are used in the actual implementation to check the consistency of the computations.

The formulas (13), (15) and (18) corresponding to the formulas (12), (14) and (17) for the $\wp$-function can be used to compute interval bounds for the derivative $\wp'(z)$ given by (6) in combination with (5). This can be done in the same way as described above for $\wp(z)$.

7. Numerical Example

We choose $g_2 := 1 - i$ and $g_3 := 0.5 + 1.5i$. The implemented routine computes the following enclosures for the corresponding zeros of the characteristic polynomial:

$$e_1 \in [\quad 7.304057751708907\mathrm{E}-001, \quad 7.304057751708914\mathrm{E}-001]$$
$$[\quad 1.44710998047266\mathrm{E}-001, \quad 1.44710998047267\mathrm{E}-001]$$
$$e_2 \in [\quad 1.92944575628788\mathrm{E}-002, \quad 1.92944575628790\mathrm{E}-002]$$
$$[-6.036291582723068\mathrm{E}-001, -6.036291582723064\mathrm{E}-001]$$
$$e_3 \in [-7.497002327337702\mathrm{E}-001, -7.497002327337697\mathrm{E}-001]$$
$$[\quad 4.589181602250396\mathrm{E}-001, \quad 4.589181602250401\mathrm{E}-001]$$

Enclosures for the values of the half-periods are computed as

$$\omega \in [\quad 1.41130965659680\mathrm{E}+000, \quad 1.41130965659682\mathrm{E}+000]$$
$$[\ -1.9614738513906\mathrm{E}-001, \ -1.9614738513904\mathrm{E}-001]$$
$$\omega' \in [-3.75227790676332\mathrm{E}-001, -3.75227790676329\mathrm{E}-001]$$
$$[\quad 1.28041044507776\mathrm{E}+000, \quad 1.28041044507777\mathrm{E}+000]$$

Enclosures for some function values are as follows:

$$\wp(0.5 + 0.5i) \in$$
$$([\quad 2.038384630688478\mathrm{E}-002, \quad 2.038384630688480\mathrm{E}-002],$$
$$[-1.988368861817518\mathrm{E}+000, -1.988368861817517\mathrm{E}+000])$$

$\wp(1 + i) \in$
$$([\quad 2.6037907231644E-002, \quad 2.6037907231646E-002],$$
$$[-6.12842236035515E-001, -6.12842236035513E-001])$$

$\wp(1 - i) \in$
$$([-1.4002133413E-001, -1.4002133411E-001],$$
$$[\quad 2.0023738186E-001, \quad 2.0023738189E-001])$$

$\wp(-0.125 + 0.5i) \in$
$$([-3.342240369084528E+000, -3.342240369084526E+000],$$
$$[\quad 1.780254108234627E+000, \quad 1.780254108234628E+000])$$

$\wp(110.5 + 151.25i) \in$
$$([\quad 2.63719260E-001, \; 2.63719263E-001],$$
$$[1.359433767E+000, 1.359433769E+000])$$

The argument $z = 1011 + 10000i$ leads to the following enclosure for the corresponding reduced Argument

$$\text{reduced arg} \in ([1.2190835456E+000, 1.2190835458E+000],$$
$$[2.2810770827E+000, 2.2810770828E+000])$$

which in turn gives the final result

$\wp(1011 + 10000i) \in$

$$([1.353529E+000, 1.353530E+000],$$

$$[3.248334E-001, 3.248337E-001])$$

Even in the last case (the argument is large in magnitude) six decimal digits of the result can be guaranteed.

As usual, complex intervals (rectangles in the complex plane) are written as pairs of real intervals.

References

[1] Abramowitz, M., Stegun, I. A. (eds.): Handbook of mathematical functions, with formulas, graphs and mathematical tables. Dover Publications, 9th printing, 1970.

[2] Adams, E., Kulisch, U. (eds.): Scientific computing with automatic result verification. San Diego: Academic Press 1992.

[3] Barth, B.: Eine verifizierte Einschließung von Werten der Weierstraßschen $\wp$-Funktion. Diplomarbeit, Universität Karlsruhe, 1991.

[4] Bromwich, T. J.: An introduction to the theory of infinite series, 2nd rev. edn. London: Macmillan 1959.

[5] Chrapan, J.: Weierstrass $\wp$-function. Aplikace matematiky 4, 16 (1971).

[6] Chritchfield, Ch. L.: Computation of elliptic functions. J. Math. Phys. 30 (2), 295–297 (1989).

[7] Davis, H. T.: Tables of the mathematical functions, vol. II. Texas: Principia Press 1963.

[8] Eckhardt, U: Zur Berechnung der Weierstraßschen Zeta- und Sigma-Funktion. Berichte der KFA Jülich, Jül-964-MA, Juni 1973.

[9] Eckhardt, U.: A rational approximation to Weierstraß' $\wp$-function. Mathematics of Computing 30 (136), 818–826 (1976).

[10] Eckhardt, U.: A rational approximation to Weierstraß' $\wp$-function (II. The Lemniscatic Case). Computing 18, 341–349 (1977).

[11] Emersleben, O.: Erweiterung des Konvergenzbereiches einer Potenzreihe durch Herausnahme von Singularitäten, insbesondere zur Berechnung einer Zetafunktion zweiter Ordnung. Math. Nachr. v. *31*, 195–220 (1966).

[12] Erdélyi, A. (ed.), Magnus, W., Oberhettinger, F., Tricomi, F. G. (Research Associates): Higher transcendental functions, Volume II, Chapter XIII (elliptic functions and integrals). New York: McGraw-Hill Book Company 1953.

[13] Graeser, E.: Einführung in die Theorie der elliptischen Funktionen und deren Anwendungen. München: R. Oldenbourg 1950.

[14] Klatte, R., Kulisch, U., Neaga, M., Ratz, D., Ullrich, Ch.: PASCAL-XSC language reference with examples. Berlin, Heidelberg: Springer 1991.

[15] Krämer, W.: Computation of verified bounds for elliptic integrals. Proceedings of the International Symposium on Computer Arithmetic and Scientific Computation, Oldenburg 1991 (SCAN91). Edited by J. Herzberger and L. Atanassova; Elsevier Science Publishers (North-Holland).

[16] Kulisch, U. W., Miranker, W. L. (eds.): A new approach to scientific computation. New York: Academic Press 1983.

[17] Milne-Thomson, L. M.: Elliptic integrals. In: [1], chap. 17, p. 587–607.

[18] F. Reutter, F.: Geometrische Darstellung der Weierstraßschen $\wp$-function. ZAMM *41* (1/2), 54–65 (Sonderdruck) (1961).

[19] Reutter, F., Haupt, D.: Zur Berechnung Jacobischer elliptischer Funktionen mittels elektronischer Rechenautomaten, mtw (Mathematik, Technik, Wirtschaft), 9. Jahrgang 1962, Heft 1 (Sonderdruck).

[20] Southard, T. H.: Weierstrass elliptic and related functions. In: [1], chap. 18, p. 627–671.

[21] Southard, T. H.: Approximation and table of the Weierstrass $\wp$-function in the equianharmonic case for real argument. Math. Tables Aids Comp. *11*, 99–100 (1957).

[22] Tricomi, F. G.: Elliptische Funktionen. Leipzig: Akademische Verlagsgesellschaft Kap. 1, 1948.

[23] Uhde, K.: Spezielle Funktionen der mathematischen Physik. Tafeln *11*, 105–108, 1964.

[24] Weierstrass, K.: Formeln und Lehrsätze zum Gebrauch der elliptischen Funktionen. Nach Vorlesungen und Aufzeichnungen dsslb. bearbeitet und herausgegeben von H. A. Schwarz, Göttingen, 1885.

[25] Whittaker, E. T., Watson G. N.: A course of modern analysis, ch. XX, XXI, 4th edn. Cambridge: Cambridge University Press 1952.

Walter Krämer
Institut für Angewandte Mathematik
Universität Karlsruhe
D-W-7500 Karlsruhe 1
Federal Republic of Germany

Bertram Barth
Institut für Industrielle Bauproduktion
Universität Karlsruhe
D-W-7500 Karlsruhe 1
Federal Republic of Germany

Computing, Suppl. 9, 161–173 (1993)

Solving Nonlinear Elliptic Problems with Result Verification Using an H^{-1} Type Residual Iteration*

Mitsuhiro T. Nakao, Fukuoka

Dedicated to Professor U. Kulisch on the occasion of his 60th birthday

Abstract — Zusammenfassung

Solving Nonlinear Elliptic Problems with Result Verification Using an H^{-1} Type Residual Iteration. In this paper, we consider a numerical technique to verify the solutions with guaranteed error bounds for nonlinear elliptic boundary value problems. Using the C^0 finite element solution and explicit error estimates for the Poisson equation, we construct, in a computer, a set of functions which satisfies the hypothesis of Sadovskii's fixed point theorem for confdensing map on a certain Sobolev space. Particularly, we propose an H^{-1} type residual iteration method which improves the ability of verification. A numerical example which confirms the usefulness of the method is presented.

AMS Subject Classification: 65N15, 65N30

Key words: Numerical verifications, elliptic problems, finite element methods, error estimates.

Zur Lösungseinschliessung bei nicht linearen elliptischen Randwertproblemen unter Verwendung von H^{-1}-artiger Residneniteration. In der vorliegenden Arbeit betrachten wir eine numerische Technik zur Verifikation von Lösungen durch Angabe von garantierten Fehlerschranken bei elliptischen Randwertproblemen. Unter Verwendung von C^0-Lösungen nach der Methode der finiten Elemente und expliziten Fehlerschranken für die Poissongleichung konstruieren wir auf einem Computer eine Funktionenmenge, die die Voraussetzungen des Fixpunktsatzes von Sadovskii für kondensierende Abbildungen auf einem gewissen Sobolevraum erfüllt. Insbesondere schlagen wir eine H^{-1}-artige Residueniteration vor, durch welche die Möglichkeit zur Verifikation verbessert wird. Ein numerisches Beispiel zeigt die Nützlichkeit der Methode.

1. Introduction

In this paper, we consider a numerical technique to enclose the solution with guaranteed error bounds for nonlinear elliptic equations of second order. Especially, we emphasize that our method needs no assumptions on the existence of the exact solutions for the original problems. That is, the numerical result also means the numerical proof for the existence of the solution. In recent years, several such kinds of numerics with result verification have been proposed for differential equations. However, there are not so many works for partial differential equations (PDEs) up to now. From this point of view, the enclosure methods based upon the monotonicity assumption on the operator, e.g. Collatz [3], are essentially different from our approach, because they need some hypotheses for the existence of solutions.

* Received September 25, 1992; revised December 23, 1992.

The author has studied for years the numerical verification of the solutions of PDEs using the finite element method (FEM) and the constructive error estimates. Our verification method means that one can use computers in proving existence and/or uniqueness of exact solutions for PDEs as well as one can get an a posteriori error bounds for the finite element solution. The basic approach of this method consists of the fixed point formulation of PDE and the construction of the function set, in a computer, satisfying the validation condition of a certain infinite dimensional fixed point theorem. In order to get such a set, we divide the verification procedure into two phases; one is the computation of a projection ($=$rounding) into some finite dimensional subspace, usually finite element space, the other is the estimation of the error ($=$rounding error) for the projection. Combining them with some iterative technique, the exact solution can be enclosed by the sum of the rounding part, which is a subset in the finite dimensional space, and the rounding error, which is indicated by a nonnegative real number. This iterative process is quite different from the monotone iteration technique such as Sattinger [19] which puts the solution between upper- and lower-solutions. The rounding procedure means to solve a linear system of finite dimensional equations with interval right-hand side. And, in the case of elliptic problems, the rounding error estimation implies the numerically constructive error estimates of the approximate solution by C^0-FEM for Poisson's equation. These two concepts enable us to treat the infinite dimensional problems as finite procedures, i.e., by computer.

Now, we briefly survey our works on the numerical verification method for PDEs. The concepts of rounding and rounding error in the Soboleve space were first introduced in [6], and we proposed a verification procedure by using Schauder's fixed point theorem with the successive iteration method for the linear elliptic problems. In [7], an extension for the nonlinear problems was considered. In order to overcome the difficulty caused by the large spectral radius of the operator, a Newton-like iterative technique combining with Sadovskii's fixed point theorem was developed in [8] for linear problems. A more general verification algorithm based on the Newton-like method was considered in [9] for one dimensional problem with more realistic nonlinear numerical examples. In principle, this formulation can also be applied to the higher dimensional problems, i.e., the elliptic PDEs. Actually, in [21], we applied this method to the verification of solutions for a nonlinear equation appeared in the mathematical biology.

Moreover, the verification for the elliptic equations with nonlinearity stronger than the polynomial order is considered in [12], and in nonsmooth-noncovex domains in [22], respectively. Furthermore, [20] describes the enclosing the solutions for parametrized equations with turning points in one dimensional cases. As for evolutional problems, in [10], [11] and [13], some prototype formulations based on the same verification principle are presented.

Recently, Plum [14]–[17] proposed other verification techniques by the use of the C^1-FEM for elliptic equations which are different from our method. His method is essentially based upon the numerical enclosure of eigenvalues using the homotopy method for the linearized elliptic operator.

In this paper, we propose some improvements on the ability of verification and the efficiency of computation for the elliptic problems. The first refinement is the use of the residual error estimation in the sense of H^{-1} norm which improves the efficiency of the Newton-like iteration. Actually, this enables us to verify some problems which could not be verified by the existing approaches. The second one is a simplification of the computation in the rounding procedure by excluding the auxiliary parameter in the verification process. Further, we also present some detailed illustrations of the interval evaluation which is essential in the computation of the rounding and the rounding error estimates.

We restrict the arguments in the paper to the plane polygonal domains. The extension to the three dimensional case, however, is straightforward for certain restricted problems, while some additional considerations will be needed for more general cases.

2. Reformulation of the Problem by the H^{-1} Residual Form

Since we use the fixed point theorem to enclose a solution of the problem, we need the reformulation of the equation as the fixed point form.

We consider the following nonlinear elliptic boundary value problem:

$$\begin{cases} \Delta u = f(x, u, \nabla u), & x \in \Omega, \\ u = 0, & x \in \partial\Omega, \end{cases} \tag{1}$$

where Ω is a convex polygonal domain in $\mathbf{R}^2$.

We assume that

A1. f is a bounded and continuous map from $H^1(\Omega)$ to $L^2(\Omega)$.

Here, for an integer m, let $H^m(\Omega) \equiv H^m$ denote L^2-Sobolev space of order m on Ω. And set $H_0^1 \equiv \{\phi \in H^1 | \mathrm{tr}(\phi) = 0 \text{ on } \partial\Omega\}$ with the inner product $\langle \phi, \psi \rangle \equiv (\nabla\phi, \nabla\psi)$, where $(\cdot, \cdot)$ means the inner product on $L^2(\Omega)$.

For example, $f(\cdot, u, \nabla u) = g_1 \cdot \nabla u + g_2 u^p$ satisfies above assumption, where $g_1 = (g_1^1, g_1^2)$ and g_2 are in $L^\infty(\Omega)$, and p an arbitrary nonnegative integer.

In the below, we use the abbreviation: $f(u) \equiv f(x, u, \nabla u)$ and simply denote $\|\cdot\|_{L^2}$ by $\|\cdot\|$.

Now let S_h be a finite dimensional subspace of H_0^1 dependent on h $(0 < h < 1)$. Usually, S_h is taken to be a finite element subspace with mesh size h. And let $P_h: H_0^1 \to S_h$ denote the H_0^1-projection defined by

$$(\nabla u - \nabla(P_h u), \nabla\hat{\phi}) = 0, \qquad \hat{\phi} \in S_h. \tag{2}$$

Then suppose that

A2. An approximate solution $u_h \in S_h$ which is fixed throughout the paper satisfies

$$(\nabla u_h, \nabla\hat{\phi}) = (-f(u_h), \hat{\phi}), \qquad \hat{\phi} \in S_h. \tag{3}$$

It is rather natural for our situation of the verification to assume the existence of the approximation solution satisfying A2, for it is a finite dimensional problem. Furthermore, see the remark 3 in the end of the section 6.

We now suppose the following approximation property of P_h:

A3. For any $\phi \in H^2 \cap H_0^1$,

$$\|\phi - P_h\phi\|_{H_0^1} \le C_1 h |\phi|_{H^2}, \tag{4}$$

where

$$|\phi|_{H^2}^2 \equiv \sum_{i,j=1}^{2} \left\| \frac{\partial^2 \phi}{\partial x_i \partial x_j} \right\|_{L^2}^2.$$

Here, C_1 is a positive constant numerically determined and independent of h. This assumption holds for many finite element subspace of piecewise linear polynomials with quasi-uniform partition ([2]). It will be easily seen, by arguments in the below, that the dependency of C_1 on h is not essential for our purpose.

On the other hand, it is well known [4], that for arbitrary $\psi \in L^2(\Omega)$ there exists a unique solution $\phi \in H^2 \cap H_0^1$ of the following Poisson's equation:

$$\begin{cases} \Delta\phi = \psi, & x \in \Omega, \\ \phi = 0, & x \in \partial\Omega. \end{cases} \tag{5}$$

When we denote the solution of (5) by $\phi \equiv A\psi$, the map $A: L^2 \to H_0^1$ is compact. Furthermore, there exists a constant C_2 satisfying

$$|\phi|_{H^2} \le C_2 \|\psi\|. \tag{6}$$

It is seen that, by the result in [4], we can take as $C_2 = 1$ for the present case. Setting $v = u - u_h$, we have by (1)

$$-\Delta v = \Delta u_h - f(u_h) - f(u_h + v) + f(u_h). \tag{7}$$

Note that, since S_h is not a subset of H^2, Δu_h does not belong to L^2 in the distributional sense, but $\Delta u_h \in H^{-1}$, where H^{-1} means the dual space of H_0^1.

We now take an element $v_0 \in H_0^1$ as the Riesz representation of $\Delta u_h - f(u_h) \in H^{-1}$.

Proposition 1. *The following estimates hold for v_0 defined above:*

$$\|v_0\|_{H_0^1} \le Ch\|f(u_h)\| \quad and \quad \|v_0\|_{L^2} \le C^2 h^2 \|f(u_h)\|, \tag{8}$$

where $C = C_1 C_2$.

Proof. v_0 is the solution of the following variational equation.

$$(\nabla v_0, \nabla \phi) = -(\nabla u_h, \nabla \phi) - (f(u_h), \phi), \qquad \phi \in H_0^1, \tag{9}$$

which can also be rewritten as the weak boundary value problem:

$$\begin{cases} -\Delta v_0 = \Delta u_h - f(u_h), & x \in \Omega, \\ v_0 = 0, & x \in \partial\Omega. \end{cases} \tag{10}$$

Notice that u_h coincides with the projection $P_h u_0 \in S_h$ of the solution u_0 to the Poisson equation: $\Delta u_0 = f(u_h)$ with homogeneous boundary condition. Consequently, these facts imply $v_0 = u_0 - P_h u_0$ which proves the proposition by using (4), (6) and the well known Aubin-Nitsche trick (e.g. [2]).

Now, setting $u = u_h + v_0 + w$, we obtain the new equation for w instead of (1) or (7):

$$\begin{cases} \Delta w = f(u_h + v_0 + w) - f(u_h), & x \in \Omega, \\ w = 0, & x \in \partial\Omega. \end{cases} \tag{11}$$

Note that there is no longer any H^{-1} term in the right-hand side of (11). From now on, v_0 is considered as the element in H_0^1 having the norm bounded by Proposition 1, though its explicit form is unknown. Thus, using the following compact map on H_0^1

$$F(u_h + v_0 + \cdot) \equiv A(f(u_h + v_0 + \cdot) - f(u_h)),$$

we have the fixed point equation:

$$w = F(u_h + v_0 + w). \tag{12}$$

Now, according to the similar manner in [8], [9] etc., we introduce a Newton-like operator for the Eq. (12).

Let $F'(u_h): H_0^1 \to H_0^1$ be the Fréchet derivative of F at u_h. Suppose that

A.4. Restriction to S_h of the operator $P_h[I - F'(u_h)]: H_0^1 \to S_h$ has an inverse

$$[I - F'(u_h)]_h^{-1}: S_h \to S_h, \qquad \text{where } I \text{ means the identity map on } H_0^1.$$

This assumption corresponds to the unique existence of the finite element solution in S_h to the linearlized equation of the original problem.

Next, for a fixed positive parameter ε such that $0 < \varepsilon < 1$, we define a nonlinear operator $T_\varepsilon: H_0^1 \to H_0^1$ as follows:

$$T_\varepsilon w \equiv w - ([I - F'(u_h)]_h^{-1} P_h + \varepsilon I)(w - F(u_h + v_0 + w)). \tag{13}$$

Then T_ε becomes a condensing map ([23]) on H_0^1, i.e., T_ε can be decomposed as the sum of contraction and compact operators. It is seen that if $-\varepsilon^{-1}$ does not coincide with any eigenvalue of the operator $P_h[I - F'(u_h)]$ on S_h, then $w = F(u_h + v_0 + w)$ is equivalent to $w = T_\varepsilon w$.

3. Computable Verification Condition

In this section, as in [6], [7] etc., we define the concepts, rounding and rounding error, which enable us to treat the infinite dimensional problem by finite procedures.

Let $\{\phi_j\}_{j=1,\ldots,M}$ be a fixed basis of S_h, where $M = \dim S_h$. And let $\mathscr{S}_{h,I}$ be the set of all linear combinations of $\{\phi_j\}$ with interval coefficients. That is,

$$\mathscr{S}_{h,I} = \left\{ W_h \subset S_h \,|\, W_h = \sum_{j=1}^{M} [\underline{a}_j, \bar{a}_j]\phi_j, \quad \underline{a}_j, \bar{a}_j \in \mathbf{R}^1 \right\} \subset 2^{S_h}.$$

Then, we define for any bounded set $W \subset H_0^1$ as follows:

Rounding:

$$R(W) \equiv \bigcap \{W_h \in \mathscr{S}_{h,I} | P_h(W) \subset W_h\}, \tag{14}$$

where $\bigcap$ means the intersection for all elements in $\mathscr{S}_{h,I}$ satisfying the condition. Note that $R(W)$ is also an element in $\mathscr{S}_{h,I}$.

Rounding error:

$$RE(W) \equiv \{\phi \in S_h^\perp \mid \|\phi\|_{H_0^1} \leq \alpha, \|\phi\|_{L^2} \leq C h \alpha\}, \tag{15}$$

where $S_h^\perp$ means the orthogonal complement of S_h in H_0^1,

$$\alpha \equiv \sup_{w \in W} \|w - P_h w\|_{H_0^1}$$

and C is the same constant as in (8).

Now we have the following computable verification condition.

Lemma 1. *Let $W \subset H_0^1(\Omega)$ be a non-empty, bounded, convex, and closed subset such that*

$$R(T_\varepsilon W) \oplus RE(T_\varepsilon W) \overset{\circ}{\subset} W, \tag{16}$$

then there exists a solution of $w = F(u_h + v_0 + w)$ in W.

Here, $\oplus$ denotes the direct sum in $H_0^1(\Omega)$, and $M_1 \overset{\circ}{\subset} M_2$ implies $\overline{M}_1 \subset \mathring{M}_2$ for any sets M_1, M_2.

Proof. The argument is essentially similar to that in [9], but, in order to keep the paper self-contained, we give a proof. By the definition, for arbitrary $w \in W$, we have $P_h(T_\varepsilon w) \in R(T_\varepsilon W)$ and $T_\varepsilon w - P_h(T_\varepsilon w) \in S_h^\perp$. Thus, by using the Aubin-Nitsche trick, we get

$$T_\varepsilon W \subset R(T_\varepsilon W) \oplus RE(T_\varepsilon W).$$

Therefore, from the assumption of the lemma and Sadovskii's fixed point theorem ([23]), there exists $w \in W$ such that $w = T_\varepsilon w$. Then we have

$$([I - F'(u_h)]_h^{-1} P_h + \varepsilon I)(w - F(u_h + v_0 + w)) = 0. \tag{17}$$

Taking account of continuous dependency of the operator T_ε on ε, the same inclusion as (16) also holds for some ε' which does not coincide with any eigenvalue of $-[I - F'(u_h)]_h^{-1}$. Hence, we obtain the desired result.

4. Verification Procedures by Computer

In order to construct the set W satisfying the verification condition (16) in a computer, we use an iterative technique similar to that in [6], [7], etc.

Let $u_h \in S_h$ and $v_0 \in H_0^1$ be the same elements as in the section 2. For any $\alpha \in R^+$, the set of all nonnegative real numbers, we set

$$[\alpha] \equiv \{\phi \in S_h^{\perp} \mid \|\phi\|_{H_0^1} \leq \alpha, \|\phi\|_{L^2} \leq Ch\alpha\}.$$

We now generate the following residual iteration sequence $\{(w_h^{(n)}, \alpha_n)\}_{n=0,1\ldots}$, where $(w_h^{(n)}, \alpha_n) \in \mathscr{S}_{h,I} \times R^+$. For $n = 0$, set $w_h^{(0)} = \{0\} \in \mathscr{S}_{h,I}, \alpha_0 = 0 \in R^+$ and $W^{(0)} = w_h^{(0)}$.

For $n \geq 1$, first, for a given $0 < \delta \ll 1$, define the δ-inflation (cf. [18]) of $(w_h^{(n-1)}, \alpha_{n-1})$ by

$$\begin{cases} \overline{w}_h^{(n-1)} \equiv w_h^{(n-1)} + \sum_{j=1}^{M} [-1,1]\delta\phi_j, \\[2mm] \overline{\alpha}_{n-1} \equiv \alpha_{n-1} + \delta. \end{cases} \qquad (18)$$

Next, for the set $\overline{W}^{(n-1)} = \overline{w}_h^{(n-1)} + [\overline{\alpha}_{n-1}]$, define $w_h^{(n)} \in \mathscr{S}_{h,I}$ and $\alpha_n \in R^+$ by

$$\begin{cases} w_h^{(n)} \equiv R(T_0 \overline{W}^{(n-1)}), \\[2mm] \alpha_n \equiv Ch \displaystyle\sup_{w \in \overline{W}^{(n-1)}} \|f(u_h + v_0 + w) - f(u_h)\|_{L^2}. \end{cases} \qquad (19)$$

Remark 1.

It should be noted that these calculations are independent of the parameter ε. The former half of (19) means that $w_h^{(n)}$ is determined by the interval vector solution for the M dimensional linear system of equations with interval right-hand side. The more detailed and concrete procedure of the computation of (19) will be described in the next section. Of course, in the actual calculation on a computer, each quantity is computed in the over-estimated sense.

Now we have the following verification condition in a computer.

Theorem 1. *If, for an integer N,*

$$w_h^{(N)} \overset{\circ}{\subset} \overline{w}_h^{(N-1)} \qquad and \qquad \alpha_N < \overline{\alpha}_{N-1},$$

then there exists an element w in $w_h^{(N)} \oplus [\overline{\alpha}_{N-1}]$ satisfying $w = F(u_h + v_0 + w)$. Here, the fomer condition implies that each coefficient interval in $w_h^{(N)}$ is strictly included by the corresponding interval in $\overline{w}_h^{(N-1)}$.

Proof. From the lemma 1, it is sufficient to prove for sufficiently small $\varepsilon > 0$

$$R(T_\varepsilon \overline{W}^{(N-1)}) \oplus RE(T_\varepsilon \overline{W}^{(N-1)}) \overset{\circ}{\subset} \overline{W}^{(N-1)}. \qquad (20)$$

First, the definition of rounding and the condition of the theorem imply

$$P_h(T_0 \overline{W}^{(N-1)}) \overset{\circ}{\subset} \overline{w}_h^{(N-1)}.$$

But the continuity arguments on T_ε in ε yields that for sufficiently small ε

$$P_h(T_\varepsilon \overline{W}^{(N-1)}) \overset{\circ}{\subset} \overline{w}_h^{(N-1)}. \qquad (21)$$

By a simple calculation, we have for arbitrary $w \in \overline{W}^{(N-1)}$

$$T_\varepsilon w - P_h(T_\varepsilon w) = (1 - \varepsilon)(w - P_h w) + \varepsilon(F - P_h F)(u_h + v_0 + w).$$

Therefore, we obtain the estimates

$$\|T_\varepsilon w - P_h(T_\varepsilon w)\|_{H_0^1} \le (1 - \varepsilon)\bar{\alpha}_{N-1} + \varepsilon \sup_{w \in \overline{W}^{(N-1)}} \|(F - P_h F)(u_h + v_0 + w)\|_{H_0^1}$$

$$\le (1 - \varepsilon)\bar{\alpha}_{N-1} + \varepsilon C h \sup_{w \in \overline{W}^{(N-1)}} \|f(u_h + v_0 + w) - f(u_h)\|_{L^2}$$

$$= (1 - \varepsilon)\bar{\alpha}_{N-1} + \varepsilon\alpha_N < \bar{\alpha}_{N-1}.$$

Combining the above with (21), we get the desired inclusion (20).

5. Computational Algorithm of the Rounding Procedure

In this section, we will mention about the concrete procedure for computation of (19), i.e., the rounding and the rounding error.

First, observe that, for $\overline{W}^{(n-1)} = \overline{w}_h^{(n-1)} + [\bar{\alpha}_{n-1}]$

$$P_h(T_0 \overline{W}^{(n-1)}) = \bigcup_{w \in \overline{W}^{(n-1)}} (I - [I - F'(u_h)]_h^{-1})P_h w + [I - F'(u_h)]_h^{-1} P_h F(u_h + v_0 + w).$$

$$(22)$$

Using the interval vector $(B_j^n)_{j=1\ldots M}$, we enclose the right-hand side of (22) as

$$w_h^{(n)} = \sum_{j=1}^{M} B_j^n \phi_j \in \mathscr{S}_{h,I}. \tag{23}$$

We define the $M \times M$ matrix $G := (g_{ii})$ by

$$g_{ji} \equiv (\nabla\phi_j, \nabla\phi_i) - \left(\frac{\partial f}{\partial u}(u_h)\phi_j + \frac{\partial f}{\partial u_x}(u_h)\cdot\nabla\phi_j, \phi_i\right), \tag{24}$$

where $u_x \equiv \nabla u$. Then G is invertible by the assumption A4.

Now, under some appropriate assumption on the smoothness of f, we have for each $w_h \in \overline{w}_h^{(n-1)}$ and $\hat{\alpha} \in [\bar{\alpha}_{n-1}]$

$$f(u_h + v_0 + w_h + \hat{\alpha}) - f(u_h) = \frac{\partial f}{\partial u}(u_h)w_h + \frac{\partial f}{\partial u_x}(u_h)\cdot\nabla w_h + \kappa(u_h, v_0, w_h, \hat{\alpha}),$$

where $\kappa(u_h, v_0, w_h, \hat{\alpha})$ means the higher order terms in w_h.

Therefore, setting $w_h^{(n-1)} = \sum_{j=1}^{M} A_j^{(n-1)}\phi_j$ we get

$$(\nabla F(u_h + v_0 + w_h + \hat{\alpha}), \nabla\phi_i) \in \sum_{j=1}^{M} A_j^{(n-1)}\left(\frac{\partial f}{\partial u}(u_h)\phi_j + \frac{\partial f}{\partial u_x}(u_h)\cdot\nabla\phi_j, \phi_i\right) + (\kappa_{n-1}, \phi_i),$$

$$(25)$$

where $\kappa_{n-1} \equiv \kappa(u_h, v_0, \overline{w}_h^{(n-1)}, [\bar{\alpha}_{n-1}])$.

Thus from (22)–(25) and the definition of the Fréchet derivative, we have

$$(\nabla P_h[I - F'(u_h)]w_h, \nabla\phi_i) \in (\kappa_{n-1}, \phi_i).$$

This proves the following proposition which is a simplified version of the corresponding expression in [21].

Proposition 2. *Interval coefficients B_j^n in (23) are determined by*

$$(B_j^n) = G^{-1}(\hat{K}_j^{n-1}).$$ (26)

Here, the interval vector $(\hat{K}_j^{n-1})_{j=1\ldots M}$ is defined as

$$\hat{K}_j^{n-1} = (\kappa_{n-1}, \phi_j), \qquad 1 \le j \le M,$$

where

$$(\kappa_{n-1}, \phi_j) \equiv \left(f(u_h + v_0 + \overline{w}_h^{(n-1)} + [\bar{\alpha}_{n-1}]) - f(u_h) - \frac{\partial f}{\partial u}(u_h)\overline{w}_h^{(n-1)} \right.$$
$$\left. - \frac{\partial f}{\partial u_x}(u_h) \cdot \nabla\overline{w}_h^{(n-1)}, \phi_j \right).$$ (27)

Next, we give some computational examples of the right-hand side of (27) and the norm estimation

$$\sup_{w \in \overline{W}^{(n-1)}} \| f(u_h + v_0 + w) - f(u_h)\|_{L^2}$$ (28)

which appears in the rounding error computation of (19).

In the below, we only consider the case $f(u) = u^m$, for more strong nonlinear case, e.g., $f(u) = e^u$, see [22]. For simplicity, we use the generic symbol w_h and $[\alpha]$ as $\overline{w}_h^{(n-1)}$ and $[\bar{\alpha}_{n-1}]$, respectively. Further, we set $\gamma_0 = \|v_0\|_{H_0^1}$.

First, observe that

$$\kappa_{n-1} = f(u_h + v_0 + w_h + [\alpha]) - f(u_h) - \frac{\partial f}{\partial u}(u_h)w_h - \frac{\partial f}{\partial u_x}(u_h) \cdot \nabla w_h$$

$$= (u_h + v_0 + w_h + [\alpha])^m - u_h^m - m u_h^{m-1} w_h$$

$$= m u_h^{m-1}(v_0 + [\alpha]) + \sum_{r=2}^{m} \binom{m}{r} u_h^{m-r}(v_0 + w_h + [\alpha])^r.$$

We now illustrate some practical calculations of (27) for this example (*cf.* [6], [21]). Each integral for the set is evaluated by intervals such as, using the Proposition 1 and the definition of $[\alpha]$,

$$(u_h^{m-1}(v_0 + [\alpha]), \phi_j) \subset [-1, 1]Ch(\gamma_0 + \alpha)\|u_h\|_{L^\infty(\Omega)}^{m-1}\|\phi_j\|$$

or

$$(u_h^{m-1}(v_0 + [\alpha]), \phi_j) \subset [-1, 1]Ch(\gamma_0 + \alpha)\|u_h^{m-1}\|_{L^2}\|\phi_j\|_{L^\infty(\Omega)}.$$

The higher order terms with respect to $(v_0 + w_h + [\alpha])$ are generally small and it is sufficient to calculate them in the over-estimated sense by the similar manner described later.

Next, we show some examples on the estimation of the norm:

$$\| f(u_h + v_0 + w_h + [\alpha]) - f(u_h) \|$$

$$= \left\| m u_h^{m-1}(v_0 + [\alpha]) + m u_h^{m-1} w_h + \sum_{r=2}^{m} \binom{m}{r} u_h^{m-r}(v_0 + w_h + [\alpha])^r \right\|$$

which has to be evaluated in computing the latter half of (19). This is usually evaluated as the sum of separated norms by applying the triangle inequality.

For example, we use Hölder's inequality to get for any $q, r \geq 1$

$$\| v_0^q [\alpha]^r \| \leq \| v_0 \|_{L^4}^q \| [\alpha] \|_{L^4}^r . \tag{29}$$

Further, by the use of Sobolev's imbedding theorem with explicit constant (e.g. [5]), we have for any $\phi \in H_0^1$ and $p > 1$,

$$\| \phi \|_{L^p} \leq \frac{p}{4} |\Omega|^{1/p} \| \phi_x \|_{L^2}^{1/2} \| \phi_y \|_{L^2}^{1/2}$$

$$\leq \frac{p\sqrt{2}}{8} |\Omega|^{1/p} \| \phi \|_{H_0^1} , \tag{30}$$

where $|\Omega|$ implies the measure of Ω.

Therefore, from the definition of $[\alpha]$ we obtain the following estimates

$$\| v_0^q [\alpha]^r \| \leq 2^{-(q+r)/2} q^q r^r |\Omega|^{(q+r)/4qr} \gamma_0^q \alpha^r .$$

Also we have an alternate estimation using the L^2 estimates for v_0 as follows:

$$\| v_0^q [\alpha]^r \| \leq \| v_0 \|_{L^2}^{1/2} \| v_0 \|_{L^{4(2q-1)}}^{q-1/2} \| [\alpha] \|_{L^{8r}}^r$$

$$\leq 2^{(2q+2r-1)/4} (q - 1/2)^{q-1/2} r^r \sqrt{Ch} |\Omega|^{1/4} \gamma_0^q \alpha^r .$$

The latter estimates will be better for sufficiently small h provided that $|\Omega|$ is not so small. Further, $\| v_0^q \|$, $\| [\alpha]^r \|$ etc., are similarly estimated.

Additionally, we need usual interval techniques (e.g. [1]) to enclose the ranges of various formulae as narrow as possible.

6. A Numerical Example

We now give an example which was verified by the present method but the existing techniques failed to verify.

We consider the following so-called Emden's equation:

$$\begin{cases} -\Delta u = u^2 & \text{in } \Omega, \\ \quad u = 0 & \text{on } \partial\Omega. \end{cases} \tag{31}$$

Let Ω be a rectangular domain $(0,1) \times (0,1) \subset \mathbf{R}^2$ and let $S_h(x)$ denote the set of continuous and piecewise linear polynomials on $(0,1)$ with uniform mesh and homogeneous boundary conditions. We set $S_h \equiv S_h(x) \otimes S_h(y)$. Then, $M = dim\, S_h =$

$(L - 1)^2$, where L denotes the number of partitions for the interval $(0, 1)$, i.e., $h = \dfrac{1}{L}$.
The constants previously appeared can be taken as $C_1 = \dfrac{1}{\pi}$, $C_2 = 1$, respectively ([9], [21]).

Because the magnitude of the solution of this problem is very large, i.e., $\|u_h\|_{L^\infty(\Omega)} \approx$ 30, it is difficult to apply the usual approaches without any residual iteration. This equation is also treated in [21] using the residual technique based on the pseudo-Hermite interpolation in $H^2(\Omega)$. However, some considerable efforts are needed to obtain such a smooth interpolation.

By using the present method, we could verify the same problem without such an additional procedures. The solution is enclosed by $u_h + v_0 + w_h^N + [\bar{\alpha}_{N-1}]$ in the below.

Conditions:

Number of partition: $L = 60$ ($h = \frac{1}{60}$, $M := dim\, S_h = 3481$)
$\delta = 10^{-3}$

Results:

Initial approximate solution u_h defined in (3) was computed by
the pseudo-spectral method as in [21].
Initial residual error in Proposition 1: $\|v_0\|_{H_0^1} \leq 1.474$, $\|v_0\|_{L^2} \leq 0.00782$
Iteration number for verification: $N = 4$
Maximum width of coefficient intervals in $w_h^{(N)} = 2.738$
Additional H_0^1 error: $\bar{\alpha}_{N-1} = 0.1848$, L^2 error: $Ch\alpha_{N-1} = 0.000981$

Fig. 1 shows the outline of the shape of the interval function $u_h + w_h^{(N)}$ along the line $y = 0.5$. That is, there exists an exact solution to (31) between those two curves with additional L^2-error: 0.00782.

The computation was executed by the FACOM VP2600/10 at the Computer Center in Kyushu University with the total CPU time 449 sec.

Remark 2.
We could also verify the above problem even for the more coarse meshes, e.g., $L = 30$. These facts show that the above H^{-1} residual method yields the significant improvement on the ability of verification. However, the error bound obtained is rather large compared with the residual iteration method by smoothing in [21]. This will be inevitable so long as we use the less smooth element, i.e., C^0-class.

Remark 3.
In general, it is difficult to take the approximate solution u_h satisfying the equation (3) in the mathematically rigorous sense. But it is not so difficult that we strictly enclose the solution of (3) by some interval methods (e.g. [1], [18]). Then, our arguments in the above are still valid if we consider u_h as an element in $\mathscr{S}_{h,I}$ in the verification process (19). Further, in order for the calculation of $F'(u_h)$, it will be sufficient to take a point element $\hat{u}_h \in u_h$ instead of u_h itself.

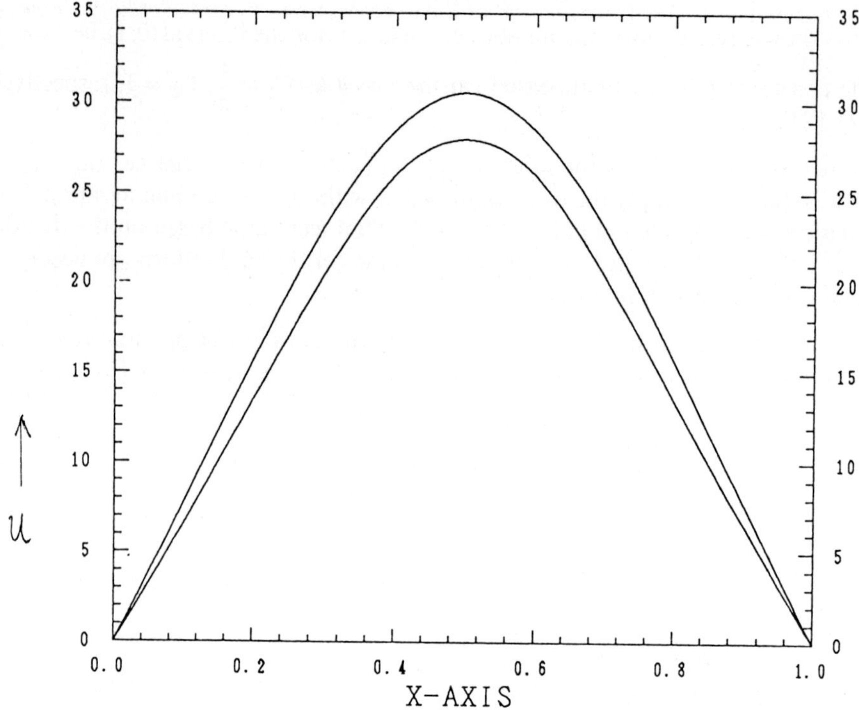

Figure 1. Range of nontrivial solution of Emden's equation for $Y = 0.5$

Remark 4.

In the above calculations, we used usual double precision floating point arithmetic instead of strict interval computations (e.g. ACRITH, PASCAL-SC etc.). Therefore, it means that we neglected the round-off error. The reason is that the main purpose of our numerical experiments is the estimation of the truncation errors which usually, roughly speaking, are over 10^{10} times larger than the round-off errors. Of course, we will have to use those verification software systems for the more delicate problems, e.g., for the case that we require the mathematical proof of the problem whose solutions are mathematically unknown.

Acknowledgement

The author is greatly indebted to Nobito Yamamoto and Yoshitaka Watanabe, Kyushu University, for their assistance in programming and computing the numerical example in the paper.

References

[1] Alefeld, G., Herzberger, J.: Introduction to interval computations. New York: Academic Press 1983.
[2] Ciarlet, P. G.: The finite element method for elliptic problems. Amsterdam: North-Holland 1978.

[3] Collatz, L.: Guaranteed inclusions of solutions of some types of boundary value problems. In: Ullrich, C. (ed.) Computer arithmetic and self-validating numerical methods. Boston: Academic Press 1990.

[4] Grisvard, P.: Elliptic problems in nonsmooth domain. Boston: Pitman 1985.

[5] Johnson, C., Schatz, A. H., Wahlbin, L. B.: Crosswind smear and pointwise errors in stream diffusion finite element methods. Math. Comp. *49*, 25–38 (1987).

[6] Nakao, M. T.: A numerical approach to the proof of existence of solutions for elliptic problems. Japan Journal of Applied Mathematics *5*, 313–332 (1988).

[7] Nakao, M. T.: A computational verification method of existence of solutions for nonlinear elliptic equations, Lecture Notes in Num. Appl. Anal., 10, (1989) 101–120. In proc. Recent Topics in Nonlinear PDE 4, Kyoto, 1988, North-Holland/Kinokuniya, 1989.

[8] Nakao, M. T.: A numerical approach to the proof of existence of solutions for elliptic problems II. Japan Journal of Applied Mathematics *7*, 477–488 (1990).

[9] Nakao, M. T.: A numerical verification method for the existence of weak solutions for nonlinear boundary value problems. Journal of Mathematical Analysis and Applications *164*, 489–507 (1992).

[10] Nakao M. T.: Solving nonlinear parabolic problems with result verification: Part I. Journal of Computational and Applied Mathematics *38*, 323–334 (1991).

[11] Nakao, M. T. & Watanabe, Y.: Solving nonlinear parabolic problems with result verification: Part II, Several space dimensional case, Research Report of Mathematics of Computation, Kyushu University, RMC 67-05 (1992), 10 pages.

[12] Nakao, M. T., Yamamoto, N.: Numerical verifications of solutions for elliptic equations with strong nonlinearity. Numerical Functional Analysis and Optimization *12*, 535–543 (1991).

[13] Nakao, M. T.: Numerical verifications of solutions for nonlinear hyperbolic equations, Research Report of Mathematics of Computation, Kyushu University, RMC 66-07 (1991), 13 pages.

[14] Plum, M.: Computer-assisted existence proofs for two-point boundary value problems. Computing *46*, 19–34 (1991).

[15] Plum, M.: Existence proofs in combination with error bounds for approximate solutions of weakly nonlinear second-order elliptic boundary value problems. ZAMM *71* (6), T660–T662 (1991).

[16] Plum, M.: Explicit H_2-estimates and pointwise bounds for solutions of second-order elliptic boundary value problems. Journal of Mathematical Analysis and Applications *165*, 36–61 (1992).

[17] Plum, M.: Numerical existence proofs and explicit bounds for solutions of nonlinear elliptic boundary value problems. Computing *49*, 25–44 (1992).

[18] Rump, S. M.: Solving algebraic problems with high accuracy. In: Kulisch, U., Miranker, W. L. (eds.) A new approach to scientific computation. New York: Academic Press 1983.

[19] Sattinger, D. H.: Topics in stability and bifurcation theory. Berlin, Heidelberg: Springer 1973 (Lecture Notes in Mathematics, 309).

[20] Tsuchiya, T., Nakao, M. T.: Numerical verification of solutions of parametrized nonlinear boundary value problems with turning points, Research Report of Mathematics of Computation, Kyushu University, RMC 67-02 (1992), 17 pages.

[21] Watanabe, Y., Nakao, M. T.: Numerical verifications of solutions for nonlinear elliptic equations. Japan Journal of Industrial and Applied Mathematics (in press).

[22] Yamamoto, N., Nakao, M. T.: Numerical verifications of solutions for elliptic equations in nonconvex polygonal domains, Numerische Mathematik (in press).

[23] Zeidler, E.: Nonlinear functional analysis and its applications I. New York: Springer 1986.

Mitsuhiro T. Nakao
Department of Mathematics
Faculty of Science
Kyushu University 33
Fukuoka 812, Japan

Computing, Suppl. 9, 175–190 (1993)

© Springer-Verlag 1993
Printed in Austria

The Wrapping Effect, Ellipsoid Arithmetic, Stability and Confidence Regions*

A. Neumaier, Freiburg

Dedicated to Prof. U. Kulisch on the occasion of his 60th birthday

Abstract — Zusammenfassung

The Wrapping Effect, Ellipsoid Arithmetic, Stability and Confidence Regions. The wrapping effect is one of the main reasons that the application of interval arithmetic to the enclosure of dynamical systems is difficult. In this paper the source of wrapping is analyzed algebraically and geometrically. A new method for reducing the wrapping effect is proposed, based on an interval ellipsoid arithmetic.

Applications are given to the verification of stability regions for nonlinear discrete dynamical systems and to the computation of rigorous confidence regions for nonlinear functions of normally distributed random vectors.

AMS Subject Classification: 65G10, 34D20, 62G15

Key words: Wrapping effect, ellipsoid arithmetic, stability of dynamical systems, rigorous confidence region.

Der Verpackungseffekt, Ellipsoidarithmetik, Stabilität und Konfidenzbereiche. Der Verpackungseffekt ist eine der Hauptursachen dafür, daß die Anwendung von Intervallverfahren auf die Einschließung dynamischer Systeme schwierig ist. In dieser Arbeit wird dieser Effekt algebraisch und geometrisch analysiert. Um den Verpackungseffekt zu reduzieren, wird eine neue Methode vorgestellt, die auf einer Intervall-Ellipsoidarithmetik basiert.

Als Anwendungen werden die Verifikation von Stabilitätsbereichen nichtlinearer diskreter dynamischer Systeme und die Berechnung von rigorosen Konfidenzbereichen für nichtlineare Funktionen normalverteilter Zufallsvariablen skizziert.

1. Introduction

Consider the simple triangular linear system $Ax = b$, where

$$A = \begin{bmatrix} 1 & & & & & \\ 1 & 1 & & & & \\ 1 & 1 & 1 & & & 0 \\ & 1 & 1 & 1 & & \\ & & \ddots & \ddots & \ddots & \\ 0 & & & 1 & 1 & 1 \end{bmatrix}, \quad b = \begin{bmatrix} \beta \\ 0 \\ 0 \\ 0 \\ \vdots \\ 0 \end{bmatrix}. \tag{1}$$

---

* Received September 21, 1992; revised December 16, 1992.

Clearly

$$x_1 = \beta, \qquad x_2 = -\beta, \tag{2}$$

$$x_l = -x_{l-1} - x_{l-2} \qquad \text{for } l > 2, \tag{3}$$

so that

$$x_l = \begin{cases} -\beta & \text{if } l = 3k - 1, \\ 0 & \text{if } l = 3k, \\ \beta & \text{if } l = 3k + 1. \end{cases}$$

Suppose we only know that $\beta \in [-\varepsilon, \varepsilon]$. If we use naive interval arithmetic to solve the triangular system we get

$$x_1 \in [-\varepsilon, \varepsilon], \qquad x_2 = -x_1 \in [-\varepsilon, \varepsilon],$$

$$x_3 = -x_1 - x_2 \in -[-\varepsilon, \varepsilon] - [-\varepsilon, \varepsilon] = [-2\varepsilon, 2\varepsilon],$$

and inductively

$$x_l = [-a_l\varepsilon, a_l\varepsilon],$$

with the Fibonacci sequence $a_1 = a_2 = 1, a_{l+1} = a_l + a_{l-1}$, hence $a_l > \text{const.} \cdot 1.618^l$. The interval bounds grow exponentially, and compared with the optimal bounds

$$x_l \in \begin{cases} [0, 0] & \text{if } l = 3k, \\ [-\varepsilon, \varepsilon] & \text{otherwise}, \end{cases}$$

the overestimation is excessive for large n. The same kind of overestimation persists when (1) is replaced by a system with wider bands.

The reason for the overestimation is the dependence of the x_l on the *same* vaguely known number β, while interval arithmetic—due to its memory-less nature— assumes that all the x_l vary independently over their enclosing interval.

Of course, there are interval methods which are more reliable, and in this case even optimal, namely those which explicitly precondition the system (1) by an approximate (midpoint) inverse. However, the inverse of A is a full lower triangular matrix which means that work for forming and solving the preconditioned system is of order $O(n^2)$, which is an order of magnitude larger than the work for solving (1) approximately (and in our case exactly since the matrix entries are so simple). For bounded systems arising in the solution of time-dependent problems over many time steps, this increase in work needed for realistic enclosures limits the scope of current interval methods. Thus it is very important to understand this overesti- mation problem in detail, and to device ways of reducing the complexity while still obtaining reasonable bounds. In this paper we only consider methods which con- serve the time-like recurrent structure of banded equations. Other methods (Gambill and Skeel [5], Alvarado [1], [2]) which use divide and conquer strategies will not be discussed here.

We obtain an intuitive geometric interpretation of the mechanism underlying the overestimation if we rewrite the difference Eq. (3) as

$$\begin{pmatrix} x_l \\ x_{l+1} \end{pmatrix} = \begin{pmatrix} 0 & 1 \\ -1 & -1 \end{pmatrix} \begin{pmatrix} x_{l-1} \\ x_l \end{pmatrix}. \tag{4}$$

If we change notation we can view this as the particular case

$$A = \begin{pmatrix} 0 & 1 \\ -1 & -1 \end{pmatrix}, \qquad b = \begin{pmatrix} 0 \\ 0 \end{pmatrix}, \qquad x_l \in \mathbb{R}^2 \tag{5}$$

of the discrete dynamical system

$$x_{l+1} = Ax_l + b. \tag{6}$$

Geometrically, (6) describes an affine transformation of x_l to x_{l+1}. The set of allowed positions of the initial value $x_0 = \begin{pmatrix} \beta \\ -\beta \end{pmatrix} (\beta \in [-\varepsilon, \varepsilon])$—corresponding to $\begin{pmatrix} x_1 \\ x_2 \end{pmatrix}$ in (2)—is a line segment, and, as affine transforms of x_0, all x_l are line segments, too.

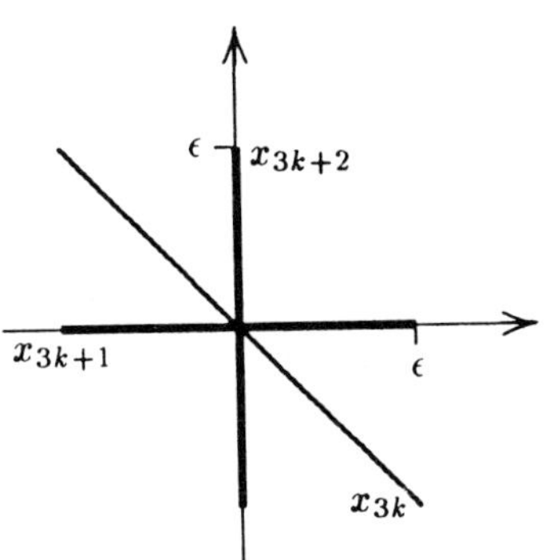

Figure 1. Optimal solution sets

In the first step, the effect of interval arithmetic is the replacement of the initial line segment by the smallest box $x_0 = \begin{pmatrix} [-\varepsilon, \varepsilon] \\ [-\varepsilon, \varepsilon] \end{pmatrix}$ containing it. In the next step, this box is transformed into a parallelogram, and the smallest box containing it is $x_1 = \begin{pmatrix} [-\varepsilon, \varepsilon] \\ [-2\varepsilon, 2\varepsilon] \end{pmatrix}$. In the next step, x_1 transforms into another parallelogram whose interval enclosure is the box $x_2 = \begin{pmatrix} [-2\varepsilon, 2\varepsilon] \\ [-3\varepsilon, 3\varepsilon] \end{pmatrix}$, etc. One sees that the tiny precious birthday present x_l is wrapped into layer after layer of wrapping paper until a very conspicuous present x_l results, whose size has no longer anything to do with its contents. This is the so-called *wrapping effect*, already observed in the early days of interval calculations (Moore [16], [17]).

In general, the behavior of the iteration (6) depends on the spectrum of A. If all eigenvalues of A have absolute values <1, the x_l tend to a limit point x^*

which is the solution of the linear system $x^* = Ax^* + b$. If some eigenvalues have absolute values >1 then, for most starting points, the iteration will diverge, $\|x_l\| \to \infty$ for $l \to \infty$.

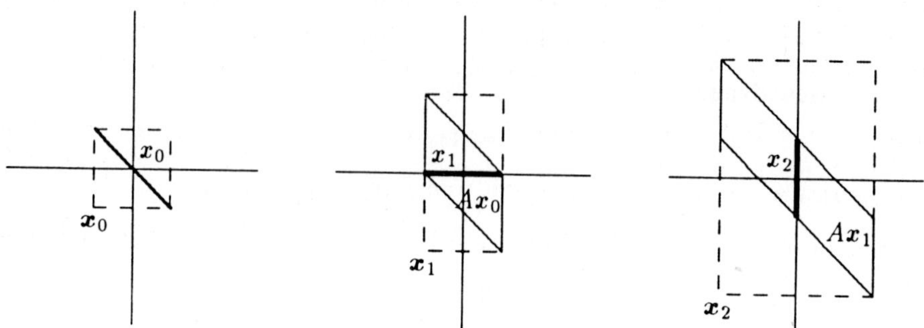

Figure 2. The wrapping effect

The volume of a set changes under the transformation (6) by a factor of $|\det A|$, the product of the absolute values of the eigenvalues of A. Therefore we can assess the amount of overestimation per step by monitoring the *overestimation factor*

$$q = \operatorname{vol} x_{l+1} / |\det A| \operatorname{vol} x_l,\tag{7}$$

where $\operatorname{vol}[\underline{x}, \bar{x}] = (\bar{x}_1 - \underline{x}_1) \ldots (\bar{x}_n - \underline{x}_n)$ is the volume of a box $x = [\underline{x}, \bar{x}]$ in $\mathbb{R}^n$.

2. Simplices, Parallelepipeds, Hyperoctahedra and Ellipsoids

It is clear that interval boxes are not sufficiently variable in shape to model all affine transforms of an initial box; to eliminate the wrapping effect for the case of the iteration (6) or any iteration

$$x_{l+1} = A_l x_l + b_l\tag{8}$$

we need to use a class of enclosure sets which is closed under affine transformations and still simple enough to be described with little effort.

A natural class of such sets is the class of all polytopes, but working with general polytopes is time-consuming. The simplest polytopes, the simplices, work and have indeed be used for enclosures (Conradt [3], Jansson [10], Rump [22]). Parallelepipeds are another natural affinely closed class of polytopes, and have been used in all recent enclosure methods for ordinary differential equations (Eijgenraam [4], Lohner [14], [15]); their special advantage is that the boxes belongs to this class, hence they can be handled easily by interval methods. Hyperoctahedra form another simple affinely closed class, but have received no attention so far. See also the survey article Nickel [20].

A very important affinely closed class consists of the set of (hyper-)ellipsoids. Since ellipsoids arise naturally in many applications, notably as confidence regions of stochastic variables, and since they are invariant under a much bigger symmetry group than the other classes mentioned, they have received considerable attention around 1968 (Kahan [11], Jackson [8], [9]). But in the absence of a simple way of calculating with ellipsoids—the Minkowski sum of two ellipsoids has a complicated shape and is expensive to enclose by another ellipsoid—they have not been used in actual codes for rigorous computation. However, as pointed out by a referee, *approximate* calculations with ellipsoids have found successful applications to several problems in linear control theory, see Kurzhanski & Vályi [13], Ovseevich & Chernousko [21] and the references there. In particular, the discrete case of the *reachable set problem* amounts to finding enclosures of (8) where A_l is known exactly and x_0 and the b_l vary in specified sets.

As we shall see, the use of ellipsoids *in combination with* intervals allows the efficient handling of ellipsoids and makes them again an interesting competitor for the best representation of multivariate enclosure sets.

In this section we represent ellipsoids in a form which, for different norms, would give other families of affine closed classes of convex sets which can be easily represented on a computer; parallelepipeds and hyperoctahedra become cases of this concept when the 2-norm is replaced by the ∞-norm or the 1-norm, respectively. We show how the chosen representation allows a very simple implementation of the iteration (8) without overestimation. But there are problems of numerical stability which often lead to a blow-up of the enclosures when implemented in (outward rounded) finite precision arithmetic. The next section is therefore devoted to the derivation of a stable version of the enclosure method.

In the following, we use freely interval arithmetical extensions of vector and matrix operations (cf. Neumaier [19]), writing interval quantities in boldface types. $\|\cdot\|$ denotes the Euclidian vector and matrix norm.

An *ellipsoid* is a set of the form

$$E(z, L, r) := \{z + L\xi \mid \xi \in \mathbb{R}^n, \|\xi\| \le r\}, \tag{9}$$

with fixed $z \in \mathbb{R}^n$, $L \in \mathbb{R}^{n \times n}$ (lower triangular), $r \in \mathbb{R}_+$. Thus ellipsoids are the images of a ball $\{\xi \in \mathbb{R}^n \mid \|\xi\| \le r\}$ under affine mappings $\xi \to z + L\xi$. Since $\|\xi\|$ is invariant under orthogonal transformations, lower triangular matrices are sufficient to represent all ellipsoids (including degenerate cases); for other norms, L would have to be unrestricted.

Proposition. *Suppose* (8) *holds with nonsingular* A_l. *Then*

$$x_l \in E(z_l, L_l, r) \quad \Rightarrow \quad x_{l+1} \in E(z_{l+1}, L_{l+1}, r), \tag{10}$$

where

$$z_{l+1} = A_l z_l + b_l, \qquad L_{l+1} = A_l L_l. \tag{11}$$

Proof. If $x_l = z_l + L_l \xi$ then $x_{l+1} = A_l(z_l + L_l \xi) + b_l = (A_l z_l + b_l) + (A_l L_l)\xi = z_{l+1} + L_{l+1}\xi.$ $\qquad\square$

While this proposition seems to handle completely all linear dynamical systems, this only holds under the assumption of exact arithmetic. In finite precision arithmetic, one must account for the rounding errors in the formation of z_{l+1} and L_{l+1}.

The simplest way to do this is to allow the z_l and L_l themselves to have interval components, and to compute (11) by outward rounding. But the recurrence for the z_l is precisely the same as for the x_l, so the wrapping effect appears in the z_l and magnifies rounding errors exponentially; and the same holds for each column of the L_l. Thus the proposition is useless for actual computation. To avoid excessive overestimation we must therefore keep z_l and L_l real, and account for the rounding errors made by increasing the ellipsoid radius r.

3. The Enclosure of Affine Transforms of Ellipsoids

We study the adaption of r in a slightly more general setting motivated by the fact that in many realistic situations the entries of the A_l are not precisely known and are allowed to vary in intervals. Thus we want to study the transformation of an ellipsoid under all affine mappings

$$x \to Ax + b \qquad (A \in A, b \in b), \tag{12}$$

where A is an $n \times n$ interval matrix and b an n-dimensional interval vector.

Theorem 1. *Suppose that $x \in E(z, L, r)$. Select arbitrary $\bar{z} \in \mathbb{R}^n$ and nonsingular $\bar{L} \in \mathbb{R}^{n \times n}$, and let*

$$\tilde{r} := \|\bar{L}^{-1}(Az + b - \bar{z})\| + \|\bar{L}^{-1}AL\|r. \tag{13}$$

Then

$$Ax + b \in E(\bar{z}, \bar{L}, \tilde{r}) \qquad \text{for all } A \in A, b \in b. \tag{14}$$

Proof. By assumption, $x = x + L\xi$ for some ξ with $\|\xi\| \le r$. Therefore

$$Ax + b = A(z + L\xi) + b = Az + b + AL\xi$$

$$= \bar{z} + (Az + b - \bar{z} + AL\xi) = \bar{z} + \bar{L}\bar{\xi},$$

where

$$\bar{\xi} = \bar{L}^{-1}(Az + b - \bar{z}) + \bar{L}^{-1}AL\xi \in \bar{L}^{-1}(Az + b - \bar{z}) + (\bar{L}^{-1}AL)\xi$$

satisfies

$$\|\bar{\xi}\| \le \|\bar{L}^{-1}(Az + b - \bar{z})\| + \|\bar{L}^{-1}AL\|\,\|\xi\| \le \tilde{r}. \qquad \square$$

Of course, if we want to have a *good* enclosure we must choose $\bar{z}$ and $\bar{L}$ properly. When A and b are precisely known (so that we can use ordinary arithmetic) we see that the choice $\bar{z} = Az + b$, $\bar{L} = AL$, $\tilde{r} = r$ is consistent with (13) and recovers the previous proposition. In the presence of roundoff error or other uncertainties, this suggests the choice

$$\bar{z} = \text{mid}(Az + b), \qquad \bar{L} = \text{mid}(AL), \tag{15}$$

where the interval expressions are computed with outward rounding to take correctly care of rounding errors. However, (13) shows that this will be disastrous when $\bar{L}$ is ill-conditioned since it blows up the radius $\bar{r}$ to a very large number. And, unfortunately, repeated iteration of (12) using (15) yields always, sooner or later, such ill-conditioned $L = (\text{mid } A)^l L_0$, unless all eigenvalues of mid A have the same absolute value. Therefore, we must look for a regularized version which keeps $\bar{r}$ reasonably small. Let us introduce some notation. For a vector x, we denote by $|x|$ the vector with components $|x_i|$. $A_{i\cdot}$ denotes the i-th row of a matrix A, and $v(A)$ denotes the *hybrid norm* of A (Neumaier [18]), i.e. the vector with components $v_i(A) := \|A_{i\cdot}\|$. It is easy to see that

$$|Ax| \leq v(A)\|x\|,$$

and

$$\|v(A)\| = \sqrt{\text{tr } A^T A} = \|A\|_{\text{F}} \qquad (\geq \|A\|)$$

is the *Frobenius* (or *Schur*) norm of A. We write

$$\bar{z} := \text{mid}(Az + b), \qquad d := |Az + b - \bar{z}|, \tag{16}$$

$$B := \text{mid}(AL), \qquad d' := v(AL - B). \tag{17}$$

From (13) and monotony, we get, for any nonsingular diagonal matrix D,

$$\bar{r} \leq \|\bar{L}^{-1}D\|\,\|D^{-1}(Az + b - \bar{z})\| + (\|\bar{L}^{-1}B\| + \|\bar{L}^{-1}D\|\,\|D^{-1}(AL - B)\|_{\text{F}})r$$

$$\leq \|\bar{L}^{-1}D\|\,\|D^{-1}d\| + (\|\bar{L}^{-1}B\| + \|\bar{L}^{-1}D\|\,\|D^{-1}d'\|)r,$$

hence

$$\bar{r} \leq \|\bar{L}^{-1}B\|r + \|\bar{L}^{-1}D\|q \tag{18}$$

where

$$q = \|D^{-1}d\| + \|D^{-1}d'\|r. \tag{19}$$

In (18) and (19) all intervals are eliminated, making it more suitable for analysis. We want to choose L such that the volume of the new ellipsoid is small. Now this volume is proportional to $\bar{r}^n \det \bar{L}$, and the next result shows how a nearly optimal $\bar{L}$ can be found. Since a change of $\bar{r}$ is equivalent to rescaling L we aim at a radius $\bar{r} \approx 1$, thus having small $\bar{L}$ when rL and hence rB are small.

Theorem 2. *Suppose that*

$$r^2 BB^T + q^2 DD^T = \bar{L}\bar{L}^T, \tag{20}$$

$$\bar{r} = \|\bar{L}^{-1}B\|r + \|\bar{L}^{-1}D\|q. \tag{21}$$

Then $\bar{r} \leq 2$, and for arbitrary nonsingular $\tilde{L}$ we have

$$(\|\tilde{L}^{-1}B\|r + \|\tilde{L}^{-1}D\|q)^n |\det \tilde{L}| \geq |\det \bar{L}|. \tag{22}$$

In particular, choosing $\bar{L}$ by (20) implies optimality of $\bar{L}$ within a factor of 2 for the radius.

Proof. We use the abbreviations $U = \bar{L}^{-1}B$, $V = \bar{L}^{-1}D$ to rewrite (20), (21) as

$$r^2 UU^T + q^2 VV^T = I,$$

$$\|U\|r + \|V\|q = \bar{r}.$$

With $W = \tilde{L}^{-1}\bar{L}$, the left hand side of (22) becomes

$$(\|WU\|r + \|WV\|q)^n |\det \bar{L}|/|\det W|.$$

Now

$$\|W(rU, qV)\|^2 = \|W(rU, qV)(rU, qV)^T W^T\| = \|WW^T\| = \|W\|^2,$$

hence

$$|\det W| \leq \|W\|^n \leq \|W(rU, qV)\|^n \leq (r\|(WU, 0)\| + q\|(0, WV)\|)^n$$

$$= (r\|WU\| + q\|WV\|)^n$$

so the left hand side of (22) is $\geq |\det \bar{L}|$.

On the other hand,

$$\|U\| \leq r^{-1}, \qquad \|V\| \leq q^{-1},$$

hence $\bar{r} \leq 2$. $\qquad\qquad\qquad\qquad\qquad\qquad\qquad\qquad\qquad\qquad\qquad\qquad\square$

Of course, this optimality result is based on the upper bound (18) which is not exact, but which is easily computable. The diagonal matrix D can still be chosen freely, and its optimal choice is unsettled. However, a natural choice for D comes from balancing the contributions of the components to q in (19), and suggests that we take

$$D = \text{Diag}(d_1 + d_1'r, \ldots, d_k + d_k'r). \tag{23}$$

(This forces $D^{-1}d + D^{-1}d'r = (1, \ldots, 1)^T$, leading to $\sqrt{n} \leq q \leq \sqrt{2n}$.) As one can see from (16) and (17), D will have entries of the order of the radii of A and b, so that the contributions in (20), (21) to $\bar{L}$ and $\bar{r}$ remain small as long as B is well conditioned. Thus $\bar{L}\bar{L}^T \approx r^2 BB^T$ remains small when rB was small and $Q = r\bar{L}^{-1}B$ satisfies $Q^TQ \approx I$ so that $\bar{r} \approx \|Q\| \approx 1$ and everything is stable.

The matrix $\bar{L}$ is determined by (20) only upto an orthogonal transformation, and is best chosen as a Cholesky factor of the left hand side of (20). One sees that because of the regularizing diagonal term in (20), $\bar{L}$ will be well-conditioned even when $B = \text{mid}(AL)$ is ill-conditioned or singular; so the instability mentioned earlier has been removed successfully. However, the formation of $\bar{L}$ by (20) directly is numerically unstable when B is ill-conditioned, and we must proceed in a slightly different way. The first possibility is to add to the matrix $N = r^2 BB^T + q^2 DD^T$ extra diagonal terms to force it positive definite (e.g. $N_{ii} \leftarrow N_{ii}(1 + \varepsilon^{1/2})$). Another possibility uses the fact that the matrix $M := (rB, qD) \in \mathbb{R}^{n \times 2n}$ satisfies $MM^T = r^2 BB^T + q^2 DD^T = \bar{L}\bar{L}^T$; hence we can obtain $\bar{L}$ from an LQ-factorization ($=$ transposed QR factorization) of M. Small diagonal entries in $\bar{L}$ due to roundoff can be corrected by replacing the diagonal entries $\bar{L}_{ii}$ with $\bar{L}_{ii} + \text{sgn}\,\bar{L}_{ii} \cdot \eta \cdot v_i(M)$, where $\eta = n^{3/2}\varepsilon$ and ε denotes the machine accuracy.

Collecting together the various formulas we find the following

Algorithm. (*Ellipsoid propagation algorithm*)

Purpose: Enclose the transformation of the ellipsoid $\{z + L\xi \mid \|\xi\| \leq r\}$
(L lower triangular) by $x \to Ax + b$ within the ellipsoid
$\{\bar{z} + \bar{L}\bar{\xi} \mid \|\bar{\xi}\| \leq \bar{r}\}$ ($\bar{L}$ lower triangular)

! The rounding mode for computing each left hand side is indicated by
! $\supseteq$ (outward), $\approx$ (approximate), $\geq$ (upwards)
$\underset{\sim}{\bar{z}} \supseteq Az + b$, $\bar{z} \approx \operatorname{mid} \underset{\sim}{\bar{z}}$, $d \geq |\bar{z} - \underset{\sim}{\bar{z}}|$
$\underset{\sim}{B} \supseteq AL$, $B \approx \operatorname{mid} \underset{\sim}{B}$, $d' \geq v(\underset{\sim}{B} - B)$
$D \approx \operatorname{Diag}(d_1 + d'_1 r, \ldots, d_n + d'_n r)$
$q \geq \|D^{-1}d\| + \|D^{-1}d'\|r$
$M \approx (rB, qD)$
$\bar{L}\bar{Q} \approx M$
For $i = 1, \ldots, n$, change $\bar{L}_{ii}$ to $\bar{L}_{ii} + \operatorname{sgn} \bar{L}_{ii} \cdot \eta \cdot v_i(M)$
Enclose the solution of $\bar{L}C = B$ by C and compute an upper bound γ
  on the largest singular value of C
Enclose the solution of $\bar{L}C = D$ by C and compute an upper bound δ
  on the largest singular value of C
$\bar{r} \geq \gamma r + \delta q$.

4. Bounds for the Range over an Ellipsoid

The wrapping effect does not only occur for linear transformations, but even more when one transforms a set by a nonlinear transformation. In this case, even exact arithmetic does not allow optimal enclosures since nonlinear mappings generally distort the shapes of ellipsoids (or simplices, parallelepipeds and hyperoctahedra), and clearly the amount of unavoidable wrapping increases with the amount of nonlinearity present in the set. In particular, this implies that one will be able to obtain realistic enclosures of general nonlinear transformations only for sufficiently narrow sets (where nonlinearities contribute in the order of the squared diameter only). On the other hand, realistic enclosures of an image of a big set can be obtained only for mappings which are nearly linear.

We now show that the techniques of the previous section suffice to enclose nonlinear images of narrow ellipsoids by another ellipsoid. The key is the observation that for any C^1-function $F: \mathbb{R}^n \to \mathbb{R}^m$ (the dimensions need not be equal!) which is defined by a Lipschitz expression (a notion explained in Neumaier [19]), any bounded set $E \subseteq \mathbb{R}^n$ and all *centers* $z \in E$, one can find a *slope matrix* A such that we can represent F as *centered form*

$$x \in E \quad \Rightarrow \quad F(x) = F(z) + \tilde{A}(x - z) \quad \text{for some } \tilde{A} \in A. \tag{24}$$

Thus if we know an enclosure

$$b \supseteq F(z) \tag{25}$$

of $F(z)$ which accounts for roundoff, we find that

$$F(x) \in \bigcup \{Ax + b \,|\, x \in E\},$$

so that the theory of the previous section applies when E is an ellipsoid.

The only difficulty is the computation of the slope matrix A. An optimal computation of A using the full ellipsoid information seems difficult; therefore we compute A using an interval enclosure for the ellipsoid E by a box x; then recursive techniques (Krawczyk & Neumaier [12], Neumaier [19]) allow the computation of the slope matrix A for the box x, and this is obviously also a slope matrix for the subset E of x. The optimal box is given by

Theorem 3. *The smallest box containing the ellipsoid $E(z, L, r)$ is*

$$x := \Box E(z, L, r) = z + [-r, r]v(L). \tag{26}$$

Proof. The i-th component of x is given by the hull of all $x_i = (z + L\xi)_i$ with $\|\xi\| \le r$. Now, by the Cauchy-Schwarz inequality,

$$|x_i - z_i| = |(L\xi)_i| = |L_{i\cdot}\xi| \le \|L_{i\cdot}\|_2 \|\xi\|_2 = v_i(L)\|\xi\|_2 \le v_i(L)r,$$

and clearly the bound can be attained with either sign of $x_i - z_i$. Hence formula (26).
$$\Box$$

In practice, when doing a sequence of nonlinear transformations, it is important to update the box containing the new ellipsoid $E(\bar{z}, \bar{L}, \bar{r})$ by

$$\bar{x} = (A(x - z) + b) \cap (\bar{z} + [-\bar{r}, \bar{r}]v(\bar{L})). \tag{27}$$

This often eliminates the ends of long and thin ellipsoids which (due to overestimation) no longer contain points of the (iterated) image of the original set; these ends would inflate the box (26) considerably.

This is particularly relevant when $A \ge 0$, where it is well known that the formula $\bar{x} = A(x - z) + b$ gives optimal boxes (though large volume overestimation).

Another trick is often important to guarantee reasonable results when the nonlinear transformation $F(x, \lambda)$ depends on a parameter vector λ which varies in a box λ (or an ellipsoid). Treating the λ_i as interval constants in the centered form (24) often leads to significant wrapping, especially after a sequence of several transformations

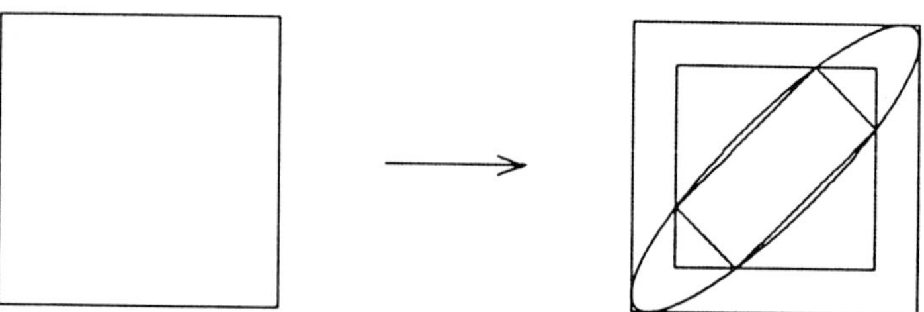

Figure 3. Updating the box by (27)

involving λ. In this case, the correct way to treat these parameters is by extending the state vector x to $x' = \begin{pmatrix} x \\ \lambda \end{pmatrix}$ and enclosing it by higher-dimensional ellipsoids. While this incurs in the first step a volume overestimation factor of the unit ball volume, this factor will often have been gained after a few more steps, due to reduced wrapping. For wide intervals in λ, however, this need no longer be the case, and one may have to resort to methods of global optimization (Hansen [6]).

Examples. We iterate (27) for a two-dimensional discrete dynamical system

$$x_{l+1} = A_l x_l + b_l \tag{28}$$

where $A_l \in A$, $b_l \in b = 10^{-12}(\begin{bmatrix} -1;1 \\ -1;1 \end{bmatrix})$, $x_0 \in x_0 = (\begin{bmatrix} -1;1 \\ -1;1 \end{bmatrix})$, and various matrices A. Thus we start with a square of side 2, and assume small vagueness in the vectors b_l but (except in case 4) large vagueness in the coefficients of A_l.

In each case we list the side β_l of a cube with the same volume as x_l and the side γ_l of a cube with the same volume as the ellipsoid. For comparison we also iterate in naive interval arithmetic

$$y_{l+1} = Ay_l + b_l,$$

starting with $y_0 = x_0$, and record the quotient β_l'/β_l, where β_l' is the side of a cube with the same volume as y_l. The computations were done with the CALCULUS system [7] of Siegfried Rump, whose help with the examples is gratefully acknowledged. On the machine used, $\varepsilon = 10^{-17}$.

Case 1: $\quad A = \begin{pmatrix} p & p \\ -p & p \end{pmatrix}, \quad p = [\tfrac{4}{10}, \tfrac{5}{10}].$

l	β_l	γ_l	β_l'/β_l
10	1.33E$-$01	6.64E$-$02	1.51E$+$01
20	6.22E$-$03	3.11E$-$03	3.22E$+$02
30	2.92E$-$04	1.46E$-$04	6.86E$+$03
40	1.37E$-$05	6.84E$-$06	1.47E$+$05
50	6.42E$-$07	3.21E$-$07	3.12E$+$06
60	3.01E$-$08	1.51E$-$08	6.65E$+$07
70	1.43E$-$09	7.11E$-$10	1.41E$+$09
80	7.69E$-$11	3.85E$-$11	2.61E$+$10
90	1.39E$-$11	6.92E$-$12	1.45E$+$11
100	1.09E$-$11	5.44E$-$12	1.84E$+$11

The ellipsoids contract until about the size of rad b; naive interval arithmetic does not contract.

Case 2: $\quad A = \begin{pmatrix} 0 & 1 \\ -1 & -p \end{pmatrix}, \quad p = [1, \tfrac{10}{9}].$

l	β_l	γ_l	β'_l/β_l
10	1.21E+01	5.79E+00	3.02E+01
20	5.60E+01	2.42E+01	1.31E+03
30	2.05E+02	9.79E+01	7.19E+04
40	7.09E+02	3.69E+02	4.18E+06
50	2.69E+03	1.35E+03	2.21E+08
60	1.09E+04	5.02E+03	1.11E+10
70	3.99E+04	1.89E+04	6.03E+11
80	1.36E+05	6.86E+04	3.56E+13
90	4.85E+05	2.45E+05	2.01E+15
100	1.83E+06	8.76E+05	1.08E+17

All matrices in A have determinant 1, so the dynamical system is volume preserving. However, the ellipsoid volume grows exponentially, with an average factor of $(8.76 \cdot 10^5)^{1/100} \approx 1.147$, but much less than the boxes in naive calculation.

Case 3: $A = \begin{pmatrix} 0 & 1 \\ 1 & p \end{pmatrix}$, $\quad p = [1, \frac{10}{9}]$.

l	β_l	γ_l	β'_l/β_l
10	3.64E+02	7.78E+03	1.00E+00
20	7.32E+04	8.98E+08	1.00E+00
30	1.48E+07	1.04E+14	1.00E+00
40	2.96E+09	1.20E+19	1.00E+00
50	5.95E+11	1.39E+24	1.00E+00
60	1.20E+14	1.60E+29	1.00E+00
70	2.41E+16	1.84E+34	1.00E+00

Again the system is volume preserving but we get boxes exploding with a factor ≈ 1.714 per iteration and ellipsoids exploding even faster. The intersection in (27) is here very effective. Since $A \geq 0$, naive interval calculation gives optimal boxes (but not optimal volumes), explaining $\beta'_l = \beta_l$. In this example, the ellipsoids are clearly not useful.

Case 4: $A = \begin{pmatrix} 8.0 & -2.6 \\ 9.0 & -2.8 \end{pmatrix}$.

a) with direct computation of $\bar{L}$ from (20):

l	β_l	γ_l	β'_l/β_l
10	3.96E+07	2.36E+05	1.05E+03
20	3.87E+14	2.30E+12	2.53E+06
30	3.78E+21	2.25E+19	6.07E+09
40	3.69E+28	2.20E+26	1.46E+13
50	3.60E+35	2.15E+33	3.52E+16

Again the system is volume preserving, but with eigenvalues 0.2 and 5 which cause the ellipsoids to flatten rapidly. The growth rate of the box volumes is about 2/3 of that of the naive calculation, a consequence of the diagonal needle shape of the ellipsoids.

b) with $\bar{L}$ computed as stated in the algorithm:

l	β_l	γ_l	β_l'/β_l
1	2.24E+01	1.42E+00	1.00E+00
2	1.30E+02	1.42E+00	1.88E+00
3	6.51E+02	1.42E+00	4.08E+00
4	3.26E+03	1.42E+00	8.89E+00
5	1.63E+04	1.42E+00	1.94E+01
6	8.14E+04	1.42E+00	4.22E+01
7	4.07E+05	1.42E+00	9.19E+01
8	2.04E+06	1.42E+00	2.00E+02
9	1.03E+07	1.43E+00	4.34E+02
10	5.42E+07	1.51E+00	8.92E+02
20	1.27E+18	1.28E+10	8.99E+02
30	2.97E+28	5.79E+20	8.99E+02

Here the ellipsoid volume hardly grows initially, but after a while the volumes grow at the same rate as in the naive calculation.

The rapid blow-up of the ellipsoids in case 4, where the ambiguity in (28) is only small, $O(10^{-12})$, suggests that the method proposed does not yet choose nearly optimal enclosing ellipsoids, so that there is further room for improvement.

5. Application: Stability Regions

A discrete dynamical system

$$x_{l+1} = F(x_l) \tag{29}$$

is called *stable* in a region E when, for any initial state $x_0 \in E$, the state vectors x_l defined by (29) remains bounded for all l. A sufficient condition for this is—when E is bounded—that

$$F(x) \in E \qquad \text{for all } x \in E. \tag{30}$$

If F is also continuous, then Brouwer's fixed point theorem guarantees a fixed point of F under these conditions.

Clearly we can verify (30) by the techniques of the present paper when $E = (z, L, r)$ is an ellipsoid. Since the image is to be enclosed by the same ellipsoid, the natural choice here is $\bar{z} = z$ and $\bar{L} = L$. Then, using (24) and (13), a sufficient condition for (30), hence for stability, is

$$\|L^{-1}(Az + b - z)\| + \|L^{-1}AL\|r \le r. \tag{31}$$

Since an initial enclosure is here not given we are free to choose z and L arbitrarily, and we want to choose it to make the verification of (31) most likely. Since we know that E must contain a fixed point, we choose z as an approximation to such a fixed point; then $F(z) \approx z$ implies that the first norm in (31) will be small.

The second norm in (31) must be made <1. Since in the limit $r \to 0$, where the ellipsoid shrinks to the point z, the slope matrix tends to $F'(z)$ (assuming F to be differentiable), we have $F'(z) \in A$. Thus we try to make $\|L^{-1}F'(z)L\|$ small, and in particular <1.

Since the norm of a matrix is an upper bound for the spectral radius, and the latter is invariant under similarity transformations, $F'(z)$ must have spectral radius $\rho(F'(z)) < 1$, i.e., z must be an attractive fixed point. In this case we can transform $F'(z)$ to a block-diagonal form by the modal matrix S whose columns are eigenvectors of real eigenvalues, or real and complex part of eigenvectors to complex eigenvalues. (This assumes that $F'(z)$ is nondefective. In the nearly defective case one must consider higher-dimensional invariant subspaces.) The diagonal blocks of $S^{-1}F(z)S$ will have the form (λ) for real eigenvalues and $\begin{pmatrix} \operatorname{Re}\lambda & \operatorname{Im}\lambda \\ -\operatorname{Im}\lambda & \operatorname{Re}\lambda \end{pmatrix}$ for complex eigenvalues. Thus if we take for L an approximation to S (silently dropping the assumption of L being lower triangular) we find that $L^{-1}F'(z)L$ and hence $L^{-1}AL$ (for small r; the entries of A have radius $O(r)$) are approximately block-diagonal. Now let

$$C \supseteq L^{-1}(AL), \qquad C \approx \operatorname{mid} C, \qquad R \geq |C - C|. \tag{32}$$

By construction, C is nearly block-diagonal, and $R = O(r)$. The form of the diagonal blocks implies that CC^T is approximately diagonal, with diagonal entries $\approx |\lambda|^2 < 1$. Thus $\|CC^T\|_\infty$ will approximately equal the square of the spectral radius of $F'(z)$, and since

$$\|L^{-1}AL\| \leq \|C\| \leq \|C\| + \|R\| \leq \sqrt{\|C^TC\|_\infty} + \sqrt{\operatorname{tr} R^TR}$$

we get the *stability condition*

$$\|L^{-1}(Az + b - z)\|_2 + (\sqrt{\|C^TC\|_\infty} + \sqrt{\operatorname{tr} R^TR})r \leq r \tag{33}$$

as verifiable sufficient condition for stability in $E(z, L, r)$. Moreover, by our analysis, the left hand side is $\rho(F'(z))r + O(r^2) < r$ for sufficiently small r, so that (33) verifies some stability region around any attractive fixed point. By checking (33) for various r (using a bisection procedure in $[0, r]$) one can find the maximal radius in which stability can be verified.

6. Application: Confidence Regions

Let x be an n-dimensional random vector with mean z, covariance matrix Σ and Gaussian distribution. We want to find a confidence region $\bar{E}$ for the transformed random vector

$$\bar{x} = F(x)$$

which contains $\bar{x}$ with specified probability α (or higher).

Since $(x - z)^T \Sigma^{-1}(x - z)$ is $\chi^2(n)$ distributed, one can compute a radius r_α such that

$$(x - z)^T \Sigma^{-1}(x - z) \leq r_\alpha^2 \qquad \text{with probability } \alpha. \tag{34}$$

If we use a Cholesky factorization LL^T of Σ and introduce $\xi = L^{-1}(x - z)$ we find $x = z + L\xi$ and $\xi^T\xi = (x - z)^T L^{-T} L^{-1}(x - z) = (x - z)^T \Sigma^{-1}(x - z)$. Hence (34) can be rewritten as

$$x \in E(z, L, r_\alpha) \qquad \text{with probability } \alpha. \tag{35}$$

Therefore, the image of $E(z, L, r_\alpha)$ under the transformation F will contain $\bar{x}$ with probability α, and the enclosing ellipsoid will therefore satisfy

$$\bar{x} \in \bar{E}(\bar{z}, \bar{L}, \bar{r}_\alpha) \qquad \text{with probability } \geq \alpha. \tag{36}$$

Thus we have a rigorous confidence region for $\bar{x}$ to the confidence level α, and since the probability is at least α, overestimation leads to an error on the safe side.

References

[1] Alvarado, F. L.: Sparse W-matrix for interval arithmetic. Manuscript (1990).
[2] Alvarado, F. L.: Sparsity preservation in matrix interval solutions. Manuscript (1990).
[3] Conradt, J.: Ein Intervallverfahren zur Einschließung des Fehlers einer Näherungslösung bei Anfangswertaufgaben für Systeme von gewöhnlichen Differentialgleichungen. Freiburger Intervall Berichte 80/1, Inst. f. Angew. Math., Univ. Freiburg (1980).
[4] Eijgenraam, P.: The solution of initial value problems using interval arithmetic. Math. Centre Tracts 144, Amsterdam (1981).
[5] Gambill, T. N., Skeel, R. D.: Logarithmic reduction of the wrapping effect with applications to ordinary differential equations. SIAM J. Numer. Anal. 25, 153–162 (1988).
[6] Hansen, E.: Global optimization using interval analysis. New York: Marcel Dekker 1992.
[7] Husung, D., Rump, S. M.: CALCULUS. In: Kulisch, U. (ed.) Wissenschaftliches Rechnen mit Ergebnisverifikation. Berlin: Vieweg 1989.
[8] Jackson, L. W.: A comparison of ellipsoidal and interval arithmetic error bounds, numerical solutions of nonlinear problems (notice). SIAM Rev. 11, 114 (1969).
[9] Jackson, L. W.: Interval arithmetic error-bounding algorithms. SIAM J. Numer. Anal. 12, 223–238 (1975).
[10] Jansson, C.: A geometric approach for computing a posteriori error bounds for the solution of a linear system. Computing 47, 1–9 (1991).
[11] Kahan, W. M.: Circumscribing an ellipsoid about the intersection of two ellipsoids. Can. Math. Bull. 11, 437–441 (1968).
[12] Krawczyk, R., Neumaier, A.: Interval slopes for rational functions and associated centered forms. SIAM J. Numer. Anal. 22, 604–616 (1985).
[13] Kurzhanski, A. B., Vályi, I.: Ellipsoidal techniques for dynamic systems: The problem of control synthesis. Dynamics and Control 1, 357–378 (1991).
[14] Lohner, R.: Enclosing the solutions of ordinary initial- and boundary-value problems. In: Kaucher, E., Kulisch, U., Ullrich, Ch. (eds.) Computerarithmetic, pp. 255–286. Stuttgart: Teubner 1987.
[15] Lohner, R.: Einschließung der Lösung gewöhnlicher Anfangs- und Randwertaufgaben und Anwendungen. Dissertation, Univ. Karlsruhe (1988).
[16] Moore, R. E.: Interval arithmetic and automatic error analysis in digital computing. PhD thesis. Appl. Math. Statist. Lab. Rep. 25, Stanford University (1962).
[17] Moore, R. E.: Interval analysis. Englewood Cliffs, NJ: Prentice-Hall 1966.

[18] Neumaier, A.: Hybrid norms and bounds for overdetermined linear systems. Linear Algebra Appl. (To appear).
[19] Neumaier, A.: Interval methods for systems of equations. Cambridge: Cambridge University Press 1990.
[20] Nickel, K.: Using interval methods for the numerical solution of ODE's. Freiburger Intervall-Berichte *83/10*, 13–44 (1983).
[21] Ovseevich, A. I., Chernousko, F. L.: On optimal ellipsoids approximating reachable sets. Problems of Control and Information Theory *16*, 125–134 (1987).
[22] Rump, S. M.: On the solution of interval linear systems. Computing *47*, 337–353 (1992).

Institut für Angewandte Mathematik
Universität Freiburg
Hermann-Herder-Strasse 10
D-W-7800 Freiburg
Federal Republic of Germany

Computing

© Springer-Verlag 1993
Printed in Austria

Validated Solution of Large Linear Systems*

S. M. Rump, Hamburg

Dedicated to Professor U. Kulisch on the occasion of his 60th birthday

Abstract — Zusammenfassung

Validated Solution of Large Linear Systems. Some new methods will be presented for computing verified inclusions of the solution of large linear systems. The matrix of the linear system is typically of sparse or band structure. There are no prerequisites for the matrix, such as being M-matrix, symmetric, positive definite or diagonally dominant. For general band matrices of lower, upper bandwidth p, q of dimension n the computing time is $n \cdot (pq + p^2 + q^2)$. Examples with up to 1.000.000 unknowns will be presented.

AMS Subject Classification: 65G10, 65F05

Key words: Sparse linear systems, banded linear systems, automatic result verification, interval methods.

Lösung großer linearer Gleichungssysteme mit Ergebnisverifikation. Es werden neuartige Methoden vorgestellt zur Berechnung sicherer Schranken der Lösung großer linearer Gleichungssysteme. Die Matrix des Gleichungssystems ist typischerweise spärlich besetzt oder hat Bandstruktur. Es werden keinerlei Voraussetzungen an die Matrix gestellt wie etwa M-Matrix, symmetrisch, positiv definit oder diagonal dominant. Für Bandmatrizen von oberer bzw. unterer Bandbreite p bzw. q der Dimension n ist die Rechenzeit $n \cdot (pq + p^2 + q^2)$. Es werden Beispiele bis Dimension 1.000.000 diskutiert.

0. Notation

Let $\mathbb{R}$ denote the set of real numbers, $\mathbb{R}^n$ vectors and $\mathbb{R}^{n \times n}$ matrices over those. The letter n is only used for the dimension of vectors and matrices, others than n-vectors and $n \times n$-matrices do not occur in this paper.

$\mathbb{P}T$ denotes the power set over T, $\mathbb{I}T$ the interval extension for $T \in \{\mathbb{R}, \mathbb{R}^n, \mathbb{R}^{n \times n}\}$. Usually hyperrectangles are used but others are not excluded. It should be stressed that interval operations producing validated bounds are rigorously and very efficiently implementable on digital computers, see [26], [1], [5], [29] for details.

Intervals are written in brackets; $a \pm b$ denotes the interval $[a - b, a + b]$; for some interval $[X]$, the absolute value $|[X]|$ is defined by $|[X]| := \max\{|x| \mid x \in [X]\}$, $\mathrm{mid}([X])$ and $\mathrm{rad}([X])$ denote the midpoint and the radius of an interval $[X]$, respectively. Those terms apply to vectors and matrices componentwise.

The interior of a set is denoted by int, ρ denotes the spectral radius of a matrix and $\rho([A]) := \max\{\rho(A) \mid A \in [A]\}$ for $[A] \in \mathbb{I}\mathbb{R}^{n \times n}$. An interval linear system is some-

---

* Received September 16, 1992; revised December 17, 1992.

times written in short notation $[A] \cdot x = [b]$, solving it means to compute bounds for

$$\sum ([A], [b]) := \{x \in \mathbb{R}^n | \exists A \in [A], b \in [b] \quad \text{with} \quad Ax = b\}.$$

$\sigma_1, \ldots, \sigma_n$ denote the singular values of a matrix in nonincreasing order such that $\sigma_1 = \|A\|_2$. If not stated otherwise, all operations are real or floating-point operations. We use operations $\triangle$ with upwardly directed rounding, $* \in \{+, -, \cdot, /\}$ having the property

$$a \triangle b \geq a * b$$

where the latter operation $*$ is the real operation. In case a, b are vectors or matrices the $\leq$-sign applies componentwise.

1. Introduction

Few papers are known dealing with the problem of finding validated inclusions for the solution of sparse linear systems without calculating an approximate inverse of the system matrix. All of the papers known to the author not using an approximate inverse require special properties of the system matrix, which is essentially being an M-matrix. The approximate inverse of a sparse matrix is in general full thus limiting the size of the tractable problems significantly. This is because of limitations in memory and because for sparse or banded systems the computing time depends quadratically on n. Our goal is to go to large sizes, that is 100.000 unknowns and beyond and to keep the computing time for sparse or banded systems linearly dependent on n.

There are very interesting papers for condition estimation of sparse matrices (cf. [4], [13]). However, these are yielding estimations rather than verified bounds.

In this paper we describe our method in detail for banded linear systems. We give numerical examples for banded and for sparse systems. For sparse systems the techniques for reducing bandwidth, symbolic pivoting and others (cf. [13]) can be applied. The resulting linear system of reduced bandwidth or reduced profile can be treated by our methods.

There is one yet unpublished method without using an approximate inverse and without prerequisites on the matrix by Jansson [21]. Besides that there are essentially two different approaches known in the literature. The first is the direct extension of some numerical decomposition algorithm by means of replacing every real operation by the corresponding interval operation. It has been shown that, for example, the interval version of Gaussian elimination is executable in this way for diagonally dominant matrices or M-matrices. In the general case intervals tend to grow in diameter rapidly due to data dependencies such that soon a pivot column only consists of intervals containing zero and the algorithm stops prematurely. This effect depends mainly on the dimension, not on the condition number. As a rule of thumb for general matrices with floating-point input data, for example for random matrices, the range of application of this approach is roughly limited to dimension 50 when calculating in double precision which is equivalent to 17 decimals. The dimension is even more limited for interval input data.

The other approach uses fixed point methods. We briefly describe this ansatz because it gives insight in the problems we have to deal with.

Let a linear system $Ax = b$ with matrix $A \in \mathbb{R}^{n \times n}$ and right hand side $b \in \mathbb{R}^n$ be given together with some $\tilde{x} \in \mathbb{R}^n$, $R \in \mathbb{R}^{n \times n}$. $\tilde{x}$ is considered to be an approximate solution to the linear system, R an approximate inverse of A. Krawczyk [22], [23] defines for $X \in \mathbb{IR}^n$ the following operator

$$K(X) := \tilde{x} - R \cdot (A\tilde{x} - b) + (I - RA) \cdot (X - \tilde{x}). \tag{1.1}$$

He shows that

$$\|I - RA\| < 1 \quad \text{and} \quad K(X) \subseteq X \quad \text{implies} \ \exists \hat{x} \in X : A\hat{x} = b$$

(see also [27], [28]). In [31] it has been shown that the assumption $\|I - RA\| < 1$ can be replaced by $K(X) \subseteq int(X)$. Algorithms were designed to compute validated inclusions for the solution of general nonlinear systems [32]. There are a number of specializations to specific problems such as polynomial zeros [7], algebraic eigenproblems [31], evaluation of arithmetic expressions [8] and others taking advantage of the special situation. A basic theorem for linear systems is as follows.

Theorem 1.1. *Let $\mathscr{A} \in \mathbb{PR}^{n \times n}$, $\mathscr{B} \in \mathbb{PR}^n$ be given and let $\tilde{x} \in \mathbb{R}^n$, $R \in \mathbb{R}^{n \times n}$, $\varnothing \neq X \in \mathbb{PR}^n$, X being compact. Define*

$$\mathscr{Z} := R \cdot (\mathscr{B} - \mathscr{A}\tilde{x}) \quad \text{and} \quad \mathscr{C} := I - R \cdot \mathscr{A}, \tag{1.2}$$

$$L(X) := \mathscr{Z} + \mathscr{C} \cdot X, \tag{1.3}$$

all operations being power set operations. If

$$L(X) \subseteq int(X) \tag{1.4}$$

then R and every $A \in \mathbb{R}^{n \times n}$, $A \in \mathscr{A}$ is nonsingular and for every $b \in \mathscr{B}$ the unique solution $\hat{x} := A^{-1}b$ satisfies

$$\hat{x} \in \tilde{x} + L(X). \tag{1.5}$$

The proof consists of three basic steps. First take fixed but arbitrary $A \in \mathscr{A}$, $b \in \mathscr{B}$ thus reducing the problem to a point problem. Second, show that $C := I - RA \in \mathscr{C}$ is convergent ($\rho(C) < 1$) and therefore A and R are nonsingular. Moreover, the iteration $x^{k+1} := R(b - A\tilde{x}) + C \cdot x^k$ has a unique fixed point $\hat{x} \in X$. Third show that this fixed point is the (unique) solution of $Ax = b$.

Thus theorem 1 already verifies the solvability of the linear system and gives a sufficient criterion for some $X \in \mathbb{PR}^n$ for including the solution. To devise an algorithm for *finding* a validated inclusion $[X]$ we have to solve two problems. First the operations have to become executable on the computer and second we need a constructive way to obtain a suitable $[X]$. The first problem is solved by using interval operations rather than power set operations. On the computer floating-point bounds for the intervals are used. Then systems with $[A] \in \mathbb{IR}^{n \times n}$, $[b] \in \mathbb{IR}^n$ can be attacked. This includes for example point matrices the entries of which not being exactly representable on the computer by replacing those by the smallest enclosing machine interval (see [1], [28], [5], [29]).

For the second problem we use an iteration with a so-called ε-inflation (see [31], [33]). In this technique for a starting interval $[X] := [Z] := R \cdot ([b] - [A] \cdot \tilde{x})$ the iterated interval is made "fatter" in every step. This is used in combination with an Einzelschrittverfahren. It can be shown [33] that a validated inclusion will be found

- for point system $Ax = b$ and *power set operations*
  iff $\rho(I - R \cdot A) < 1$
- for an interval linear system $[A]x = [b]$ and *interval operations*
  iff $\rho(|I - R \cdot [A]|) < 1$.

All of the fixed point methods known in the literature basically use theorem 1, especially (1.2)–(1.4), in one or the other way. Thus in our discussions for sparse linear systems we may concentrate on how to satisfy those conditions.

For simplicity let a point linear system $Ax = b$, $A \in \mathbb{R}^{n \times n}$, $b \in \mathbb{R}^n$ be given. We do not impose restrictions on A or b. For large banded or sparse linear systems the original approach cannot be used because it needs an approximate inverse R of A which is in general full. We may omit this by using some decomposition of A. For $A = LU$ and $R = U^{-1}L^{-1}$ we obtain for $x \in \mathbb{R}^n$

$$R \cdot (b - A\tilde{x}) + (I - RA)x = U^{-1}L^{-1} \cdot (b - A\tilde{x} + (LU - A) \cdot x). \qquad (1.6)$$

L and U preserve a banded structure of A. In a practical application we would think of replacing U^{-1} and L^{-1} by an efficient algorithm for solving triangular systems. From a mathematical point of view L and U are arbitrary; in the practical application they are the floating-point approximations of triangular factors of an LU, LDL^T, LDM^T, Cholesky or some other decomposition. If for some $L, U \in \mathbb{R}^{n \times n}$ and $[X] \in \mathbb{IR}^n$ we can show that

$$M([X]) := U^{-1}L^{-1}(b - A\tilde{x} + (LU - A) \cdot [X]) \subseteq \text{int}([X]) \qquad (1.7)$$

then theorem 1 implies that A is nonsingular and the unique solution $\hat{x} = A^{-1} \cdot b$ satisfies $\hat{x} \in \tilde{x} + M([X])$. $LU - A$ can be estimated during the decomposition of A, most simple and without additional cost for example using Crout's variant. Thus we have reduced our problem to computing a validated inclusion of the solution of a linear system with triangular point matrix and interval right hand side.

In (1.7) $b - A\tilde{x}$ is of order $\varepsilon \cdot \|A\| \cdot \|\tilde{x}\|$ if $\tilde{x}$ is a reasonable approximate solution, for example the one computed by floating-point Gaussian elimination. Also, numerical error analysis tells us that $LU - A$ will be of the order $\varepsilon \cdot \|A\|$. $[X]$ shall contain the error of the approximate solution $\tilde{x}$ which means that $(LU - A) \cdot [X]$ will be an interval vector of small magnitude. Thus we would not loose too much accuracy going to intervals being symmetric to the origin. This saves us half of the storage per interval vector. Clearly, for $0 < x \in \mathbb{R}^n$

$$|U^{-1}L^{-1}|\cdot(|b - A\tilde{x}| + |LU - A|\cdot x) < x \qquad (1.8)$$

implies A being nonsingular and $A^{-1}b \in \tilde{x} \pm x$.

This reduces our problem to solving a triangular system with right hand side $[b]$ symmetric to the origin and we may further simplify it by setting $[b] := [-1, 1]$. In other words find

$$\text{validated bounds for } S := \{L^{-1}\cdot b | -\mathbf{1} \le b \le \mathbf{1}\}, L \in \mathbb{R}^{n \times n} \text{ lower triangular.} \qquad (1.9)$$

All of the papers [11], [12], [24] using the fixed point approach solve (1.9) using interval forward substitution:

$$\text{for } i = 1: n \text{ do } [x]_i = \left([-1, +1] - \sum_{j=1}^{i-1} L_{ij}\cdot[x]_j\right)\bigg/ L_{ii} \qquad (1.10)$$

all operations in (1.10) being interval operations. Thus the intervals $[x]_j$ are symmetric to the origin and (1.10) can be written using absolute values $x_j := |[x]_j|$

$$\text{for } i = 1: n \text{ do } x_i = \left(1 + \sum_{j=1}^{i-1} |L_{ij}|\cdot x_j\right)\bigg/ |L_{ii}| \qquad (1.11)$$

yielding a true inclusion $S \subseteq [-x, +x]$, where $x = (x_i)$. The overestimation can be estimated observing $x = \langle L \rangle^{-1}\cdot e$ where $e \in \mathbb{R}^n$, $e_i = 1$ for $1 \le i \le n$ and $\langle L \rangle$ is Ostrowski's comparison matrix (see [29]):

$$\langle L \rangle_{ij} := \begin{cases} |L_{ii}| & \text{for } i = j \\ -|L_{ij}| & \text{otherwise.} \end{cases}$$

For our special right hand side the maximal overestimation is equal to the ratio

$$\|\langle L \rangle^{-1}\|_\infty / \|L^{-1}\|_\infty. \qquad (1.12)$$

If we could estimate $\|L^{-1}\|_\infty$ then our problem (1.9) would be solved. In practical applications the ratio (1.12) is exponentially increasing with n unless L has special properties. Such properties are L and U being M-matrices in which case $L = \langle L \rangle$, $U = \langle U \rangle$. This is the reason why the set of solutions of linear systems with M-Matrix as coefficient matrix can be enclosed without overestimation when using interval Gaussian elimination. To further illustrate the effect consider the following example due to Neumaier (cf. [30]):

$$L = \begin{bmatrix} 1 & & & & & \\ 1 & 1 & & & & \\ 1 & 1 & 1 & & & \\ & 1 & 1 & 1 & & \\ & & & \ddots & & \\ & & & 1 & 1 & 1 \end{bmatrix}, \qquad [b]_i = [-1, +1]. \qquad (1.13)$$

Using interval forward substitution we obtain with $E := [-1, 1]$

$$[x]_1 = E$$
$$[x]_2 = E - [x]_1 = 2 \cdot E$$
$$[x]_3 = E - [x]_1 - [x]_2 = 4 \cdot E$$
$$[x]_4 = E - [x]_2 - [x]_3 = 7 \cdot E$$

with exponentially growing diameter of $[x]_i$. This can also be seen from $\langle L \rangle^{-1}$ which we show for $n = 7$:

$$\langle L \rangle^{-1} = \begin{pmatrix} 1 & 0 & 0 & 0 & 0 & 0 & 0 \\ 1 & 1 & 0 & 0 & 0 & 0 & 0 \\ 2 & 1 & 1 & 0 & 0 & 0 & 0 \\ 3 & 2 & 1 & 1 & 0 & 0 & 0 \\ 5 & 3 & 2 & 1 & 1 & 0 & 0 \\ 8 & 5 & 3 & 2 & 1 & 1 & 0 \\ 13 & 8 & 5 & 3 & 2 & 1 & 1 \end{pmatrix} .$$

Thus $[x]$ computed by (1.10) is a huge overestimation of the true solution set which computes to

$$(L^{-1} \cdot [b])_i = |L^{-1}| \cdot E = (i - [i/3]) \cdot E \subseteq n \cdot E,$$

where $[i/3]$ denotes the largest integer not exceeding $i/3$.

This can be seen from

$$\begin{pmatrix} 1 & 0 & 0 & 0 & 0 & 0 & 0 \\ -1 & 1 & 0 & 0 & 0 & 0 & 0 \\ 0 & -1 & 1 & 0 & 0 & 0 & 0 \\ 1 & 0 & -1 & 1 & 0 & 0 & 0 \\ -1 & 1 & 0 & -1 & 1 & 0 & 0 \\ 0 & -1 & 1 & 0 & -1 & 1 & 0 \\ 1 & 0 & -1 & 1 & 0 & -1 & 1 \end{pmatrix} .$$

Unfortunately, this behaviour is typical for practical examples with matrices without special properties.

Methods based on the first approach (replacing floating-point operations by their corresponding interval operations in some numerical decomposition algorithm) are by their nature restricted to matrices with special properties (see for example [1], [29]). See also [35] for an interval version of Bunemann's algorithm for the Poisson equation.

As we have just seen the fixed point approach as described in the literature is also restricted to matrices having special properties. This approach is used in [3], [11], [12], [24].

Using a coded version [3] of this algorithm the effect can be demonstrated. We used algorithm DSSSB with IWK = 5 which means the maximum possible amount of work is invested. We used $A = 0.1 \cdot LL^T$ with L from (1.13) and right hand side

$(1,0,\ldots,0)^T$. The factor 0.1 is used to make the factors of A not exactly representable on the computer. Then using double precision floating-point format which is approximately 17 decimal digits the algorithm fails for $n \geq 41$. For $n = 41$ we have $\mathrm{cond}_2(A) = 2.3e3$, $\mathrm{cond}_1(A) = \mathrm{cond}_\infty(A) = 3.7e3$. Taking the matrix (4.20) from [17] with $a = 1$ and the same right hand side $(1,0,\ldots,0)^T$ the algorithm fails for $n \geq 48$. For $n = 48$ we have $\mathrm{cond}_2(A) = 42$, $\mathrm{cond}_1(A) = \mathrm{cond}_\infty(A) = 63$.

The amount of overestimation (1.12) is displayed in the following table.

Table 1.1. Overestimation of interval Gaussian elimination for L from (1.13)

n	10	20	30	40	50
$\|\langle L\rangle^{-1}\|_\infty/\|L^{-1}\|_\infty$	20.4	1265	1.0e5	9.9e6	9.7e8

The figures demonstrate the exponential behaviour of the overestimation.

2. The Method

In order to bound (1.9) we may look for the singular values of L. Let U_r be the n-dimensional disk of radius r. Then $\|L^{-1} \cdot u\|_2, u \in U_r$ is bounded by $\sigma_n(L)^{-1} \cdot \|u\|_2 = \sigma_n(L)^{-1} \cdot r$. Thus a validated lower bound on the smallest singular value of a triangular matrix would solve the problem. This, in turn, would also yield a validated condition estimator. The problem of finding fast and reliable (although not validated) condition estimators has been attacked by many authors ([9], [10], [16], [18], [19], [2], [6]).

Given an approximation $\tilde{\lambda}$ of $\sigma_n(L)$, $\tilde{\lambda}^2$ is an approximate eigenvalue of LL^T. If for some $\kappa \in \mathbb{R}$ being slightly less than one we could prove that $LL^T - \kappa\tilde{\lambda}^2 \cdot I$ is positive semidefinite then $\kappa^{1/2} \cdot \tilde{\lambda}$ proved to be a lower bound of $\sigma_n(L)$.

L is a Cholesky factor of LL^T. The change of the Cholesky factor L into G with $GG^T = LL^T - \tilde{\lambda}^2 I$ is given by the following formulas:

$$\sum_{v=1}^{i} G_{iv}^2 = \sum_{v=1}^{i} L_{iv}^2 - \tilde{\lambda}^2 \qquad \text{for } i = j$$

$$\sum_{v=1}^{j} G_{iv} G_{jv} = \sum_{v=1}^{j} L_{iv} L_{jv} \qquad \text{for } i > j. \tag{2.1}$$

We need, however, a validation for the fact that $LL^T - \tilde{\lambda}^2 I$ is positive semidefinite. When performing an exact Cholesky factorization of $LL^T - \tilde{\lambda}^2 I$ this is true if the algorithm is executable, i.e. if the diagonal elements stay nonnegative. Using floating-point operations we have to estimate the rounding errors during the computation. Rather than estimating them a priori by replacing the floating-point operations by the corresponding interval operations we estimate them a posteriori by estimating the difference of GG^T and $LL^T - \tilde{\lambda}^2 I$ for the computed Cholesky factor G and by using perturbation theory.

For the diagonal elements this means

$$\text{computing} \quad G_{ii} := \left(\sum_{v=1}^{i} L_{iv}^2 - \sum_{v=1}^{i-1} G_{iv}^2 - \tilde{\lambda}^2 \right)^{1/2} \quad \text{approximatively and}$$

$$\text{estimating} \quad |(LL^T - \tilde{\lambda}^2 I - GG^T)_{ii}| = \left| \sum_{v=1}^{i} L_{iv}^2 - \sum_{v=1}^{i} G_{iv}^2 - \tilde{\lambda}^2 \right| \quad \text{rigorously.}$$

For off-diagonal elements this means

$$\text{computing} \quad G_{ij} := \left(\sum_{v=1}^{j} L_{iv} L_{jv} - \sum_{v=1}^{j-1} G_{iv} G_{jv} \right) \Big/ G_{jj} \quad \text{approximatively and}$$

$$\text{estimating} \quad |(LL^T - \tilde{\lambda}^2 I - GG^T)_{ij}| = \left| \sum_{v=1}^{j} L_{iv} L_{jv} - \sum_{v=1}^{j} G_{iv} G_{jv} \right| \quad \text{rigorously.}$$

The computation and the estimation can essentially be done in one step. First the
common part of both sums, resp. is evaluated *with* error estimation, then the midpoint is used for the floating-point components G_{ii}, G_{ij} of G, resp. and the interval
part for the error estimation. If only the four basic interval operations, that is IEEE
754 [20] arithmetic, is available that is the best we can do. If a precise scalar product
[25], [26] is available then we can do better. For the diagonal elements we compute
the exact value

$$dot := \sum_{v=1}^{i} L_{iv}^2 - \sum_{v=1}^{i-1} G_{iv}^2 - \tilde{\lambda}^2$$

and for S being the value of dot rounded to nearest we get $G_{ii} := fl(\sqrt{S})$, that is G_{ii}
is the floating-point square root of S. Then we use the accumulating feature of the
scalar product and compute the exact value of $dot - G_{ii}^2$. This value rounded to the
smallest enclosing interval provides a very sharp bound for the error $(LL^T - \tilde{\lambda}^2 I -
GG^T)_{ii}$. For the off-diagonal elements we proceed in a similar way. To avoid to
formulate the algorithm twice we simply state in the diagonal case

Compute S, ΔS such that

$$\sum_{v=1}^{i} L_{iv}^2 - \sum_{v=1}^{i-1} G_{iv}^2 - \tilde{\lambda}^2 \in S \pm \Delta S.$$

For basic interval operations this means S being the midpoint, ΔS the radius of the
left hand side computed in naive interval arithmetic. With the precise scalar product
we proceed as described before. The off-diagonal elements are treated similarly.

Having an estimation on $E := LL^T - \tilde{\lambda}^2 I - GG^T$ and assuming the diagonal of G
being nonnegative implies that $LL^T - \tilde{\lambda}^2 I - E$ is positive semidefinite. Hence perturbation theory tells us that the eigenvalues of $LL^T - \tilde{\lambda}^2 I$ are not smaller than
$-\rho(E)$ (cf. [15], Corollary 8.1.3) and those of LL^T not smaller than $\tilde{\lambda}^2 - \rho(E)$. Now
$\rho(E)$ can be estimated conveniently by $\|E\|_\infty$ which is done in the following algorithm. There the ith row sum is stored in e_i. When computing the ij-th component
of G the error $(LL^T - \tilde{\lambda}^2 I - GG^T)_{ij}$ contributes to e_i and e_j due to symmetry. To
obtain an upper bound on $\|E\|_\infty$ upward directed rounding is used in the computa-
tion of the e_i and e_{max}.

We give the algorithm for a triangular matrix L with full lower triangle. It can be altered for band matrices in a straightforward manner. Pivoting is omitted because $LL^T - \tilde{\lambda}^2 I$ is (hopefully) positive definite.

Given nonsingular lower triangular $L \in \mathbb{R}^{n \times n}$ and $\tilde{\lambda} \in \mathbb{R}$ do

$e_{\max} := 0$
for $i = 1: n$ do $e_i := 0$;
for $i = 1: n$ do
   for $j = 1: i - 1$ do
      Compute S, ΔS such that

$$\sum_{v=1}^{j} L_{iv} L_{jv} - \sum_{v=1}^{j-1} G_{iv} G_{jv} \in S \pm \Delta S;$$

     $G_{ij} := fl(S/G_{jj})$;
     Compute ΔT such that

$$|S - G_{ij} G_{jj}| \leq \Delta T;$$

     $d := \Delta S \triangle \Delta T$; $e_i := e_i \triangle d$; $e_j := e_j \triangle d$;
     Compute S, ΔS such that

$$\sum_{v=1}^{i} L_v^2 - \sum_{v=1}^{i-1} G_{iv}^2 - \tilde{\lambda}^2 \in S \pm \Delta S;$$

     $G_{ii} := fl(\sqrt{S})$;
     Compute ΔT such that

$$|S - G_{ii}^2| \leq \Delta T;$$

     $e_i := e_i \triangle \Delta S \triangle \Delta T$;
$e_{\max} := \max_i e_i$;

Algorithm 2.1. *Cholesky factorization of $LL^T - \tilde{\lambda}^2 I$ with lower bound for $\sigma_n(L)$*

In precise computation, ΔS and ΔT as well as T would be zero according to (2.1). The main effort in the algorithm goes into the two inner products for computing S together with a validated bound. If L is a lower triangular band matrix of bandwidth p then the vector e needs only to be of length $p + 1$ storing the values cyclically. Also, G needs only $(p + 1) * (p + 1)$ elements of storage.

It should be stressed that G is computed in floating-point arithmetic without presumptions on its accuracy. If the algorithm finishes successfully, i.e. the radicands are nonnegative, then the G_{ii} are nonnegative and therefore GG^T is positive semidefinite with

$$\|(LL^T - \tilde{\lambda}^2 I) - GG^T\|_\infty \leq e_{\max}. \tag{2.2}$$

The eigenvalues of LL^T are the squared singular values of L and are bounded from below by $\tilde{\lambda}^2 - e_{\max}$. This establishes the following theorem.

Theorem 2.1. *If algorithm 2.1 finishes successfully (all square roots real) then $LL^T - (\tilde{\lambda}^2 - e_{\max})I$ is positive semidefinite. If $\tilde{\lambda}^2 \geq e_{\max}$ then*

$$\sigma_n(L) \geq (\tilde{\lambda}^2 - e_{\max})^{1/2}.$$

The computing time for L with lower bandwidth p ($L_{ij} = 0$ for $i > j + p$) is less than $n \cdot p^2 + O(np)$ multiplications and additions plus $n(p + 1)$ divisions and n square roots.

Proof. The first part has been proved above, the computing time is a straightforward operation count. ∎

In our applications we are particularly interested in sparse linear systems. This fact should be taken into account when implementing algorithm 2.1. For example, in case of a band matrix L the scalar products become very short compared to n.

Theorem 2.1 can be applied as follows. Consider some decomposition of A, for example $\tilde{L}\tilde{U} \approx A$ with $\tilde{A} := \tilde{L}\tilde{U}$. Then traditional norm estimates can be used to compute validated bounds for the solution together with theorem 2.1.

Theorem 2.2. *Let $A \in \mathbb{R}^{n \times n}$, $b \in \mathbb{R}^n$ be given as well as nonsingular $\tilde{A} \in \mathbb{R}^{n \times n}$ and $\tilde{x} \in \mathbb{R}^n$. Define $\Delta A := \tilde{A} - A$ and suppose $\sigma_n(\tilde{A}) > n^{1/2} \cdot \|\Delta A\|_\infty$.*

Then A is not singular and for $\hat{x} := A^{-1}b$ holds

$$\|\hat{x} - \tilde{x}\|_\infty \leq \frac{n^{1/2} \cdot \|b - A\tilde{x}\|_\infty}{\sigma_n(\tilde{A}) - n^{1/2} \cdot \|\Delta A\|_\infty}. \tag{2.3}$$

Proof. Since $\|\tilde{A}^{-1} \cdot \Delta A\|_2 \leq \sigma_n(\tilde{A})^{-1} \cdot \|\Delta A\|_2 \leq n^{1/2} \cdot \sigma_n(\tilde{A})^{-1} \cdot \|\Delta A\|_\infty < 1$ the matrix $I - \tilde{A}^{-1} \cdot \Delta A = \tilde{A}^{-1} \cdot A$ and hence A is invertible. Now

$$(I - \tilde{A}^{-1} \cdot \Delta A)(\hat{x} - \tilde{x}) = \tilde{A}^{-1} \cdot A \cdot (\hat{x} - \tilde{x}) = \tilde{A}^{-1} \cdot (b - A\tilde{x}).$$

Using $\|(I - F)^{-1}\| \leq (1 - \|F\|)^{-1}$ for convergent $F \in \mathbb{R}^{n \times n}$ this implies

$$\|\hat{x} - \tilde{x}\|_\infty \leq \frac{\|\tilde{A}^{-1} \cdot (b - A\tilde{x})\|_\infty}{1 - \|\tilde{A}^{-1} \cdot \Delta A\|_\infty} \leq \frac{\|\tilde{A}^{-1}\|_\infty \cdot \|b - A\tilde{x}\|_\infty}{1 - \|\tilde{A}^{-1}\|_\infty \cdot \|\Delta A\|_\infty} \tag{2.4}$$

and using $\|B\|_\infty \leq n^{1/2} \cdot \|B\|_2$ for $B \in \mathbb{R}^{n \times n}$ yields

$$\|\hat{x} - \tilde{x}\|_\infty \leq \frac{n^{1/2} \cdot \sigma_n(\tilde{A})^{-1} \cdot \|b - A\tilde{x}\|_\infty}{1 - n^{1/2} \cdot \sigma_n(\tilde{A})^{-1} \cdot \|\Delta A\|_\infty}$$

proving the theorem. ∎

In a practical application $\tilde{A}$ is some floating-point decomposition of A, for example $\tilde{A} = \tilde{L}\tilde{U}$. Then the application of theorem 2.2 runs as follows. The nonsingularity of $\tilde{A}$ is obvious. Compute an approximate solution $\tilde{x}$ of $\tilde{A}x = b$ thru forward and backward substitution and a lower bound for $\sigma_n(\tilde{A})$ by

$$\sigma_n(\tilde{A}) = \sigma_n(\tilde{L}\tilde{U}) \geq \sigma_n(\tilde{L}) \cdot \sigma_n(\tilde{U}) \tag{$*$}$$

and theorem 2.1. Then check for $\sigma_n(\tilde{A}) > n^{1/2} \cdot \|\Delta A\|_\infty$ using ($*$) to satisfy the conditions of theorem 2.1.

In the estimation (2.3) one may try to avoid or diminish the factor $n^{1/2}$. If some $B \in \mathbb{R}^{n \times n}$ is sparse with at most $\tau(B)$ elements per row, then it is not difficult to prove

$$\|B\|_\infty \leq \tau(B)^{1/2} \cdot \|B\|_2.$$

One may hope that given B with $\tau(B^{-1})$ small then also the factor $n^{1/2}$ may be decreased. Unfortunately, this is not true. Consider

$$
B^{-1} = \begin{pmatrix}
1 & & & & & & \\
& -1 & & & & & \\
& 1 & -1 & & & & \\
& & 1 & \ddots & & & \\
& & & \ddots & -1 & \\
& & & & 1 & \varepsilon
\end{pmatrix}.
$$

Then for small ε

$$
\|B\|_2 \approx \varepsilon^{-1} \cdot n^{1/2}, \quad \|B\|_\infty \approx \varepsilon^{-1} \cdot n \qquad \text{with } \|B\|_\infty / \|B\|_2 \lesssim n^{1/2}.
$$

To apply theorem 2.1 in order to obtain a lower bound on the smallest singular value of a triangular matrix L we need an approximation $\tilde{\lambda} \approx \sigma_n(L)$. There are two ways to obtain such an approximation. First, we could use our favourite condition estimator. This is fast and, according to our experimental results, works fine. The second method would be to apply inverse power iteration to LL^T using forward and backward substitution. The heuristic is that an L occuring in practice has a well separated smallest singular value. Due to our practical results in most cases 2 or 3 iterations sufficed to generate 3 correct decimal digits. This is far more than necessary.

In the applications we have in mind we can do better using a 2-norm estimate and taking advantage of the structure of A. Let $A \in \mathbb{R}^{n \times n}$ be a band matrix of lower, upper bandwidth p, q, that is

$$
A_{ij} = 0 \qquad \text{for } i > j + p \qquad \text{and for } j > i + q.
$$

Let $\alpha := \max|A_{ij}|$. Then both $\|A\|_1$ and $\|A\|_\infty$ are bounded by $(p + q + 1) \cdot \alpha$ and using $\|A\|_2^2 \le \|A\|_1 \cdot \|A\|_\infty$ yields

$$
\|A\|_2 \le (p + q + 1) \cdot \alpha. \tag{2.5}
$$

Hence $\|A\|_2 \le (\|A\|_1 \cdot \|A\|_\infty)^{1/2}$ will in general be significantly smaller than $n^{1/2} \cdot \|A\|_\infty$. Applying this to ΔA in (2.4) with the 2-norm instead of ∞-norm we obtain the following result.

Theorem 2.3. *Let $A \in \mathbb{R}^{n \times n}$, $b \in \mathbb{R}^n$ be given as well as nonsingular $\tilde{A} \in \mathbb{R}^{n \times n}$ and $\tilde{x} \in \mathbb{R}^n$. Define $\Delta A := \tilde{A} - A$ and suppose $\sigma_n(\tilde{A}) > (\|\Delta A\|_1 \cdot \|\Delta A\|_\infty)^{1/2}$.*

Then A is not singular and for $\hat{x} := A^{-1}b$ holds

$$
\|\hat{x} - \tilde{x}\|_\infty \le \|\hat{x} - \tilde{x}\|_2 \le \frac{\|b - A\tilde{x}\|_2}{\sigma_n(\tilde{A}) - (\|\Delta A\|_1 \cdot \|\Delta A\|_\infty)^{1/2}}. \tag{2.6}
$$

Theorem 2.3 also follows by a fixed point argument. Using $\tilde{A} = LU$ and a disk U_r of radius r instead of $[x]$ in (1.7) gives according to theorem 1.1

$$
\tilde{A}^{-1} \cdot (b - A\tilde{x} + (\tilde{A} - A) \cdot U_r) \subseteq \text{int}(U_r) \quad \Rightarrow
$$
$$
A \text{ is not singular and } A^{-1}b \in \tilde{x} + U_r. \tag{2.7}
$$

The inclusion in (2.7) is satisfied if

$$\sigma_n(\tilde{A})^{-1} \cdot (\|b - A\tilde{x}\|_2 + r \cdot \|\Delta A\|_2) < r.$$

This yields a lower bound on r and with a continuity argument (2.6).

The heuristic is that the elements of ΔA are roughly of the same size, namely $\varepsilon \|A\|$. In the application of (2.6) we have to check $\sigma_n(\tilde{A}) > (\|\Delta A\|_1 \cdot \|\Delta A\|_\infty)^{1/2}$ to verify $\rho(\tilde{A}^{-1} \cdot \Delta A) < 1$ which is according to (2.5) more likely to happen than $\sigma_n(\tilde{A}) > n^{1/2} \cdot \|\Delta A\|_\infty$. Moreover, computing $\tilde{x}$ by Gaussian elimination we know that the residual $\|b - A\tilde{x}\|$ will be of the order $\varepsilon \cdot \|A\| \cdot \|\tilde{x}\|$ (cf. [15]).

In the following we add some computational hints for specific cases being relevant in practice.

I) *A is M-matrix.* Apply [34].

If A is symmetric positive definite we can use algorithm 2.1 to calculate a lower bound for $\sigma_n(A)$ directly. When replacing

$$\sum_{v=1}^{j} L_{iv}L_{jv} \qquad \text{by } A_{ij} \text{ in row 7 and}$$

$$\sum_{v=1}^{i} L_{iv}^2 \qquad \text{by } A_{ii} \text{ in row 13,}$$

then obviously

$$\sigma_n(A) \geq (\tilde{\lambda}^2 - \varepsilon_{max})^{1/2}.$$

Replacing $\tilde{A}$ by A in theorem 2.3 then yields

$$\|\hat{x} - \tilde{x}\|_\infty \leq \|\hat{x} - \tilde{x}\|_2 \leq (\tilde{\lambda}^2 - e_{max})^{-1/2} \cdot \|b - A\tilde{x}\|_2. \tag{2.8}$$

II) *A is symmetric positive definite.* Compute a floating-point Cholesky decomposition $A \approx \tilde{G}\tilde{G}^T$ and an approximation $\tilde{\sigma}$ of the smallest singular value of A. Apply algorithm 2.1 altered as described above with $\tilde{\lambda} = 0.9 \cdot \tilde{\sigma}$ to compute a lower bound on $\sigma_n(A)$ and apply (2.8).

In case A is not symmetric positive definite one may use the following method. Having some approximate decomposition $A \approx \tilde{F} \cdot \tilde{G}$ compute an approximation $\tilde{\lambda}$ to the smallest singular value of A by inverse power method applied to $\tilde{F}\tilde{G} \cdot (\tilde{F}\tilde{G})^T$. If $\tilde{F}, \tilde{G}$ are triangular this is inexpensive. Then apply theorem 2.1 with some obvious modifications to $A^T A - \tilde{\lambda}^2 I$ to bound $\sigma_n(A^T A)$.

This approach is working only for moderate condition numbers because the condition number of $A^T A$ is that of A squared. For working precision ε this limits the scope of application to $\text{cond}(A) < \varepsilon^{-1/2}$ rather than $\text{cond}(A) < \varepsilon^{-1}$.

In contrast we estimate the smallest singular value of the factors of the decomposition separately. We have to take provision that the condition numbers of the factors are of the same order, namely $\text{cond}(A)^{1/2}$. In this case the square of the condition number of the factors is still of the order of $\text{cond}(A)$ and no additional restrictions are imposed on A.

In the case A is symmetric we can do a little bit better than using LDL^T. Instead, let D_1, D_2 be diagonal such that $D_1 D_2 = D$, $|D_1| = |D_2| = |D|^{1/2}$. Then $LDL^T = (LD_1) \cdot (LD_2)^T$ and the usual LDL^T decomposition can be modified in an obvious way to compute $L_1 := L \cdot D_1$ and $L_2 := L \cdot D_2$ directly instead of L and D. Furthermore $D_1 = Q \cdot D_2$ with Q being a diagonal matrix with $+1$ or -1 in the diagonal thus being orthogonal. Therefore LD_1 and LD_2 have the same singular values and lower bound for $\sigma_n(LD_1)$ suffices for our purposes.

Despite saving computing time the heuristic is that $\sigma_n(LD_1)^2$ provides a better lower estimate for $\sigma_n(LDL^T)$ than $\sigma_n(L)^2 \cdot \sigma_n(D)$. Practical examples support this heuristic to a certain point. The same heuristic applies to general nonsymmetric matrices.

III) *A is symmetric.* Compute an approximate $\tilde{L}_1 \cdot \tilde{L}_2^T$ decomposition as described above and an approximation σ of the smallest singular value of $\tilde{L}_1$. Apply algorithm 2.1 with $\tilde{\lambda} = 0.9 \cdot \tilde{\sigma}$ to compute a lower bound σ on $\sigma_n(\tilde{L}_1)$ and apply theorem 2.2 or 2.3 with $\tilde{A} := \tilde{L}_1 \cdot \tilde{L}_2^T$ and $\sigma_n(\tilde{A}) \geq \sigma^2$.

In the general case we may apply an LU-decomposition. However, L tends to be fairly well-conditioned whereas the condition of A moves into U. Thus we may run into difficulties trying to estimate $\sigma_n(U^T U)$. On the other hand the LDM^T-decomposition can be altered in an obvious way to distribute $D = D_1 \cdot D_2$, $|D_1| = |D_2| = |D|^{1/2}$ both in L and M as we did in the $L_1 L_2^T$-decomposition in the symmetric case. This yields an LM^T-decomposition, L and M no longer being unit lower triangular. The heuristic is that then L and M are more or less equally conditioned, the condition number not being much bigger than the square root of the condition number of A.

IV) *A is general nonsymmetric.* Compute an approximate $\tilde{L} \cdot \tilde{M}^T$-decomposition of A and approximations $\tilde{\sigma}_1$, $\tilde{\sigma}_2$ for the smallest singular value of $\tilde{L}$, $\tilde{M}$, respectively. Apply algorithm 2.1 with $\tilde{\lambda}_1 = 0.9 \cdot \tilde{\sigma}_1$, $\tilde{\lambda}_2 = 0.9 \cdot \tilde{\sigma}_2$ to compute a lower bound σ_1, σ_2 on $\sigma_n(\tilde{L})$, $\sigma_n(\tilde{M})$ and apply theorem 2.2 or 2.3 with $\tilde{A} = \tilde{L} \cdot \tilde{M}^T$ and $\sigma_n(\tilde{A}) \geq \sigma_1 \cdot \sigma_2$.

It should be pointed out that the heuristic for cases III) and IV) works for many examples but also has its drawbacks. In the moment we do not know a general strategy for choosing a decomposition $A \approx \tilde{F}\tilde{G}$ which maximizes $\sigma_n(\tilde{F}) \cdot \sigma_n(\tilde{G})$. In case of symmetric positive definite A the method of choice is of course the Cholesky decomposition $A = GG^T$ with $\sigma_n(A) = \sigma_n(G)^2$.

Let $L \in \mathbb{R}^{n \times n}$ be of lower triangular of bandwidth p. Then approximations of the smallest singular value of L are either computed by

- inverse power iteration for LL^T at the cost of $2np$ ops per iteration or
- using some condition estimator at the cost of $c \cdot np$ ops, c small.

As has been pointed out before this is small against np^2. Thus the total computing time for either of the algorithms for a linear system $Ax = b$ with A of lower, upper bandwidth p, q, respectively, $p \ll n$, $q \ll n$ is

I) *A is an M-matrix:*	$n \cdot pq$	ops
II) *A is symmetric positive definite:*	$n \cdot p^2$	ops
III) *A is symmetric indefinite:*	$\frac{3}{2} n \cdot p^2$	ops
IV) *A is general matrix:*	$n \cdot (pq + p^2 + q^2)$	ops.

Finally we want to mention how to use our methods in an interval setting, that is to solve $[A]x = [b]$, $[A] \in \mathbb{IR}^{n \times n}$, $[b] \in \mathbb{IR}^n$ which means computing an inclusion of $\Sigma([A], [b]) := \{x \in \mathbb{R}^n | \exists A \in [A] \exists b \in [b]: Ax = b\}$. Theorem 2.3 extends as follows.

Theorem 2.4. *Let $[A] \in \mathbb{IR}^{n \times n}$, $[b] \in \mathbb{IR}^n$ be given as well as nonsingular $\tilde{A} \in \mathbb{R}^{n \times n}$ and $\tilde{x} \in \mathbb{R}^n$. Define $\Delta A := |[A] - \tilde{A}|$ and suppose $\sigma_n(\tilde{A}) > (\|\Delta A\|_1 \cdot \|\Delta A\|_\infty)^{1/2}$.*

Then every $A \in [A]$ is nonsingular and for every $\hat{x} := A^{-1}b$, $A \in [A]$, $b \in [b]$ holds

$$\|\hat{x} - \tilde{x}\|_\infty \leq \|\hat{x} - \tilde{x}\|_2 \leq \frac{\| |[b] - [A] \cdot \tilde{x}| \|_2}{\sigma_n(\tilde{A}) - (\|\Delta A\|_1 \cdot \|\Delta A\|_\infty)^{1/2}}. \tag{2.9}$$

The **proof** follows by applying theorem 2.3 to each $A \in [A]$, $b \in [b]$. ∎

We shortly describe an algorithm for solving a general interval linear system. We use the property $A \in [A] \Rightarrow \|A\|_p \leq \|mid([A])\|_p + \|rad([A])\|_p$ for $p \in \{1, \infty\}$.

Let $[A] \in \mathbb{IR}^{n \times n}$, $[b] \in \mathbb{IR}^n$ be given.

1) For $mA := mid([A])$ compute an approximate decomposition $\tilde{L} \cdot \tilde{M}^T =: \tilde{A} \approx mA$ (see IV)) in floating-point arithmetic together with estimates ζ_1, ζ_∞ on $\|\tilde{A} - mA\|_1$, $\|\tilde{A} - mA\|_\infty$, resp.
2) Solve $\tilde{L} \cdot \tilde{M}^T \cdot \tilde{x} = mb$, $mb := mid([b])$ by floating-point forward and backward substitution to obtain $\tilde{x}$
3) Compute approximations for the smallest singular value $\tilde{\sigma}_1$, $\tilde{\sigma}_2$ of $\tilde{L}$, $\tilde{M}$ by floating-point inverse power method applied to $\tilde{L}\tilde{L}^T$, $\tilde{M}\tilde{M}^T$, resp. or by some condition estimator
4) Apply algorithm 2.1 to compute lower bounds σ_1, σ_2 on $\sigma_n(\tilde{L})$, $\sigma_n(\tilde{M})$ using $\lambda := 0.9 \cdot \tilde{\lambda}_i$. If algorithm 2.1 does not finish successfully try smaller values for $\tilde{\lambda}_i$.
5) Calculate $z = \sup(|[b] - [A] \cdot \tilde{x}|)$ and upper bounds $\eta_1 \geq \|rad([A])\|_1$, $\eta_\infty \geq \|rad([A])\|_\infty$ using interval arithmetic (for η_1, η_∞ upwardly directed rounding suffices).
6) If $\mu := \sigma_1 \cdot \sigma_2 - ((\zeta_1 + \eta_1)(\zeta_\infty + \eta_\infty))^{1/2} > 0$ then every $A \in [A]$ is nonsingular and

$$\|\hat{x} - \tilde{x}\|_\infty \leq \|\hat{x} - \tilde{x}\|_2 \leq \mu^{-1} \cdot \|z\|_2$$

for every $\hat{x} = A^{-1}b$ with $A \in [A]$, $b \in [b]$.

Algorithm 2.2. *Inclusion of the solution of a general interval linear system*

If in case of point linear systems very high accuracy of the inclusion is desired $\tilde{x}$ may be stored in $\tilde{x}_1$ and $\tilde{x}_2$ with $\tilde{x} = \tilde{x}_1 + \tilde{x}_2$ (staggered correction, see [31], [36]). In this case $b - A\tilde{x}_1 - A\tilde{x}_2$ should be calculated in double the working precision. Using this method frequently very high or least significant bit accuracy is achieved. A simpler way is to perform a residual iteration

$$x^{k+1} := x^k + \tilde{M}^{-T}\tilde{L}^{-1}(b - Ax^k) \tag{2.10}$$

as usual. Only in the final step the addition is not executed but $\tilde{x}_1 := x^k$ and $\tilde{x}_2 := \tilde{M}^{-T}\tilde{L}^{-1}(b - Ax^k)$ are stored in separate vectors. This saves computing time and produces similar results to storing $\tilde{x}$ in two parts $\tilde{x}_1$, $\tilde{x}_2$ from the beginning.

3. Computational Results

In the following we give numerical results for three different types of our algorithm:

(1) The *symmetric positive definite* case using a Cholesky-decomposition and proceeding as described in (II).
(2) The *symmetric case* using a modified LDL^T-decomposition *without* pivoting as described in (III).
(3) The *general case* using an LU-decomposition *with* pivoting from LAPACK.

In the following tables we display

n	dimension of the matrix
$cond(A)$	approximation of the $\|\cdot\|_\infty$-condition number of A
$iter$	number of inverse power iterations to obtain an approximation for $\sigma_n(A)$
$\sigma_{min}(A)$	lower bound for the smallest singular value of A
$\|\hat{x} - \tilde{x}\|_\infty / \|\tilde{x}\|_\infty$	upper bound for the relative error of the approximate solution $\tilde{x}$.

The condition number is estimated using the vector obtained by the inverse power iteration. Working accuracy is IEEE 754 double precision (approximately 17 decimals). As described in (2.10) we split $\tilde{x}$ into $\tilde{x}_1$, x_2 and compute $b - A\tilde{x}_1 - A\tilde{x}_2$ in quadruple precision.

In all of the following examples the

- right hand side b is computed such that the solution $\hat{x}$ of $Ax = b$ is $\hat{x}_i := (-1)^{i+1} \cdot 1/i.$

This introduces different magnitudes in the solution together with some roughness.

The first example, only displayed for reference purposes, is a discretisation of a Poisson equation

$$M := \begin{bmatrix} 4 & -1 & \\ -1 & 4 & \ddots \\ & \ddots & \ddots \end{bmatrix} ; \qquad A := \begin{bmatrix} M & -I & \\ -I & M & \ddots \\ & \ddots & \ddots \end{bmatrix} \tag{3.1}$$

with I being the identity matrix. We used three different bandwidthes p.

Table 3.1. Matrices (3.1) for different bandwidthes

n	p	cond	iter	$\sigma_{min}(A)$	$\|\hat{x} - \tilde{x}\|_\infty / \|\tilde{x}\|_\infty$
200	5	2.90E+01	4	5.24E-01	1.15E-22
2000	5	2.98E+01	3	5.18E-01	1.18E-22
20000	5	2.98E+01	3	5.18E-01	1.18E-22
200	10	7.73E+01	4	3.21E-01	4.49E-22
2000	10	9.78E+01	5	2.86E-01	5.87E-22
20000	10	9.86E+01	3	2.85E-01	5.91E-22
200	20	7.73E+01	4	3.21E-01	8.81E-22
2000	20	3.41E+02	4	1.53E-01	3.94E-21
20000	20	3.57E+02	3	1.50E-01	4.12E-21

The second example is (4.16) from Gregory/Karney [17] with bandwidth 2.

$$A := \begin{bmatrix} 5 & -4 & 1 & & & & \\ -4 & 6 & -4 & 1 & & & \\ 1 & -4 & 6 & -4 & 1 & & \\ \multicolumn{7}{c}{\dotfill} \\ & & 1 & -4 & 6 & -4 & 1 \\ & & & 1 & -4 & 6 & -4 \\ & & & & 1 & -4 & 5 \end{bmatrix}$$

Example (4.16) from [17]

Here the increasing condition number limits the dimension to the same amount as for a pure floating-point algorithm.

Table 3.2. Matrix (4.16) from [17]

n	cond	iter	σ_{min}	$\|\hat{x} - \tilde{x}\|_\infty / \|\tilde{x}\|_\infty$
100	1.71E+07	2	9.67E-04	2.82E-18
200	2.68E+08	2	2.44E-04	6.26E-17
500	1.03E+10	2	3.93E-05	1.87E-15
1000	1.65E+11	2	9.85E-06	3.95E-14
2000	2.63E+12	2	2.46E-06	7.01E-13
5000	1.03E+14	2	3.95E-07	2.53E-11
10000	1.64E+15	2	9.87E-08	5.38E-10
20000	2.63E+16	2	2.47E-08	1.83E-08
50000	1.03E+18	2	4.05E-09	failed

Another example with high condition numbers are Hilbert matrices, $A_{ij} := 1/(i + j - 1)$. The following table shows the results.

Table 3.3. Hilbert matrices

n	cond	iter	$\sigma_{min}(A)$	$\|\hat{x} - \tilde{x}\|_\infty / \|\tilde{x}\|_\infty$
5	6.94E+05	3	1.81E-03	1.10E-22
6	2.26E+07	3	3.29E-04	4.44E-21
7	7.42E+08	3	5.91E-05	1.76E-19
8	2.45E+10	3	1.05E-05	1.88E-14
9	8.08E+11	3	1.87E-06	2.45E-16
10	2.68E+13	3	3.31E-07	1.86E-11
11	8.84E+14	3	5.83E-08	8.41E-10
12	2.60E+16	3	1.03E-08	2.38E-11
13	2.72E+17	2	1.21E-09	failed

Using Neumaier's example (1.13) we can show the behaviour for larger dimensions. We used $A = 10^{-1} \cdot LL^T$ producing a matrix of bandwidth 2. The factor 10^{-1} is introduced to make the factors of A not exactly representable. Otherwise a decomposition algorithm would rapidly produce the *exact* Cholesky factors. Here we observe decreasing precision of $\|\hat{x} - \tilde{x}\|_\infty / \|\tilde{x}\|_\infty$ with increasing condition number.

Table 3.4. Neumaier's example with $A = 10^{-1}LL^T$, L from (1.13)

n	cond	iter	$\sigma_{min}(A)$	$\|\hat{x} - \tilde{x}\|_\infty / \|\tilde{x}\|_\infty$
100	1.26E+04	3	2.68E-02	3.49E-21
200	4.95E+04	3	1.35E-02	2.71E-20
500	3.06E+05	3	5.43E-03	8.50E-20
1000	1.22E+06	3	2.72E-03	3.40E-19
2000	4.87E+06	3	1.36E-03	1.36E-18
5000	3.04E+07	3	5.44E-04	8.47E-18
10000	1.22E+08	3	2.72E-04	3.39E-17
20000	4.87E+08	3	1.36E-04	1.35E-16
50000	3.04E+09	3	5.44E-05	8.47E-16
100000	1.22E+10	3	2.72E-05	3.39E-15
500000	3.04E+11	3	5.44E-06	8.47E-14
1000000	1.22E+12	3	2.72E-06	3.39E-13

Next we go to the symmetric indefinite case. The first example is taken from [17], (4.20) with $a = 1$, bandwidth 2.

$$A := \begin{pmatrix} -1 & 2 & 1 & & & \\ 2 & 0 & 2 & 1 & & \\ 1 & 2 & 0 & 2 & 1 & \\ \hdotsfor{6} \\ & & 1 & 2 & 0 & 2 \\ & & & 1 & 2 & -1 \end{pmatrix}$$

Example (4.20) from [17]

The eigenvalues are $\lambda_k = \left(1 - 2\cos\dfrac{k\pi}{n+1}\right)^2 - 3$, $1 \le k \le n$. We also display the computed upper bound on $\|A - \tilde{L}_1 \cdot \tilde{L}_2^T\|_2$. It is $\sigma_n(\tilde{L}_1) = \sigma_n(\tilde{L}_2)$.

Table 3.5. Example (4.20) from [17], $a = 1$

n	cond	$\|A - \tilde{L}_1 * \tilde{L}_2^T\|_2$	iter	$\sigma_{min}(\tilde{L}_1)$	$\|\hat{x} - \tilde{x}\|_\infty / \|\tilde{x}\|_\infty$
100	5.37E+01	8.82E-14	3	1.01E-02	8.65E-21
200	9.02E+01	8.82E-14	3	7.66E-03	1.80E-20
500	3.29E+02	8.82E-14	3	2.62E-03	1.28E-19
1000	6.18E+02	2.59E-13	3	1.28E-03	1.02E-18
2000	1.29E+03	6.54E-12	3	3.27E-04	3.43E-18
5000	3.33E+03	6.54E-12	3	1.43E-04	1.82E-17
10000	5.52E+03	6.54E-12	3	9.57E-05	7.57E-17
20000	1.13E+04	6.54E-12	3	5.68E-05	2.62E-16
50000	3.24E+04	7.75E-12	3	2.04E-05	4.13E-15
100000	6.26E+04	7.75E-12	3	1.04E-05	7.62E-14

The results show that, as before, few inverse power iterations are necessary to obtain an approximation for the smallest singular value of $\tilde{L}_1$. The iteration is stopped when two successive iterates differ relatively less than 10^{-3}. Note that $\sigma_n(\tilde{L}_1)$ is fairly small in magnitude. This is due to the fact that the decomposition is performed without pivoting. Nevertheless sharp inclusions of the solution are achieved.

The next two tables show the behaviour for larger bandwidths. We use the abbreviation $M(a, b, c \ldots)$ denoting a symmetric matrix with value a in the diagonal, b in the first subdiagonal, c in the second and so forth.

Table 3.6. $M(1, -2, 3, 4, -5)$, bandwidth 4

n	cond	$\|A - \tilde{L}_1 * \tilde{L}_2^T\|_2$	iter	$\sigma_{min}(\tilde{L}_1)$	$\|\hat{x} - \tilde{x}\|_\infty / \|\tilde{x}\|_\infty$
100	2.07E+01	6.61E-14	6	2.43E-02	5.06E-20
200	2.82E+01	2.64E-13	6	1.55E-02	1.24E-19
500	1.16E+02	2.64E-13	4	6.78E-03	6.44E-19
1000	1.96E+02	2.64E-13	5	3.66E-03	2.18E-18
2000	4.18E+02	4.08E-12	5	1.12E-03	2.48E-17
5000	1.47E+03	4.46E-12	3	4.49E-04	1.42E-16
10000	1.98E+03	4.46E-12	6	2.93E-04	3.30E-16
20000	4.24E+03	2.15E-11	5	9.86E-05	3.65E-15
50000	1.14E+04	2.15E-11	5	5.06E-05	1.14E-14
100000	1.93E+04	2.26E-11	7	2.60E-05	4.61E-14
200000	4.99E+04	3.66E-11	5	1.09E-05	4.26E-13

Table 3.7. $M(1, -2, 3, 4, -5, 5, 4, 3, 2, 1)$, bandwidth 9

n	cond	$\|A - \tilde{L}_1 * \tilde{L}_2^T\|_2$	iter	$\sigma_{min}(\tilde{L}_1)$	$\|\hat{x} - \tilde{x}\|_\infty / \|\tilde{x}\|_\infty$
100	1.08E+01	1.09E-12	4	2.05E-02	1.36E-19
200	4.38E+01	1.83E-12	3	7.91E-03	8.74E-19
500	7.46E+01	1.83E-12	4	5.26E-03	2.06E-18
1000	2.23E+02	6.48E-12	5	2.11E-03	1.24E-17
2000	2.54E+02	6.48E-12	6	1.89E-03	1.56E-17
5000	9.65E+02	6.48E-12	6	6.94E-04	1.28E-16
10000	2.23E+03	1.13E-11	5	2.41E-04	1.01E-15
20000	3.12E+03	1.13E-11	6	1.93E-04	1.54E-15
50000	7.06E+03	4.97E-10	4	3.78E-05	6.49E-14
100000	9.48E+03	4.97E-10	7	3.32E-05	1.07E-13

Again the comparitatively small values of $\sigma_n(\tilde{L}_1)$ are due to the lack of pivoting.

Finally we give same examples for the general case. First we show random matrices with upper and lower bandwidth 8 and uniformly distributed entries in the interval $[-1, 1]$.

Table 3.8. Random matrices, upper and lower bandwidth 8

n	cond	$\|A - \tilde{L}\tilde{U}\|_2$	$\sigma_{min}(\tilde{L})$	$\sigma_{min}(\tilde{U})$	$\|\hat{x} - \tilde{x}\|_\infty / \|\tilde{x}\|_\infty$
100	1.2E+03	1.2E-16	7.1E-02	2.5E-02	5.5E-26
200	3.5E+03	1.4E-16	1.0E-01	5.6E-03	2.6E-25
500	9.3E+04	1.4E-16	2.6E-02	1.9E-03	4.1E-23
1000	6.3E+04	2.1E-16	1.3E-02	4.0E-04	2.1E-21
2000	4.4E+05	3.5E-16	1.4E-02	1.0E-04	7.7E-24
5000	4.9E+05	3.2E-16	7.7E-03	1.7E-04	1.3E-23
10000	8.7E+05	2.9E-16	1.8E-02	5.9E-05	6.0E-23
20000	2.2E+05	3.4E-16	2.4E-02	1.0E-04	5.5E-23
50000	1.6E+06	2.8E-16	1.0E-02	2.3E-05	7.6E-22

The LU-decomposition is performed using routines DGBTRF and DGBTRS from LAPACK with pivoting. The smallest singular values of $\tilde{L}$ and $\tilde{U}$ are not too near. This improves when distributing the diagonal of $\tilde{U}$ among $\tilde{L}$ and $\tilde{U}$. Finally we show an example being unsymmetric in upper and lower bandwidth.

Table 3.9. Random matrices, upper/lower bandwidth 8/6

n	cond	$\|A - \tilde{L}\tilde{U}\|_2$	$\sigma_{min}(\tilde{L})$	$\sigma_{min}(\tilde{U})$	$\|\hat{x} - \tilde{x}\|_\infty / \|\tilde{x}\|_\infty$
100	5.1E+02	1.1E-16	9.1E-02	2.0E-02	1.3E-26
200	1.4E+03	1.8E-16	4.5E-02	1.2E-03	8.1E-24
500	7.8E+06	2.3E-16	4.3E-02	2.7E-06	4.6E-23
1000	3.6E+07	2.2E-16	3.0E-02	3.8E-07	3.8E-22
2000	2.0E+05	5.8E-16	4.0E-02	5.9E-05	3.4E-24
5000	2.3E+06	4.3E-16	1.0E-02	7.7E-06	2.5E-22

Random matrices with symmetric upper and lower bandwidth are fairly well-conditioned. This changes when the bandwidth becomes unsymmetric. Then for moderate dimension we run into fairly ill-conditioned matrices. Again the numbers become better when distributing of $\tilde{U}$ among $\tilde{L}$ and $\tilde{U}$.

4. Sparse Matrices

Following we give some remarks and computational results concerning sparse matrices. For $A \in \mathbb{R}^{n \times n}$ let

$$\underline{m}_i(A) := \min\{j \mid A_{ij} \neq 0\} \quad \text{and} \quad \overline{m}_i := \max\{j \mid A_{ij} \neq 0\}$$

for $1 \leq i \leq n$. The envelope is defined by

$$\mathrm{env}(A) := \{(i, j) \mid \underline{m}_i(A) \leq j \leq \overline{m}_i(A), 1 \leq i, j \leq n\}.$$

The profile of A is the number of elements in $\mathrm{env}(A)$, i.e. $\sum_{i=1}^{n} (\overline{m}_i(A) - \underline{m}_i(A) + 1)$. We treat matrices with an envelope such that the profile is significantly less than n^2.

It is well-known that LU, LDL^T, LDM^T and Cholesky decomposition without pivoting produce factors which remain in the envelope of A. Algorithm 2.1 is applicable with some obvious modifications since the Cholesky factor of $LL^T - \tilde{\lambda}^2 I$ has the same envelope as L. Hence the algorithmic approaches I), II), III), and IV) described in chapter 2 can be used.

Following we list some test results from the Harwell test case library (cf. [14]). The first column gives the name of the matrix within the library. Next to the dimension we list the lower and upper bandwidth and the total number of nonzero elements k; the other data are as in the tables of the previous chapter.

Table 4.1. Harwell test cases

Matrix	n	p	q	profile	cond	$\|A - \tilde{L}\tilde{U}\|_2$	$\sigma_{\min}(\tilde{L})$	$\sigma_{\min}(\tilde{U})$	$\|\hat{x} - \tilde{x}\|_\infty / \|\tilde{x}\|_\infty$
gre_216	216	14	36	876	2.7e+02	3.1e-15	3.6e-02	1.4e-02	7.9e-27
gre_343	343	18	49	1435	2.5e+02	5.6e-15	2.9e-02	1.2e-02	2.4e-26
gre_512	512	24	64	2192	3.8e+02	7.4e-15	2.0e-02	1.6e-02	6.8e-26
west0167	167	158	20	507	2.8e+06	1.6e-16	8.3e-02	1.4e-05	4.6e-22
west0381	381	363	153	2157	2.0e+06	1.1e-15	3.6e-02	2.7e-03	8.8e-25
bcsstk08	1074	590	590	7017	6.1e+06	1.6e-16	8.3e+00	4.6e+03	6.6e-23
bcsstk14	1806	161	161	32630	4.3e+04	1.8e-15	3.4e+01	5.0e+03	1.8e-25

In contrast to the examples in the previous chapter the matrices have been equilibrated in some cases reducing the condition number by several orders of magnitude.

5. Conclusion

The presented algorithm in its different versions for symmetric positive definite, symmetric indefinite and general matrices works for high dimensions. Possible improvements and open questions are the following.

Using a condition estimator instead of inverse power iteration would eventually be cheaper but has not been tested yet. For symmetric indefinite and for general matrices it is not clear how to choose a proper decomposition $A \approx \tilde{F} \cdot \tilde{G}$ in order to minimize $\sigma_n(\tilde{F}) \cdot \sigma_n(\tilde{G})$. Using and LDL^T or LDM^T decomposition with D equally distributed among the other factors works fine in many cases but also has its drawbacks. The estimations given by the algorithm are ∞-norm estimates on the relative error of an approximate solution $\tilde{x}$. These are good estimates on the relative error of each individual component as long as those do not differ too much in magnitude. However, even then they can be improved by making $b - A\tilde{x}$ smaller thru staggered correction. For interval data usually the components of the solution are of similar magnitude anyway.

References

[1] Alefeld, G., Herzberger, J.: Introduction to interval compoutations. New York: Academic Press 1983.
[2] Anderson, E.: Robust triangular solvers for use in condition estimation. Cray Research (1991).
[3] High-Accuracy Arithmetic Subroutine Library, Program Description and User's Guide, Release 3, IBM Publications, Document Number SC 33-6164-3 (1986).
[4] Arioli, M., Demmel, J. W., Duff, I. S.: Solving sparse linear systems with backward error. SIAM J. Matrix Anal. Appl. *10* (2), 165–190 (1989).
[5] Bauch, H., Jahn, K.-U., Oelschlägel, D., Süsse, H., Wiebigke, V.: Intervallmathematik, Theorie und Anwendungen. Mathematisch-naturwissenschaftliche Bibliothek, Bd. 72. Leipzig: B. G. Teubner 1987.
[6] Bischof, Ch. H., Tang, P. T. P.: Robust incremental condition estimators. Argonne National Lab. (1992).
[7] Böhm, H.: Berechnung von Polynomnullstellen und Auswertung arithmetischer Ausdrücke mit garantierter, maximaler Genauigkeit. Ph.D. dissertation, University of Karlsruhe (1983).
[8] Böhm, H., Rump, S. M.: Least significant bit evaluation for arithmetic expressions. Computing *30*, 189–199 (1983).
[9] Cline, A. K., Moler, G. B., Stewart, G. W., Wilkinson, J. H.: An estimate for the condition number of a matrix. SIAM J. Num. Anal. *16*, 368–375 (1979).
[10] Cline, A. K., Conn, A. R., Van Loan, C.: Generalizing the LINPACK condition estimator. In: Hennart, J. P. (ed.) Numerical analysis. New York: Springer-Verlag 1982 (Lecture Notes in Mathematics, 909).
[11] Cordes, D., Kaucher, E.: Self-validating computation for sparse matrix problems. In: Kaucher, E., Kulisch, U., Ullrich, Ch. (eds.) Computerarithmetic: scientific computation and programming languages. Stuttgart: Teubner B. G. 1987.
[12] Cordes, D.: Spärlich besetzte Matrizen. In: Kulisch, U. (ed.) Wissenschaftliches Rechnen mit Ergebnisverifikation—Eine Einführung, ausgearbeitet von S. Geörg, R. Hammer und D. Ratz. Berlin: Akademie Verlag und Wiesbaden: Vieweg Verlagsgesellschaft 1989.
[13] Duff, I. S., Erisman, A. M., Reid, J. K.: Direct methods for sparse matrices. Oxford: Clarendon Press 1986.
[14] Duff, I. S., Grimes, R., Lewis, J.: Sparse matrix test problems. ACM Transactions on Math. Software *15* (1), 1–14 (1989).
[15] Golub, G., v. Loan, C.: Matrix computations, 2nd edn. Baltimore: John Hopkins University Press 1989.
[16] Grimes, R. G., Lewis, J. G.: Condition number estimation for sparse matrices. SIAM J. Sci. and Stat. Comp. *2*, 384–388 (1991).

[17] Gregory, R. T., Karney, D. L.: A collection of matrices for testing computational algorithms. New York, London: John Wiley 1969.
[18] Hager, W.: Condition estimates. SIAM J. Sci. and Stat. Comp. 5, 311–316 (1984).
[19] Higham, N. J.: Fortran codes for estimating the one-norm of a real or complex matrix, with applications to condition estimation. ACM Trans. Math. Soft. 14, 381–396 (1987).
[20] IEEE Standard for Binary Floating-Point Arithmetic, ANSI/IEEE Standard 754 (1985).
[21] Jansson, Ch.: private communication.
[22] Krawczyk, R.: Newton-Algorithmen zur Bestimmung von Nullstellen mit Fehlerschranken. Computing 4, 187–201 (1969).
[23] Krawczyk, R.: Fehlerabschätzung bei linearer Optimierung. In: Nickel, K. (ed.) Interval Mathematics. Berlin, Heidelberg: Springer 1975 (Lecture Notes in Computer Science 29).
[24] Krämer, W.: Verified solution of eigenvalue problems with sparse matrices. Proceedings of 13th World Congress on Computation and Applied Mathematics, IMACS '91, Dublin, 32–33 (1991).
[25] Kulisch, U.: Grundlagen des numerischen Rechnens, Reihe Informatik 19. Mannheim, Wien: BI-Verlag 1976.
[26] Kulisch, U., Miranker, W. L.: Computer arithmetic in theory and practice. New York: Academic Press 1981.
[27] Moore, R. E.: A test for existence of solutions for non-linear systems. SIAM J. Numer. Anal. 4, 611–615 (1977).
[28] Moore, R. E.: Methods and applications of interval analysis. SIAM, Philadelphia (1979).
[29] Neumaier, A.: Interval methods for systems of equations. Cambridge: Cambridge University Press 1990.
[30] Neumaier, A.: The wrapping effect, ellipsoid arithmetic, and stability and confidence regions. In: Albrecht, R., Alefeld, G., Stetter, H. J. (eds.) Validation Numerics, pp. 175–190. Wien New York: Springer 1993 (Computing, Suppl. 6).
[31] Rump, S. M.: Kleine Fehlerschranken bei Matrixproblemen. Dissertation, Universität Karlsruhe (1980).
[32] Rump, S. M.: Solving algebraic problems with high accuracy, Habilitationsschrift. In: Kulisch, U. W., Miranker, W. L. (eds.) A new approach to scientific computation, 51–120. New York: Academic Press, 1983.
[33] Rump, S. M.: On the solution of interval linear systems. Computing 47, 337–353 (1992).
[34] Rump, S. M.: Inclusion of the solution for large linear systems with M-matrix. In: Atanassova, L., Herzberger J. (eds.) Computer arithmetic and enclosure methods. North-Holland: Amsterdam, London: 1992.
[35] Schwandt, H.: An interval arithmetic approach for the construction of an almost globally convergent method for the solution of the nonlinear poisson equation on the unit square. SIAM J. Sci. Stat. Comp. 5 (2), 427–452 (1984).
[36] Stetter, H. J.: Sequential defect correction in high-accuracy floating-point arithmetics. In: Griffith, D. F. (ed.) Numerical analysis (Proceedings, Dundee 1983), 186–202. Berlin, Heidelberg: Springer 1984 (Lecture Notes in Math. 1066).

Siegfried M. Rump
Technische Informatik III
TU Hamburg-Harburg
Eissendorferstrasse 38
D-W-2100 Hamburg 90
Federal Republic of Germany

Computing, Suppl. 9, 213–231 (1993)

Computing
© Springer-Verlag 1993
Printed in Austria

The Interval Buneman Algorithm for Arbitrary Block Dimension*

H. Schwandt, Berlin

Dedicated to Professor U. Kulisch on the occasion of his 60th birthday

Abstract — Zusammenfassung

The Interval Buneman Algorithm for Arbitrary Block Dimension. The interval arithmetic Buneman algorithm is a "fast solver" for a class of block tridiagonal systems with interval coefficients. In the present paper, we consider a modification for arbitrary block dimension and we discuss its inclusion properties.

AMS Subject Classification (1985): 65F05, 65G10

Key words: Buneman algorithm, optimal inclusions, interval analysis.

Der Intervall-Buneman-Algorithmus für beliebige Blockdimension. Der intervallarithmetische Buneman-Algorithmus stellt einen "schnellen Löser" für eine spezielle Klasse von Blocktridiagonalsystemen mit Intervallkoeffizienten dar. In dieser Arbeit wird eine Modifikation für beliebige Blockdimension betrachtet, und es werden ihre Einschließungseigenschaften untersucht.

1. Introduction

Since the introduction of the Buneman algorithm [3], [4], a wellknown fast solver for linear systems of equations with block tridiagonal coefficient matrices of the form (-I, A, -I) which mostly result from difference methods for (elliptic) partial differential equations, several variants have been published among which the $FACR$-algorithm [15], an efficient combination with FFT-based techniques, a parallel version [14] or algorithms for arbitrary block dimension [16], [17] should be mentioned. The above papers also contain variants for related matrices resulting from different boundary conditions (Neumann, periodic) in the solution of elliptic boundary value problems. The Buneman algorithm is a stabilized form of block cyclic reduction profiting from the particular matrix structure which is preserved in a series of reduction steps resulting in an arithmetic complexity of order $O(\log_2(N)N)$ which is roughly one order of magnitude smaller than that of classical solvers.

In the present paper, we discuss the properties of a modification of the interval arithmetic Buneman algorithm for arbitrary block dimension q, abbreviated by $IBUD$ in the sequel, based on an idea for point (noninterval) systems from [17].

---

* Received October 16, 1992; revised December 18, 1992.

The original algorithm requires a number of $q = 2^{n+1} - 1$ block equations while the dimension of A is arbitrary. For many applications, in particular for large (point and interval) systems or in domain decomposition algorithms, this condition on q can be too restrictive.

The treatment of interval systems $(\mathbf{M}, \mathbf{y})$ by a suitable interval method LES yields and interval vector $\mathbf{x}$ including the following set of solutions

$$\mathbf{x} := LES(\mathbf{M}, \mathbf{y}) \supseteq SOL(\mathbf{M}, \mathbf{y}) := \{\mathbf{x} \in \mathbf{R}^N | M\mathbf{x} = \mathbf{y}, M \in \mathbf{M}, \mathbf{y} \in \mathbf{y}\}. \quad (1.1)$$

This relation is the base for the discussion of inclusion properties in the context of interval systems of linear form. Many methods for the treatment of interval systems $(\mathbf{M}, \mathbf{y})$ have been published in various contexts. To mention a few, we note [1], [5], [8], [9], [10], [13] and also related methods [6], [7]. The importance of the interval Buneman algorithm IBU—see [11], [12], e.g.—for block tridiagonal interval systems $(\mathbf{M}, \mathbf{y})$, $\mathbf{M} = (-S, \mathbf{A}, -T)$, where S, T are real point matrices and where $\mathbf{A}$ and $\mathbf{y}$ have real compact intervals as coefficients, is due to the fact that it seems to be up to now the only "fast" direct interval solver. IBU (and also $IBUD$) is mainly used for the treatment of linear interval systems in Newton-like methods for large nonlinear systems of equations arizing from the discretization of almost linear elliptic boundary value problems ([11], [12], e.g). As a typical simple example, we mention the almost linear Dirichlet problem

$$a(x)u_{xx} + bu_{yy} + c(x)u_x + du_y = f(x, y, u)$$

or $\qquad (a(x)u_x)_x + (bu_y)_y = f(x, y, u) \quad$ on a rectangle $\Omega \subset \mathbf{R}^2$.

$$u(x, y) = g(x, y) \quad \text{on } \Gamma_\Omega, a, b > 0, f_u \geq 0.$$

The standard five point discretization with central difference quotients yields a nonlinear system of the form $f(u) = 0$, $f(u) = Fu + \phi(u)$ with $F = (-S, \mathbf{A}, -T)$ and a diagonal matrix $\phi'(u) \geq o$. In appropriate Newton-like interval methods ([11], [12]), systems $(\mathbf{M}, \mathbf{y})$, $\mathbf{M} = (-S, \mathbf{A}, -T)$, have to be treated in each iteration, i.e. $IBU(D)$ can be integrated. It is also possible to treat Neumann, periodic or mixed boundary conditions, three dimensional problems, circular domains with polar coordinates, nine point discretizations of Poisson's equation. By applying matrix decompositions in Newton-like methods such that the form $(-S, \mathbf{A}, -T)$ of the coefficient matrix is preserved in each step, more general problems with more variable coefficients as in

$$a(x, y)u_{xx} + c(x, y)u_{yy} + d(x, y)u_x + e(x, y)u_y = f(x, y, u)$$

can be solved; nonrectangular domains can be treated by integrating $IBU(D)$ in a domain decomposition method. $IBU(D)$ can also be applied in the context of methods for parabolic problems.

In the discussion of $IBUD$ we focus our attention mainly on the applicability, i.e. all interval operations have to be well defined, and on the inclusion properties in view of (1.1). It is wellknown that, in general, $SOL(\mathbf{M}, \mathbf{y})$ is not an interval vector [1]. The optimum we can expect is an optimal inclusion, i.e. that $\mathbf{x}$ is the smallest interval vector which still includes the solution set. In the present paper, we discuss in particular under which conditions on the interval coefficients of **M** and **y** and on the block dimension q of **M**, optimal inclusions can be guaranteed by *IBUD*.

After the introduction of some notation in §2, we derive the algorithm in §3. In §4 we discuss the main interval arithmetic properties of *IBUD*. §5 contains some numerical results.

2. Notation

For the introduction of interval arithmetic and interval methods, we refer to [1], for example. We denote real numbers by $a, \ldots, z$ or by greek letters, real point vectors and matrices by a, ..., z and A, ..., Z, resp., intervals by $A, \ldots, Z$, real interval vectors and matrices by **a**, ..., **z** and **A**, ..., **Z**, resp. All intervals are assumed to be real and compact: $A = [i(A), s(A)]$, $i(A)$, $s(A) \in R$, $i(A) \le s(A)$. For interval vectors and matrices we use the notation $\mathbf{a} = (A_i)_{i=1}^{N} = [i(\mathbf{a}), s(\mathbf{a})] = ([i(A_i), s(A_i)])_{i=1}^{N}$ and $\mathbf{A} = (A_{i,j})_{i,j=1}^{N} = [i(\mathbf{A}), s(\mathbf{A})] = ([i(A_{i,j}), s(A_{i,j})])_{i,j=1}^{N}$, where the bounds are defined componentwise. We also use the componentwise ordering: $A \le B \Leftrightarrow \forall i, \ j \in \{1, \ldots, N\}: a_{i,j} \le b_{i,j}$. We denote by $I(R)$, $V_N(I(R))$, $M_{NN}(R)$, $M_{NN}(I(R))$ the sets of real compact intervals, N-dimensional interval vectors, $N \times N$ point and interval matrices, resp. The absolute value, midpoint, radius of an interval are defined by $|A| = \min\{|i(A)|, |s(A)|\}$, $m(A) = (i(A) + s(A))/2$, $r(A) = (s(A) - i(A))/2$.

A real point matrix A is called an M matrix if $a_{i,j} \le 0$ for $i \ne j$, if A^{-1} exists and if $A^{-1} \ge O$. An interval matrix **A** is called an interval M matrix if all $A \in \mathbf{A}$ are M matrices. An interval matrix **A** is called a H matrix if the matrix $\langle \mathbf{A} \rangle$ defined by

$$\langle A_{i,j} \rangle := \begin{cases} |m(A_{i,i})| - r(A_{i,i}) & \text{if } i = j \\ -|A_{i,j}| & \text{if } i \ne j \end{cases} \tag{2.1}$$

is an M matrix.

The following elementary properties are relevant for the use of interval methods *LES* for the treatment of systems of linear form $(\mathbf{A}, \mathbf{y})$ in other contexts, as part of Newton like methods, for example.

$$\forall \mathbf{A} \in M_{NN}(I(R)), \quad \mathbf{y}, \mathbf{z} \in V_N(I(R)): LES(\mathbf{A}, \mathbf{y} + \mathbf{z}) \subseteq LES(\mathbf{A}, \mathbf{y}) + LES(\mathbf{A}, \mathbf{z}) \tag{2.2}$$

$$\forall \mathbf{A} \in M_{NN}(R), \quad \mathbf{y}, \mathbf{z} \in V_N(I(R)): LES(\mathbf{A}, \mathbf{y} + \mathbf{z}) = LES(\mathbf{A}, \mathbf{y}) + LES(\mathbf{A}, \mathbf{z}) \tag{2.3}$$

$$\forall \mathbf{A} \in M_{NN}(I(R)), \quad \mathbf{y} \in V_N(I(R)), \quad \alpha \in R: LES(\mathbf{A}, \alpha\mathbf{y}) = \alpha LES(\mathbf{A}, \mathbf{y}) \tag{2.4}$$

$$\forall \mathbf{A}, \mathbf{B} \in M_{NN}(I(R)), \quad \mathbf{y}, \mathbf{z} \in V_N(I(R)): \mathbf{A} \subseteq \mathbf{B}, \mathbf{y} \subseteq \mathbf{z} \Rightarrow LES(\mathbf{A}, \mathbf{y}) \subseteq LES(\mathbf{B}, \mathbf{z}) \tag{2.5}$$

$$\mathbf{A}_n \to \mathbf{A}, \quad \mathbf{y}_n \to \mathbf{y} \quad (n \to \infty) \Rightarrow LES(\mathbf{A}_n, \mathbf{y}_n) \to LES(\mathbf{A}, \mathbf{y}) \quad (n \to \infty) \tag{2.6}$$

An interval vector $LES(\mathbf{A}, \mathbf{y})$ is an optimal inclusion in the sense of (1.1) if its bounds belong to the set of solutions, i.e.

$$\exists A_1, A_2 \in \mathbf{A}, \quad y_1, y_2 \in \mathbf{y}: LES(\mathbf{A}, \mathbf{y}) = [(A_1)^{-1}y_1, (A_2)^{-1}y_2] \tag{2.7}$$

The following cases are typical for optimal inclusions:
$A \in M_{NN}(I(R))$ is an interval M matrix

a) $0 \leq i(\mathbf{y}) \Rightarrow LES(\mathbf{A}, \mathbf{y}) = [(s(\mathbf{A}))^{-1}i(\mathbf{y}), (i(\mathbf{A}))^{-1}s(\mathbf{y})]$

b) $s(\mathbf{y}) \leq 0 \Rightarrow LES(\mathbf{A}, \mathbf{y}) = [(i(\mathbf{A}))^{-1}i(\mathbf{y}), (s(\mathbf{A}))^{-1}s(\mathbf{y})]$

c) $0 \in \mathbf{y} \Rightarrow LES(\mathbf{A}, \mathbf{y}) = [(i(\mathbf{A}))^{-1}i(\mathbf{y}), (i(\mathbf{A}))^{-1}s(\mathbf{y})]$

d) $\mathbf{A} \equiv A \in M_{NN}(R) \Rightarrow \forall \mathbf{y} \in V_N(I(R)): LES(\mathbf{A}, \mathbf{y}) = [A^{-1}i(\mathbf{y}), A^{-1}s(\mathbf{y})] = A^{-1}\mathbf{y}$

$$(2.8)$$

3. The Algorithm

We assume the following situation.

A system of linear form $(\mathbf{M}, \mathbf{y})$, where $\mathbf{y} \in V_N(I(R))$,
$\mathbf{M} = (-S, -\mathbf{A}, -T) \in M_{NN}(I(R))$ block tridiagonal with q block rows,
$S, T \in M_{pp}(R)$, $\mathbf{A} \in M_{pp}(I(R))$, $N = pq$, $p, q \in N$.

$$(3.1)$$

In order to derive an interval algorithm, we first consider a point system $Mx = y$
with arbitrary $A \in \mathbf{A}$, $y \in \mathbf{y}$. We then get the following basic algorithm in analogy
to [17], generalized for $S, T \neq I$. Therefore, we only summarize the main ideas.
Starting from $M^{(0)} = (-S^{(0)}, A^{(0)}, -T^{(0)})$ with $A^{(0)} = A$, $S^{(0)} = S$, $T^{(0)} = T$ and $q_0 = q$ block equations, in each of a series of $r_q = [\log_2(q)]$, the largest integer smaller
than $\log_2(q)$, reduction steps, every second of q_r block equations is preserved by
combining it appropriately with its two neighbours: add $A^{(r)} * \text{Eq. } j + S^{(r)} * \text{Eq.}$
$j - 2^r + T^{(r)} * \text{Eq. } j + 2^r$ for $j := 2^{r+1}(2^{r+1})j_r - 2^{r+1}$. The modified algorithm differs
from the standard Buneman algorithm by the treatment of the last block equation
j_r whose different form is due to the necessary distinction of the cases "q_r odd" and
"q_r even": add $A^{(r)} * \text{Eq. } j + S^{(r)} * \text{Eq. } j - 2^r + T^{(r)}A^{(r)}(B^{(r)})^{-1}C^{(r)} * \text{Eq. } j + 2^r$ if q_r is
odd else add $A^{(r)} * \text{Eq. } j + S^{(r)} * \text{Eq. } j + 2^r$ if q_r is even. The definition of the matrices
$A^{(r)}$, $B^{(r)}$, $C^{(r)}$, $S^{(r)}$, $T^{(r)}$ (see (3.3)) results from the above combination of block
equations and requires the condition $AS = SA$, $AT = TA$, $ST = TS$. The original
matrix structure can be preserved in all steps. For $0 \leq r \leq r_q$, we get reduced
systems $M^{(r)}x^r = y^r$, $M^{(r)} \in M_{pj_r, pj_r}(R)$, $x^r, y^r \in V_{pj_r}(R)$ where

$$M^{(r)} = \begin{bmatrix} A^{(r)} & -T^{(r)} & & & \\ -S^{(r)} & A^{(r)} & -T^{(r)} & & \\ & \ddots & \ddots & \ddots & \\ & & -S^{(r)} & A^{(r)} & -T^{(r)} \\ & & & -S^{(r)} & B^{(r)}(C^{(r)})^{-1} \end{bmatrix},$$

$$x^r = \begin{bmatrix} x_{2r} \\ x_{2r+1} \\ \vdots \\ x_{j_r-2r} \\ x_{j_r} \end{bmatrix}, \quad y^r = \begin{bmatrix} y^r_{2r} \\ y^r_{2r+1} \\ \vdots \\ y^r_{j_r-2r} \\ y^r_{j_r} \end{bmatrix}.$$

$$(3.2)$$

with $B^{(0)} = A$, $C^{(0)} = I$, $j_0 = q$. According to the Buneman stabilization, the computation of new righthand sides y^r is replaced by that of auxiliary vectors p^r and q^r. Assuming all vectors with index $j \notin \{1,\ldots,q\}$ to vanish, we get the

basic algorithm $\hspace{8cm}$ (3.3)

reduction phase

for $j := 1$ **to** q **do** $\quad p_j^0 = o; \ q_j^0 = y_j;$

for $r := 0$ **to** $r_q - 1$ **do**

$\quad A^{(r+1)} = (A^{(r)})^2 - 2S^{(r)}T^{(r)}; \ S^{(r+1)} = (S^{(r)})^2; \ T^{(r+1)} = (T^{(r)})^2; \ U^{(r+1)} = S^{(r)}T^{(r)};$

$\quad q_{r+1} = q_r \operatorname{div} 2;$

$\quad$ **if** q_r even **then**

$\qquad j_{r+1} = j_r; \ B^{(r+1)} = A^{(r)}B^{(r)} - C^{(r)}U^{(r+1)}; \ C^{(r+1)} = C^{(r)};$

$\qquad p_{j_{r+1}}^{r+1} = p_{j_{r+1}}^r + (B^{(r)})^{-1}C^{(r)}(S^{(r)}p_{j_{r+1}-2^r}^r + q_{j_{r+1}}^r);$

$\qquad q_{j_{r+1}}^{r+1} = S^{(r)}q_{j_{r+1}-2^r}^r + U^{(r+1)}p_{j_{r+1}}^{r+1}$

$\quad$ **else**

$\qquad j_{r+1} = j_r - 2^r; \ B^{(r+1)} = A^{(r)}(A^{(r)}B^{(r)} - C^{(r)}U^{(r+1)}) - B^{(r)}U^{(r+1)};$

$\qquad C^{(r+1)} = B^{(r)};$

$\qquad p_{j_{r+1}}^{r+1} = p_{j_{r+1}}^r + (A^{(r)})^{-1}(S^{(r)}p_{j_{r+1}-2^r}^r + T^{(r)}p_{j_{r+1}+2^r}^r + q_{j_{r+1}}^r);$

$\qquad q_{j_{r+1}}^{r+1} = S^{(r)}q_{j_{r+1}-2^r}^r + U^{(r+1)}p_{j_{r+1}}^{r+1} + (B^{(r)})^{-1}C^{(r)}A^{(r)}(T^{(r)}q_{j_{r+1}+2^r}^r + U^{(r+1)}p_{j_{r+1}}^{r+1})$

$\qquad$ **for** $j := 2^{r+1}$ **step** 2^{r+1} **to** $j_{r+1} - 2^{r+1}$ **do**

$\qquad\quad p_j^{r+1} = p_j^r + (A^{(r)})^{-1}(S^{(r)}p_{j-2^r}^r + T^{(r)}p_{j+2^r}^r + q_j^r);$

$\qquad\quad q_j^{r+1} = S^{(r)}q_{j-2^r}^r + T^{(r)}q_{j+2^r}^r + 2U^{(r+1)}p_j^{r+1}$

solution phase

$x_{2^{r_q}} = p_{2^{r_q}}^{r_q} + (B^{(r_q)})^{-1}C^{(r_q)}q_{2^{r_q}}^{r_q};$

for $r := r_q - 1$ **step** -1 **to** 0 **do**

$\quad$ **for** $j := 2^r$ **step** 2^{r+1} **to** $j_r - 2^r$ **do**

$\qquad x_j = p_j^r + (A^{(r)})^{-1}(S^{(r)}x_{j-2^r} + T^{(r)}x_{j+2^r} + q_j^r);$

$\quad$ **if** q_r odd **then** $x_{j_r} = p_{j_r}^r + (B^{(r)})^{-1}C^{(r)}(S^{(r)}x_{j_r-2^r} + q_{j_r}^r).$ $\hspace{3cm}$ #

The second idea of the Buneman algorithm and its variants is a factorization of the matrices $A^{(r)}$ and for the modification for arbitrary q also of $B^{(r)}$, $C^{(r)}$. For $0 \le r \le r_q$, we obtain

$$\mathbf{A}^{(r)} = \prod_{i=1}^{2^r} (\mathbf{A} - \alpha_i^{(r)}\mathbf{U}), \qquad \alpha_i^{(r)} = 2\cos\left(\frac{2i-1}{2^{r+1}}\pi\right),$$

$$\mathbf{B}^{(r)} = \prod_{i=1}^{k_r} (\mathbf{A} - \lambda_i^{(r)}\mathbf{U}), \qquad \lambda_i^{(r)} = 2\cos\left(\frac{i}{k_r+1}\pi\right), \tag{3.4}$$

$$\mathbf{C}^{(r)} = \prod_{i=1}^{l_r} (\mathbf{A} - \mu_i^{(r)}\mathbf{U}), \qquad \mu_i^{(r)} = 2\cos\left(\frac{i}{l_r+1}\pi\right),$$

where $k_0 = 1$, $l_0 = 0$ and

$$k_{r+1} = \begin{cases} k_r + 2^r \\ k_r + 2^{r+1} \end{cases} \qquad l_{r+1} = \begin{cases} l_r & \text{if } q_r \text{ even} \\ k_r & \text{if } q_r \text{ odd} \end{cases}. \tag{3.5}$$

For the proof of (3.4), we first observe, that for $u \neq 0$

$$\left(\prod_{i=1}^{2^r}\left(\frac{a}{u} - \alpha_i^{(r)}\right)u^{2^r}\right)^2 - 2u^{2^{r+1}} = \left(\left(\prod_{i=1}^{2^r}\left(\frac{a}{u} - \alpha_i^{(r)}\right)\right)^2 - 2\right)u^{2^{r+1}} \tag{3.6}$$

and

$$\prod_{i=1}^{2^r}(a - \alpha_i^{(r)}u)\prod_{i=1}^{k_r}(a - \lambda_i^{(r)}u) - \prod_{i=1}^{l_r}(a - \mu_i^{(r)}u)u^{2^{r+1}}$$

$$= \prod_{i=1}^{2^r}\left(\frac{a}{u} - \alpha_i^{(r)}\right)u^{2^r}\prod_{i=1}^{k_r}\left(\frac{a}{u} - \lambda_i^{(r)}\right)u^{k_r} - \prod_{i=1}^{l_r}\left(\frac{a}{u} - \mu_i^{(r)}\right)u^{l_r+2^{r+1}}$$

$$= \left(\prod_{i=1}^{2^r}\left(\frac{a}{u} - \alpha_i^{(r)}\right)\prod_{i=1}^{k_r}\left(\frac{a}{u} - \lambda_i^{(r)}\right) - \prod_{i=1}^{l_r}\left(\frac{a}{u} - \mu_i^{(r)}\right)\right)u^{k_{r+1}} \tag{3.7}$$

for even q_r and

$$\prod_{i=1}^{2^r}(a - \alpha_i^{(r)}u)\left(\prod_{i=1}^{2^r}(a - \alpha_i^{(r)}u)\prod_{i=1}^{k_r}(a - \lambda_i^{(r)}u) - \prod_{i=1}^{l_r}(a - \mu_i^{(r)}u)u^{2^{r+1}}\right)$$

$$- \prod_{i=1}^{k_r}(a - \lambda_i^{(r)}u)u^{2^{r+1}}$$

$$= \prod_{i=1}^{2^r}\left(\frac{a}{u} - \alpha_i^{(r)}\right)u^{2^r}\left(\prod_{i=1}^{2^r}\left(\frac{a}{u} - \alpha_i^{(r)}\right)u^{2^r}\prod_{i=1}^{k_r}\left(\frac{a}{u} - \lambda_i^{(r)}\right)u^{k_r}\right.$$

$$\left. - \prod_{i=1}^{l_r}\left(\frac{a}{u} - \mu_i^{(r)}\right)u^{2^{r+1}+l_r}\right) - \prod_{i=1}^{k_r}\left(\frac{a}{u} - \lambda_i^{(r)}\right)u^{2^{r+1}+k_r}$$

$$= \left(\prod_{i=1}^{2^r}\left(\frac{a}{u} - \alpha_i^{(r)}\right)\left(\prod_{i=1}^{2^r}\left(\frac{a}{u} - \alpha_i^{(r)}\right)\prod_{i=1}^{k_r}\left(\frac{a}{u} - \lambda_i^{(r)}\right) - \prod_{i=1}^{l_r}\left(\frac{a}{u} - \mu_i^{(r)}\right)\right)\right.$$

$$\left. - \prod_{i=1}^{k_r}\left(\frac{a}{u} - \lambda_i^{(r)}\right)\right)u^{k_{r+1}} \tag{3.8}$$

for odd q_r.

For nonsingular $\mathbf{U}$, this is equivalent to

$$A^{(r+1)} = (A^{(r)})^2 - 2U^{(r+1)} = ((\tilde{A}^{(r)})^2 - 2I)U^{2^{r+1}} = \tilde{A}^{(r+1)}U^{2^{r+1}}$$

$$A^{(r)}B^{(r)} - C^{(r)}U^{(r+1)} = (\tilde{A}^{(r)}\tilde{B}^{(r)} - \tilde{C}^{(r)})U^{k_{r+1}} = \tilde{B}^{(r+1)}U^{k_{r+1}} = B^{(r+1)} \tag{3.9}$$

and

$$A^{(r)}(A^{(r)}B^{(r)} - C^{(r)}U^{(r+1)}) - B^{(r)}U^{(r+1)} = (\tilde{A}^{(r)}(\tilde{A}^{(r)}\tilde{B}^{(r)} - \tilde{C}^{(r)}) - \tilde{B}^{(r)})U^{k_{r+1}}$$

$$= \tilde{B}^{(r+1)}U^{k^{r+1}} = B^{(r+1)}, \tag{3.10}$$

where the tilde denotes the corresponding matrices of the algorithm for $M = (-I, A, -I)$ [17]. (3.9), (3.10) show that we can derive the matrix factorizations for the general case $M = (-S, A, -T)$ from those for $S = T = I$. In the latter $\tilde{A}^{(r)}$, $\tilde{B}^{(r)}$, $\tilde{C}^{(r)}$ can be expressed as Tschebyscheff matrix polynomials of the first and second kind $T_n(x)$, $U_n(x)$, respec., with the substitution $x = a/2$. Due to the extraction of the factors u^n, we simply have to substitute $x = a/2u$ in the general case.

This factorization enables us to replace all partial systems with the above matrices $A^{(r)}$, $B^{(r)}$, $C^{(r)}$ by series of simpler systems with coefficient matrices $A - \alpha U$, $\alpha \in R$, preserving the original, usually simple matrix structure:

$$A^{(r)}z = y:$$

$$z_0 = y;$$

$$\textbf{for } i := 1 \textbf{ to } 2^r \textbf{ do } z_i = (A - \alpha_i^{(r)}U)^{-1}z_{i-1}; \tag{3.11}$$

$$z := z_{2^r}$$

Using the relation

$$(A - \alpha U)z = (A - \beta U)y \quad \Leftrightarrow \quad z = y + (\alpha - \beta)U(A - \alpha U)^{-1}y, \tag{3.12}$$

matrix multiplications can be avoided where $C^{(r)}$ or $A^{(r)}$ appear on the right-hand side:

$$B^{(r)}z = C^{(r)}y:$$

$$z_0 = y;$$

$$\textbf{for } i := 1 \textbf{ to } l_r \textbf{ do } z_i = z_{i-1} + (\lambda_i^{(r)} - \mu_i^{(r)})U(A - \lambda_i^{(r)}U)^{-1}z_{i-1}; \tag{3.13}$$

$$\textbf{for } i := l_r + 1 \textbf{ to } k_r \textbf{ do } z_i = (A - \lambda_i^{(r)}U)^{-1}z_{i-1};$$

$$z = z_{k_r}.$$

$$B^{(r)}z = C^{(r)}A^{(r)}y:$$

$$z_0 = y;$$

$$\textbf{for } i := 1 \textbf{ to } k_r \textbf{ do } z_i = z_{i-1} + (\lambda_i^{(r)} - \beta_i^{(r)})U(A - \lambda_i^{(r)}U)^{-1}z_{i-1}; \tag{3.14}$$

$$z = z_{k_r}.$$

In (3.14) the roots $(\mu_i^{(r)})_{i=1,\ldots,l_r}$, $(\alpha_i^{(r)})_{i=1,\ldots,2^r}$ have to be merged appropriately [15]:

$$(\beta_i^{(r)})_{i=1,\ldots,k_r} = \text{merge}\{(\mu_i^{(r)})_{i=1,\ldots,l_r}, (\alpha_i^{(r)})_{i=1,\ldots,2^r}\} \tag{3.15}$$

We now proceed to the derivation of the interval algorithm.

We have chosen an arbitrary $M \in \mathbf{M}$, $y \in \mathbf{y}$, i.e. also $A \in \mathbf{A}$. In view of (1.1) and because of the inclusion monotonicity of interval arithmetic operations [1], we get $IBUD$ by replacing real numbers by the corresponding intervals according to the just mentioned inclusion (only the matrices $S^{(r)}$, $T^{(r)}$, $U^{(r)}$ and the roots of the Tschebyscheff polynomials are kept as real noninterval terms) and by replacing all point operations by the corresponding inteval arithmetic operations:

Modified *IBU* for arbitrary block dimension (*IBUD*) $\qquad\qquad$ (3.16)

$q_0 = j_0 = q, r_q = [\log_2(q)]; \; s := \min\{r | 0 \le r \le r_q, q_r \text{ even}\}$

$S^{(0)} = S; \; T^{(0)} = T; \; U^{(0)} = U = \sqrt{ST}; \; x_0 = x_{q+1} = o;$

reduction phase

for $j := 1$ **to** q **do** $\mathbf{p}_j^0 = o; \mathbf{q}_j^0 = y_j;$

for $r := 0$ **to** $s - 1$ **do**

$\quad S^{(r+1)} = (S^{(r)})^2; \; T^{(r+1)} = (T^{(r)})^2; \; U^{(r+1)} = S^{(r)}T^{(r)};$

$\quad q_{r+1} = q_r \operatorname{div} 2; \; j_{r+1} = j_r - 2^r; \; k_{r+1} = k_r + 2^{r+1}; \; l_{r+1} = k_r;$

$\quad$ **for** $j := 2^{r+1}$ **step** 2^{r+1} **to** j_{r+1} **do**

$\quad\quad \mathbf{p}_j^{r+1} = \mathbf{p}_j^r + L\tilde{E}S(\mathbf{A}^{(r)}, S^{(r)}\mathbf{p}_{j-2^r}^r + T^{(r)}\mathbf{p}_{j+2^r}^r + \mathbf{q}_j^r);$

$\quad\quad \mathbf{q}_j^{r+1} = S^{(r)}\mathbf{q}_{j-2^r}^r + T^{(r)}\mathbf{q}_{j+2^r}^r + 2U^{(r+1)}\mathbf{p}_j^{r+1}$

for $r := s$ **to** $r_q - 1$ **do**

$\quad S^{(r+1)} = (S^{(r)})^2; \; T^{(r+1)} = (T^{(r)})^2; \; U^{(r+1)} = S^{(r)}T^{(r)}; \; q_{r+1} = q_r \operatorname{div} 2;$

$\quad$ **if** q_r even **then**

$\quad\quad j_{r+1} = j_r; \; k_{r+1} = k_r + 2^r; \; l_{r+1} = l_r;$

$\quad\quad \mathbf{p}_{j_{r+1}}^{r+1} = \mathbf{p}_{j_{r+1}}^r + L\tilde{E}S(\mathbf{B}^{(r)}, \mathbf{C}^{(r)}(S^{(r)}\mathbf{p}_{j_{r+1}-2^r}^r + \mathbf{q}_{j_{r+1}}^r));$

$\quad\quad \mathbf{q}_{j_{r+1}}^{r+1} = S^{(r)}\mathbf{q}_{j_{r+1}-2^r}^r + U^{(r+1)}\mathbf{p}_{j_{r+1}}^{r+1}$

$\quad$ **else**

$\quad\quad j_{r+1} = j_r - 2^r; \; k_{r+1} = k_r + 2^{r+1}; \; l_{r+1} = k_r;$

$\quad\quad \mathbf{p}_{j_{r+1}}^{r+1} = \mathbf{p}_{j_{r+1}}^r + L\tilde{E}S(\mathbf{A}^{(r)}, S^{(r)}\mathbf{p}_{j_{r+1}-2^r}^r + T^{(r)}\mathbf{p}_{j_{r+1}+2^r}^r + \mathbf{q}_{j_{r+1}}^r);$

$\quad\quad \mathbf{q}_{j_{r+1}}^{r+1} = S^{(r)}\mathbf{q}_{j_{r+1}-2^r}^r + U^{(r+1)}\mathbf{p}_{j_{r+1}}^{r+1} + L\tilde{E}S(\mathbf{B}^{(r)}, \mathbf{C}^{(r)}\mathbf{A}^{(r)}(T^{(r)}\mathbf{q}_{j_{r+1}+2^r}^r$

$\quad\quad\quad\quad\quad\quad\quad\quad\quad\quad\quad + U^{(r+1)}\mathbf{p}_{j_{r+1}}^{r+1}));$

$\quad$ **for** $j := 2^{r+1}$ **step** 2^{r+1} **to** $j_{r+1} - 2^{r+1}$ **do**

$\quad\quad \mathbf{p}_j^{r+1} = \mathbf{p}_j^r + L\tilde{E}S(\mathbf{A}^{(r)}, S^{(r)}\mathbf{p}_{j-2^r}^r + T^{(r)}\mathbf{p}_{j+2^r}^r + \mathbf{q}_j^r);$

$\quad\quad \mathbf{q}_j^{r+1} = S^{(r)}\mathbf{q}_{j-2^r}^r + T^{(r)}\mathbf{q}_{j+2^r}^r + 2U^{(r+1)}\mathbf{p}_j^{r+1}$

solution phase
___

if $s = r_q$ **then** $\mathbf{x}_{2^{r_q}} = \mathbf{p}^{r_q}_{2^{r_q}} + L\widetilde{E}S(\mathbf{A}^{(r_q)}, \mathbf{q}^{r_q}_{2^{r_q}})$ **else**

$$\mathbf{x}_{2^{r_q}} = \mathbf{p}^{r_q}_{2^{r_q}} + L\widetilde{E}S(\mathbf{B}^{(r_q)}, \mathbf{C}^{(r_q)}\mathbf{q}^{r_q}_{2^{r_q}});$$

for $r := r_q - 1$ **step** -1 **to** $\min\{s, r_q - 1\} + 1$ **do**

  for $j := 2^r$ **step** 2^{r+1} **to** $j_r - 2^r$ **do**

   $\mathbf{x}_j = \mathbf{p}^r_j + L\widetilde{E}S(\mathbf{A}^{(r)}, \mathbf{S}^{(r)}\mathbf{x}_{j-2^r} + \mathbf{T}^{(r)}\mathbf{x}_{j+2^r} + \mathbf{q}^r_j);$

   if q_r odd **then** $\mathbf{x}_{j_r} = \mathbf{p}^r_{j_r} + L\widetilde{E}S(\mathbf{B}^{(r)}, \mathbf{C}^{(r)}(\mathbf{S}^{(r)}\mathbf{x}_{j_r-2^r} + \mathbf{q}^r_{j_r}));$

for $r := \min\{s, r_q - 1\}$ **step** -1 **to** 0 **do**

  for $j := 2^r$ **step** 2^{r+1} **to** j_r **do**

   $\mathbf{x}_j = \mathbf{p}^r_j + L\widetilde{E}S(\mathbf{A}^{(r)}, \mathbf{S}^{(r)}\mathbf{x}_{j-2^r} + \mathbf{T}^{(r)}\mathbf{x}_{j+2^r} + \mathbf{q}^r_j).$ $\qquad\qquad$ #

For the solution of the partial systems, we profit from the above factorizations and define the abstract method $L\widetilde{E}S$ as follows. Instead of $\mathbf{A}^{(r)}$, $\mathbf{B}^{(r)}$, $\mathbf{C}^{(r)}$, we use interval extensions [1] $\mathbf{A} - \alpha\mathbf{U}$ of the matrices $A - \alpha U$. LES denotes any interval method which is applicable to systems with coefficient matrices $A - \alpha U$.

$\mathbf{z} = L\widetilde{E}S(\mathbf{A}^{(r)}, \mathbf{y})$:

  $\mathbf{z}_0 = \mathbf{y};$

  for $i := 1$ **to** 2^r **do** $\mathbf{z}_i = LES(\mathbf{A} - \alpha^{(r)}_i\mathbf{U}, \mathbf{z}_{i-1}),$ $\qquad\qquad$ (3.17)

  $\mathbf{z} := \mathbf{z}_{2^r}$

$\mathbf{z} = L\widetilde{E}S(\mathbf{B}^{(r)}, \mathbf{C}^{(r)}\mathbf{y})$:

  $\mathbf{z}_0 = \mathbf{y};$

  for $i := 1$ to l_r do $\quad \mathbf{z}_i = \mathbf{z}_{i-1} + (\lambda^{(r)}_i - \mu^{(r)}_i)\mathbf{U}\, LES(\mathbf{A} - \lambda^{(r)}_i\mathbf{U}, \mathbf{z}_{i-1});$ $\quad$ (3.18)

  for $i := l_r + 1$ to k_r do $\quad \mathbf{z}_i = LES(\mathbf{A} - \lambda^{(r)}_i\mathbf{U}, \mathbf{z}_{i-1});$

  $\mathbf{z} = \mathbf{z}_{k_r}.$

$\mathbf{z} = L\widetilde{E}S(\mathbf{B}^{(r)}, \mathbf{C}^{(r)}\mathbf{A}^{(r)}\mathbf{y})$:

  $\mathbf{z}_0 = \mathbf{y};$

  for $i := 1$ **to** k_r **do** $\quad \mathbf{z}_i = \mathbf{z}_{i-1} + (\lambda^{(r)}_i - \beta^{(r)}_i)\mathbf{U}\, LES(\mathbf{A} - \lambda^{(r)}_i\mathbf{U}, \mathbf{z}_{i-1});$ $\quad$ (3.19)

  $\mathbf{z} = \mathbf{z}_{k_r}.$

The properties of LES are prescribed more precisely in the next section. We already mention that, according to the given problem, typical examples are variants of IGA, the interval Gauss algorithm [1], of ICR, interval cyclic reduction for tridiagonal matrices [13] or of $IBU(D)$ itself if $\mathbf{A}$ has again a blocktridiagonal structure $\mathbf{A} = (-\mathbf{V}, \mathbf{B}, -\mathbf{W})$.

From the computational point of view, the modification $IBUD$ only involves the computation of the respective last blocks $\mathbf{p}^r_{j_r}$, $\mathbf{q}^r_{j_r}$, $\mathbf{x}_{j_r}$ in all steps. While in the original algorithm partial systems with the $2^{r_q+1} - 1 = q$ coefficient matrices

$$\mathbf{A} - \alpha_i^{(r)}\mathbf{U} \qquad (1 \le i \le 2^r, 0 \le r \le r_q) \qquad (3.20)$$

and a variable number of righthand sides have to be treated, the modified algorithm requires the additional treatment of partial systems with the m_q coefficient matrices

$$\mathbf{A} - \lambda_i^{(r)}\mathbf{U} \qquad (1 \le i \le k_r, 0 \le r \le r_q),$$

$$2^{r_q+1} - 1 \le m_q := \sum_{r=0}^{r_q} k_r \le 2^{r_q+2} - 2, \qquad (3.21)$$

and one righthand side. As the treatment of these additional systems may significantly contribute to the arithmetic complexity of the algorithm, we have incorporated a modification for the case that the number of these systems can be reduced. In §4, we show that $IBUD$ and IBU coincide in the steps $0 \le r \le s$, if there exists an $s \in \{0, \dots, r_q\}$ such that q_r remains odd for $0 \le r \le s$. Consequently, we do not have to treat partial systems with matrices $\mathbf{B}^{(r)}$, $\mathbf{C}^{(r)}$ in these steps. On the other hand, in step $r = r_q$ of the solution phase, one eather needs the 2^{r_q} matrix factors of $\mathbf{A}^{(r_q)}$ (if $q = 2^{n+1} - 1$) or the k_{r_q} factors of $\mathbf{B}^{(r_q)}(\mathbf{C}^{(r_q)})^{-1}$ otherwise. Only the really needed matrices should be treated as, in view of $2^{r_q} \le k_{r_q}, q < 2^{r_q+1}$, both contribute significantly to the arithmetic complexity.

Multiplications of the form $\mathbf{S}^{(r)}\mathbf{x}$ (analogously $\mathbf{T}^{(r)}\mathbf{x}$, $\mathbf{U}^{(r+1)}\mathbf{x}$) should be carried out as a series of matrix vector multiplications with $S(T, U)$ because the coefficients may dramatically increase with increasing r. If $S, T, U \ge O$, we can write

$$\mathbf{S}^{(r)}\mathbf{x} = \mathbf{S}^{2^r}\mathbf{x} = (S(\dots S(S\mathbf{x})\dots)).$$

With $T, U \ge 0$ the same relations hold for $\mathbf{T}^{(r)}$, $\mathbf{U}^{(r)}$. Note also that $IBUD$ degenerates, with the above modifications, into the algorithm from [17] for point systems, i.e. $\mathbf{M} \equiv M$, $\mathbf{y} \equiv y$, according to its construction.

4. Properties

The first properties to be discussed for any interval method LES for the treatment of (1.1) are its applicability, i.e. all interval operations have to be welldefined [1], and the inclusion of the set of solutions according to (1.1). In the sequel, we refer to matrices $\mathbf{A} - \alpha\mathbf{U}$, where α is defined as one of the roots of (3.4):

$$\alpha \in \{\alpha_i^{(r)}, \mu_j^{(r)}, \lambda_l^{(r)} | 1 \le i \le 2^r, 1 \le j \le l_r, 1 \le l \le k_r, 0 \le r \le r_q\}. \qquad (4.1)$$

4.1. Theorem: *Assume* (3.1). *Under the conditions*

a) $U = +\sqrt{ST}$, *i.e.* $U^2 = ST$, *and* U^{-1} *exist*;
b) *LES is an interval method which is defined for systems with the coefficient structure of* $(\mathbf{A} - \alpha\mathbf{U}, \mathbf{y})$;
c) *LES is applicable to interval H matrices and satisfies* (2.2)–(2.6);
d) *The point matrix* $\langle \mathbf{B} \rangle$ *defined by*

$$\langle b_{i,j} \rangle := \begin{cases} |m(A_{i,i})| - r(A_{i,i}) - \alpha_1^{(r_q)}u_{i,i} & \text{if } i = j \\ -(|A_{i,j}| + \alpha_1^{(r_q)}u_{i,j}) & \text{if } i \neq j \end{cases}$$

is an **M** *matrix;*

the following assertions hold:

1) *IBUD can be applied to* (**M, y**);
2) *SOL*(**M, y**) $\subseteq$ *LES*(**M, y**);
3) *IBUD satisfies* (2.2)–(2.6).

Proof: *IBUD* consists of simple vector additions, matrix vector multiplications and applications of *LES* to partial systems of the form $(\mathbf{A} - \alpha\mathbf{U}, \mathbf{z})$. Therefore, the applicability of *IBUD* is shown if the applicability of *LES* to all partial systems is proved. According to the conditions b) and c), we have to show that for $0 \leq r \leq r_q$ all matrices

$$\mathbf{A} - \alpha_i^{(r)}\mathbf{U} \quad (1 \leq i \leq 2^r), \qquad \mathbf{A} - \mu_j^{(r)}\mathbf{U} \quad (1 \leq j \leq l_r),$$
$$\mathbf{A} - \lambda_l^{(r)}\mathbf{U} \quad (1 \leq l \leq k_r) \tag{4.2}$$

are interval H matrices. (3.4) and (3.5) yield

$$\alpha_i^{(r)} \leq \alpha_1^{(r_q)}, \qquad \mu_j^{(r)} \leq \mu_1^{(r_q)}, \qquad \lambda_l^{(r)} \leq \lambda_1^{(r_q)}. \tag{4.3}$$

and

$$k_r = l_r + 2^r, \quad l_r < 2^r, \qquad \text{hence } k_r < 2^{r+1}. \tag{4.4}$$

These relations imply $\mu_1^{(r_q)} < \lambda_1^{(r_q)} < \alpha_1^{(r_q)}$. The definition of the roots in (3.4) further yields $\alpha_i^{(r)} = -\alpha_{2^r+1-i}^{(r)}$, $\mu_j^{(r)} = -\mu_{l_r+1-j}^{(r)}$, $\lambda_l^{(r)} = -\lambda_{k_r+1-l}^{(r)}$. The combination of the above (in-)equalities yields

$$-\alpha_1^{(r_q)} = \alpha_{2^{r_q}}^{(r_q)} \leq \alpha_i^{(r)}, \mu_j^{(r)}, \lambda_l^{(r)} \leq \alpha_1^{(r_q)} \tag{4.5}$$

From elementary rules for midpoint and radius, we deduce

$$|m(A_{i,i} - \alpha u_{i,i})| - r(A_{i,i} - \alpha u_{i,i}) = |m(A_{i,i}) - \alpha u_{i,i}| - r(A_{i,i})$$
$$\geq |m(A_{i,i})| - r(A_{i,i}) - \alpha u_{i,i}$$
$$\geq |m(A_{i,i})| - r(A_{i,i}) - \alpha_1^{(r_q)}u_{i,i} > 0 \tag{4.6}$$

with d), (4.5), $\mathbf{U} \geq 0$ and α according to (4.1).

The inclusion monotonicity of interval operations yields

$$|b_{i,j}| = |A_{i,j}| + \alpha_1^{(r_q)}u_{i,j} \geq |A_{i,j}| + \alpha u_{i,j} \geq |A_{i,j} - \alpha u_{i,j}|. \tag{4.7}$$

As $\langle \mathbf{B} \rangle$ is an M matrix, there exists a real vector $t > o$ such that $\langle \mathbf{B} \rangle t > o$ [18]. Together with (4.5)–(4.7), we then obtain

$$(|m(A_{i,i} - \alpha u_{i,i})| - r(A_{i,i} - \alpha u_{i,i}))t_i$$
$$\geq (|m(A_{i,i})| - r(A_{i,i}) - \alpha u_{i,i})t_i$$
$$\geq (|m(A_{i,i})| - r(A_{i,i}) - \alpha_1^{(r_q)}u_{i,i})t_i$$

$$= \langle b_{i,i} \rangle t_i > - \sum_{\substack{j=1 \\ j \neq i}}^{p} \langle b_{i,j} \rangle t_j = \sum_{\substack{j=1 \\ j \neq i}}^{p} (|A_{i,j}| + \alpha_1^{(r_q)} u_{i,j}) t_j$$

$$\geq \sum_{\substack{j=1 \\ j \neq i}}^{p} |A_{i,j} - \alpha u_{i,j}| t_j \tag{4.8}$$

We have shown $\langle \mathbf{A} - \alpha U \rangle t > \mathrm{o}$ for all matrices in (4.2), i.e. all matrices $\langle \mathbf{A} - \alpha U \rangle$ defined by (2.1) are M matrices, hence all $\mathbf{A} - \alpha U$ are interval H matrices. Therefore, *LES* is applicable to all partial systems and *IBUD* is welldefined. The inclusion $SOL(\mathbf{M}, \mathbf{y}) \subseteq LES(\mathbf{M}, \mathbf{y})$ follows from the inclusion monotonicity of interval arithmetic operations [1] by applying *IBUD* simultaneously to $(\mathbf{M}, \mathbf{y})$ and to a point system $M\mathbf{x} = \mathbf{y}$ with arbitrary $M \in \mathbf{M}$, $\mathbf{y} \in \mathbf{y}$.

The properties (2.2)–(2.6) are satisfied by *LES* according to c). As *IBUD* consists of simple vector additions, matrix vector multiplications and applications of *LES*, (2.2)–(2.6) also hold for *IBUD*. This can be shown by using the relations [1] $\alpha(X + Y) = \alpha X + \alpha Y$ for $\alpha \in R$, $X, Y \in I(R)$, $A(\mathbf{B} + \mathbf{C}) = A\mathbf{B} + A\mathbf{C}$ for $A \geq \mathrm{o}$, $A \in M_{NN}(R)$, $\mathbf{B}, \mathbf{C} \in M_{NN}(I(R))$), the inclusion monotonicity and the continuity of interval arithmetic operations

#

In the next theorem we discuss the quality of the inclusion. More precisely, we give a criterion for optimal inclusions which requires additional conditions on $\mathbf{M}$ and $\mathbf{y}$. Similar results for previous versions of *IBU* can be found, for example, in [11], [12].

4.2. Theorem: *Assume (3.1). Under the conditions*

a) U^{-1} *exists*;
b) *LES is an interval method which is defined for systems with the coefficient structure of* $(\mathbf{A} - \alpha U, \mathbf{y})$;
c) *LES is applicable to interval* $\mathbf{M}$ *matrices*;
d) *LES yields optimal inclusions for* $(\mathbf{A} - \alpha U, \mathbf{y})$ *according to (2.8a–d)*;
e) $\mathbf{M}$ *and* $\mathbf{A} \pm \alpha_1^{r_q} U$ *are interval* $\mathbf{M}$ *matrices*;
f) $q = 2^n(2^m + 1) - 1$, $n, m \in N$;

the following assertions hold:

1) *IBUD can be applied to* $(\mathbf{M}, \mathbf{y})$;
2) $SOL(\mathbf{M}, \mathbf{y}) \subseteq LES(\mathbf{M}, \mathbf{y})$
3) *IBUD yields optimal inclusions of (3.1) according to (2.8a–d)*;
4) *If* $\exists s \in \{0, \ldots, r_q\} \forall r \in \{0, \ldots, s\} : q_r$ *odd, then* $A^{(r)} = B^{(r)}(C^{(r)})^{-1}$ *and IBU and IBUD coincide in these steps.*

Proof: According to Th. 4.1, *IBUD* is applicable and 2) holds as an interval M matrix is also an interval H matrix. $U = \sqrt{ST}$ exists because the (interval) M matrix property of $\mathbf{M} = (-S, \mathbf{A}, -T)$ implies $S, T \geq O$ as well as $U \geq O$.

$U \geq O$ and (4.5) yield for arbitrary $A \in \mathbf{A}$ and α from (4.1)

$$i(\mathbf{A}) - \alpha_1^{(r_q)} U \leq A - \alpha U \leq s(\mathbf{A}) + \alpha_1^{(r_q)} U. \tag{4.9}$$

According to a) $\mathbf{A}$ and $\mathbf{A} \pm \alpha_1^{(r_q)}\mathbf{U}$ are interval $\mathbf{M}$ matrices, therefore $i(\mathbf{A}) - \alpha_1^{(r_q)}\mathbf{U}$ and $s(\mathbf{A}) + \alpha_1^{(r_q)}\mathbf{U}$ are $\mathbf{M}$ matrices. Then (4.9) shows that $\mathbf{A} - \alpha\mathbf{U}$ must also be an $\mathbf{M}$ matrix, i.e. all $\mathbf{A} - \alpha\mathbf{U}$ are interval $\mathbf{M}$ matrices.

For simplicity, we only discuss optimal inclusions in the case of the lower bounds of (2.8d), as the upper bounds and a), b), c) can be treated in full analogy. The question of optimal inclusions can be essentially answered by the discussion of (3.17)–(3.19) as $S, T \geq O$ imply $S^{(r)}, T^{(r)}, U^{(r)} \geq O$. We note, for example,

$$i(\mathbf{q}_j^{r+1}) = S^{(r)}i(\mathbf{q}_{j-2^r}^r) + T^{(r)}i(\mathbf{q}_{j+2^r}^r) + 2U^{(r+1)}i(\mathbf{p}_j^{r+1}); \tag{4.10}$$

and

$$i(\mathbf{p}_j^{r+1}) = i(\mathbf{p}_j^r) + i(L\tilde{E}S(\mathbf{A}^{(r)}, S^{(r)}\mathbf{p}_{j-2^r}^r + T^{(r)}\mathbf{p}_{j+2^r}^r + \mathbf{q}_j^r)). \tag{4.11}$$

The other formulas for $\mathbf{q}^r$, $\mathbf{p}^r$ and $\mathbf{x}^r$ can be treated similarly. According to the assumption on LES, we get $LES(\mathbf{W}, \mathbf{y}) = \mathbf{W}^{-1}\mathbf{y}$ for a suitable point $\mathbf{M}$ matrix $\mathbf{W} \equiv \mathbf{W}$ and arbitrary $\mathbf{y}$, hence (3.17) yields

$$i(\mathbf{z}_0) = i(\mathbf{y}); \qquad i(\mathbf{z}_i) = (\mathbf{A} - \alpha_i^{(r)}\mathbf{U})^{-1}i(\mathbf{z}_{i-1}), \quad (1 \leq i \leq 2^r), \qquad i(\mathbf{z}) := i(\mathbf{z}_{2^r}). \tag{4.12}$$

For (4.11), this implies

$$i(\mathbf{p}_j^{r+1}) = i(\mathbf{p}_j^r) + \prod_{i=1}^{2^r}(\mathbf{A} - \alpha_i^{(r)}\mathbf{U})^{-1}(S^{(r)}i(\mathbf{p}_{j-2^r}^r) + T^{(r)}i(\mathbf{p}_{j+2^r}^r) + i(\mathbf{q}_j^r)). \tag{4.13}$$

For (3.18), we note

$$\mu_i^{(r)} < \lambda_i^{(r)} \qquad (1 \leq i \leq l_r, 0 \leq r \leq r_q) \tag{4.14}$$

and therefore

$$i(\mathbf{z}_0) = i(\mathbf{y});$$
$$i(\mathbf{z}_i) = i(\mathbf{z}_{i-1}) + (\lambda_i^{(r)} - \mu_i^{(r)})\mathbf{U}(\mathbf{A} - \lambda_i^{(r)}\mathbf{U})^{-1}i(\mathbf{z}_{i-1}), \quad (1 \leq i \leq l_r);$$
$$i(\mathbf{z}_i) = (\mathbf{A} - \lambda_i^{(r)}\mathbf{U})^{-1}i(\mathbf{z}_{i-1}), \qquad (l_r + 1 \leq i \leq k_r);$$
$$i(\mathbf{z}) = i(\mathbf{z}_{k_r}). \tag{4.15}$$

The inequalities (4.5) show, that there is no way to merge $\mu_i^{(r)}$ and $\alpha_j^{(r)}$ such, that $\beta_l^{(r)} < \lambda_i^{(r)}$ $(1 \leq i, l \leq k_r, 0 \leq r \leq r_q)$. Consequently, (3.19) cannot yield optimal inclusions as $\beta_l^{(r)} > \lambda_i^{(r)}$ for at least one pair i, j implies that the upper bound $s(\mathbf{z}_i)$ is needed for the computation of $i(\mathbf{z}_{i+1})$. (3.19) is needed only for the computation of $\mathbf{q}_{j_{r+1}}^{r+1}$ in the branch "q_r is odd" in the algorithm (3.16) for $IBUD$. Looking more closely we observe, that if we start $IBUD$ with an odd q and that q_r is odd for $0 \leq r \leq s \leq r_q$, then $k_0 = 1$, $l_0 = 0$, $k_{r+1} = k_r + 2^{r+1}$, $l_{r+1} = k_r$ imply $k_r = 2^{r+1} - 1$, $l_r = 2^r - 1$, hence

$$\mu_i^{(r)} = 2\cos\left(\frac{i}{l_r + 1}\pi\right) = 2\cos\left(\frac{i}{2^r}\pi\right) = 2\cos\left(\frac{2i}{2^{r+1}}\pi\right)$$

$$= 2\cos\left(\frac{2i}{k_r + 1}\pi\right) = \lambda_{2i}^{(r)} \qquad (1 \leq i \leq l_r) \tag{4.16}$$

and similarly

$$\alpha_i^{(r)} = 2\cos\left(\frac{2i-1}{2^{r+1}}\pi\right) = 2\cos\left(\frac{2i-1}{k_r+1}\pi\right) = \lambda_{2i-1}^{(r)} \qquad (1 \le i \le 2^r). \quad (4.17)$$

In this case, $B^{(r)}$ and $C^{(r)}$ divide $A^{(r)}$ for $0 \le r \le s$, hence IBU and $IBUD$ coincide in these steps as all applications of $L\tilde{E}S$ are reduced to (3.17). The remaining critical case consists in a transition from an even q_r to an odd q_{r+1} $(r < r_q - 1)$. We can avoid this case by the condition $q = 2^n(2^m + 1) - 1$, $n, m \in \mathbb{N}$. For $m = 0$, $IBUD$ reduces to IBU; for $n = 0$, q is a power of two. For $n, m \ne 0$, q_r is an odd number for $0 \le r \le n$ and a power of two (2^m) for $n \le r \le r_q$.

Like any integer q can be expressed by its binary representation

$$q = \sum_{r=0}^{r_q} d_r 2^r, \qquad d_r \in \{0,1\}, \qquad d_{r_q} = 1. \qquad (4.18)$$

As an integer division by 2 corresponds to a rightshift, a transition from an even q_r to an odd q_{r+1} for $r < r_q - 1$ is equivalent to $d_r = 0$, $d_{r+1} = 1$. Hence,

$$d_{r_q} = 1; \qquad \exists s \in \{0,\dots,r_q - 2\}: d_r = 0 \quad (s + 1 \le r \le r_q - 2),$$
$$d_r = 1 \quad (0 \le r \le s). \qquad (4.19)$$

This is equivalent to the condition on q.

As $IBUD$ degenerates to the (here slightly generalized) point algorithm from [17] for noninterval systems, we can conclude $i(IBUD(\mathbf{M}, \mathbf{y})) = IBUD(\mathbf{M}, i(\mathbf{y})) = \mathbf{M}^{-1}i(\mathbf{y})$ under the above conditions. The proof can be repeated for the upper bounds, hence (2.8d) is complete: $IBUD(\mathbf{M}, \mathbf{y}) = [\mathbf{M}^{-1}i(\mathbf{y}), \mathbf{M}^{-1}s(\mathbf{y})]$. This inclusion is optimal as $i(\mathbf{y}), s(\mathbf{y}) \in \mathbf{y}$, i.e. $\mathbf{M}^{-1}i(\mathbf{y}), \mathbf{M}^{-1}s(\mathbf{y}) \in SOL(\mathbf{M}, \mathbf{y})$.   #

The above theorems indicates that $IBUD$ is applicable for arbitrary q, but that optimal inclusions can only be guaranteed with a restriction on q. Compared to the original requirement $q = 2^{n+1} - 1$, however, the range of admissible q has been significantly extended.

4.3. Remark: The above theorem shows that $IBUD$ and IBU can coincide in several (or even all) steps for particular values q. As mentioned in §3, $IBUD$ should be modified in this sense, in particular in view of (3.19). For $0 \le r \le s$, we then note $B^{(r)}(C^{(r)})^{-1} = A^{(r)}$. Therefore, the different treatment of block j_r can be avoided in the corresponding steps of both the reduction and the solution phase, i.e. the j-loops are extended up to $j = j_r$ and $j = j_{r+1}$, resp.

4.4. Remark: Th. 4.1 and Th. 4.2 indicate criteria for the choice of the method LES for the solution of the partial systems of the form $(\mathbf{A} - \alpha\mathbf{U}, \mathbf{z})$. As already mentioned in §3, typical direct methods which satisfy the above conditions are various versions of the interval Gauss algorithm [1], [2], e.g., of interval cyclic reduction [13] or of $IBU(D)$ itself [12].

4.5. Remark: In the point algorithm (3.3) the order of the treatment of partial systems in (3.13) and (3.14) should be modified such that the terms $\lambda_i^{(r)} - \mu_i^{(r)}$

and $\lambda_i^{(r)} - \alpha_i^{(r)}$ are replaced by $\lambda_i^{(r)} - \mu_j^{(r)}$ and $\lambda_i^{(r)} - \alpha_k^{(r)}$, where $j \equiv j(i)$, $k \equiv k(i)$ are chosen such that these terms are minimized in order to preserve maximum accuracy [17]. In contrast to that, we have to guarantee in the interval algorithm that these terms remain nonnegative in order to get optimal inclusions. According to Th. 4.2, we note $j(i) = i$ and $k(i) = i$ in (3.18). As mentioned above, we cannot obtain optimal inclusions in (3.19). The condition on q in Th. 4.2. indicates exactly the cases in which (3.19) is not involved. In the present paper, we do not further discuss the quality of nonoptimal inclusions.

4.6. Remark: The required existence of U^{-1} is not a severe restriction as in the main applications, interval systems resulting from difference methods for elliptic boundary value problems, S and T are invertible.

5. Numerical Examples

The numerical examples have been computed on a workstation IBM RS6000/560 with the FORTRAN compiler XLF 2.2. The computation times are given in CPU seconds, measured with the system routine MCLOCK with a resolution of a 1/100 second.

Define

t_{mat} CPU time for the matrix dependent parts of *IBUD*, i.e. *LU-* or similar decompositions of the matrices $\mathbf{A} - \alpha U$ for the chosen method *LES*.

t_{rhs} CPU time for those parts of *IBUD* which only depend on the righthand sides.

$t_{tot} = t_{mat} + t_{rhs}$ total CPU time for *IBUD*

$t_{rel} = t_{tot}/q$ CPU time for *IBUD* relative to the block dimension.

The quality of the inclusion is expressed by the relative distance of the computed interval vector $\mathbf{x} := LES(M, \mathbf{y})$ and a given optimal inclusion $\mathbf{x}_{opt}$.

$$q_{rel}(\mathbf{x}, \mathbf{x}_{opt}) := \max_{1 \leq i \leq N} \left\{ \left| \frac{i(X_i) - i(X_i^{opt})}{i(X_i^{opt})} \right|, \left| \frac{s(X_i) - s(X_i^{opt})}{s(X_i^{opt})} \right| \right\}. \tag{5.1}$$

This measure is appropriate because of $\mathbf{x}_{opt} \subseteq \mathbf{x}$.

The following examples shows the typical behaviour of *IBUD* as predicted by the theory. In all examples, we exclusively use $LES = IGA$, more precisely, a version of interval Gaussian elimination for tridiagonal systems. We prescribe an optimal inclusion and define the righthand side accordingly.

1) $p = 511$, $N = pq$, $\mathbf{M} \equiv M = (-I, A, -I)$, $A = (-1, 4, -1)$,
    $x := (i + j)_{i=1}^{p}{}_{j=1}^{q}$, $\mathbf{x}_{opt} := [x, x + 1]$, $\mathbf{y} := [Mi(\mathbf{x}_{opt}), Ms(\mathbf{x}_{opt})]$;
2) $p = 511$, $N = pq$, $\mathbf{M} \equiv M = (-I, A, -I)$, $A = (-1, 4, -1)$, $x := (1)_{i=1}^{p}{}_{j=1}^{q}$,
    $\mathbf{x}_{opt} := [x, x + 1]$, $\mathbf{y} := [Mi(\mathbf{x}_{opt}), Ms(\mathbf{x}_{opt})]$;

3) $p = 511$, $N = pq$, $\mathbf{M} \equiv M = (-3.1I, A, -4.2I)$,
   $A = (-2.1, [11.2, 12.2] + 0.01j, -1.9)$, $x := (i + j)_{i=1\,j=1}^{p\quad q}$, $\mathbf{x}_{opt} := [-x, x]$,
   $y := [Mi(\mathbf{x}_{opt}), Ms(\mathbf{x}_{opt})]$.

In 1), 2) $\mathbf{M} \equiv M$ is a point M matrix, in 3), $\mathbf{M}$ is an interval M matrix. $A - 2I = (-1, 2, -1)$ and $A - 2U = (-2.1, [11.2, 12.2] - \sqrt{13.02} + 0.01j, -1.9)$ are M matrices and lower bounds for $A \pm \alpha_1^{(r_q)}I$ and $A \pm \alpha_1^{(r_q)}U$, resp., i.e the latter are M matrices [18]. The theoretical inclusions $\mathbf{x}_{opt}$ are optimal according to (2.8d) for <u>all</u> q.

Table 1. CPU-times and inclusion quality for examples 1, 2, 3

q	t_{mat}	t_{rhs}	t_{tot}	t_{rel}	$q_{rel}(\mathbf{x}^1, \mathbf{x}^1_{opt})$	$q_{rel}(\mathbf{x}^2, \mathbf{x}^2_{opt})$	$q_{rel}(\mathbf{x}^3, \mathbf{x}^3_{opt})$
62	$.530_{10}+00$	$.152_{10}+01$	$.205_{10}+01$	$.331_{10}-01$	$.101_{10}+08$	$.445_{10}+02$	$.157_{10}-01$
53	$.310_{10}+00$	$.104_{10}+01$	$.135_{10}+01$	$.214_{10}-01$	$.844_{10}-12$	$.380_{10}-13$	$.513_{10}-14$
64	$.840_{10}+00$	$.136_{10}+01$	$.220_{10}+01$	$.344_{10}-01$	$.870_{10}-12$	$.380_{10}-13$	$.513_{10}-14$
65	$.840_{10}+00$	$.138_{10}+01$	$.222_{10}+01$	$.342_{10}-01$	$.907_{10}-12$	$.380_{10}-13$	$.522_{10}-14$
66	$.860_{10}+00$	$.149_{10}+01$	$.235_{10}+01$	$.356_{10}-01$	$.151_{10}+01$	$.690_{10}+00$	$.156_{10}+01$
67	$.880_{10}+00$	$.142_{10}+01$	$.230_{10}+01$	$.343_{10}-01$	$.972_{10}-12$	$.380_{10}-13$	$.550_{10}-14$
68	$.870_{10}+00$	$.163_{10}+01$	$.250_{10}+01$	$.368_{10}-01$	$.546_{10}+01$	$.101_{10}+01$	$.330_{10}-03$
125	$.118_{10}+01$	$.351_{10}+01$	$.469_{10}+01$	$.375_{10}-01$	$.333_{10}+10$	$.530_{10}+01$	$.426_{10}-03$
126	$.119_{10}+01$	$.356_{10}+01$	$.475_{10}+01$	$.377_{10}-01$	$.333_{10}+11$	$.412_{10}+02$	$.156_{10}-01$
127	$.640_{10}+00$	$.258_{10}+01$	$.326_{10}+01$	$.255_{10}-01$	$.262_{10}-11$	$.382_{10}-13$	$.534_{10}-14$
128	$.167_{10}+01$	$.318_{10}+01$	$.485_{10}+01$	$.379_{10}-01$	$.267_{10}-11$	$.382_{10}-13$	$.534_{10}-14$
129	$.167_{10}+01$	$.320_{10}+01$	$.487_{10}+01$	$.378_{10}-01$	$.272_{10}-11$	$.382_{10}-13$	$.544_{10}-14$
130	$.168_{10}+01$	$.338_{10}+01$	$.506_{10}+01$	$.389_{10}-01$	$.153_{10}+01$	$.682_{10}+00$	$.155_{10}-01$
131	$.169_{10}+01$	$.325_{10}+01$	$.494_{10}+01$	$.377_{10}-01$	$.281_{10}-11$	$.382_{10}-13$	$.556_{10}-14$
132	$.171_{10}+01$	$.362_{10}+01$	$.533_{10}+01$	$.404_{10}-01$	$.557_{10}+01$	$.993_{10}+00$	$.327_{10}-03$
133	$.174_{10}+01$	$.366_{10}+01$	$.540_{10}+01$	$.406_{10}-01$	$.860_{10}+01$	$.144_{10}+01$	$.426_{10}-03$
134	$.175_{10}+01$	$.374_{10}+01$	$.549_{10}+01$	$.410_{10}-01$	$.888_{10}+02$	$.822_{10}+01$	$.156_{10}-01$
135	$.177_{10}+01$	$.337_{10}+01$	$.514_{10}+01$	$.381_{10}-01$	$.304_{10}-11$	$.391_{10}-13$	$.540_{10}-13$
136	$.174_{10}+01$	$.380_{10}+01$	$.554_{10}+01$	$.407_{10}-01$	$.144_{10}+02$	$.248_{10}+00$	$.225_{10}-07$
137	$.177_{10}+01$	$.384_{10}+01$	$.561_{10}+01$	$.409_{10}-01$	$.241_{10}+02$	$.375_{10}+00$	$.279_{10}-07$
138	$.179_{10}+01$	$.390_{10}+01$	$.569_{10}+01$	$.412_{10}-01$	$.128_{10}+03$	$.724_{10}+00$	$.155_{10}-01$
139	$.181_{10}+01$	$.389_{10}+01$	$.570_{10}+01$	$.410_{10}-01$	$.362_{10}+02$	$.476_{10}+00$	$.283_{10}-06$
140	$.181_{10}+01$	$.397_{10}+01$	$.578_{10}+01$	$.413_{10}-01$	$.115_{10}+04$	$.547_{10}+01$	$.326_{10}-03$

Table 1 (continued)

141	$.196_{10}+01$	$.401_{10}+01$	$.597_{10}+01$	$.415_{10}-01$	$.177_{10}+04$	$.772_{10}+01$	$.425_{10}-03$
142	$.184_{10}+01$	$.401_{10}+01$	$.585_{10}+01$	$.417_{10}-01$	$.177_{10}+05$	$.408_{10}+02$	$.156_{10}-01$
143	$.186_{10}+01$	$.406_{10}+01$	$.592_{10}+01$	$.385_{10}-01$	$.344_{10}-11$	$.392_{10}-13$	$.556_{10}-14$
144	$.188_{10}+01$	$.363_{10}+01$	$.551_{10}+01$	$.412_{10}-01$	$.319_{10}+02$	$.384_{10}-02$	$.556_{10}-14$
145	$.180_{10}+01$	$.414_{10}+01$	$.594_{10}+01$	$.411_{10}-01$	$.573_{10}+02$	$.588_{10}-02$	$.363_{10}-12$
254	$.242_{10}+01$	$.886_{10}+01$	$.113_{10}+02$	$.444_{10}-01$	$.408_{10}+15$	$.395_{10}+02$	$.155_{10}-01$
255	$.134_{10}+01$	$.697_{10}+01$	$.831_{10}+01$	$.324_{10}-01$	$.120_{10}-10$	$.386_{10}-13$	$.556_{10}-14$
256	$.337_{10}+01$	$.814_{10}+01$	$.115_{10}+02$	$.450_{10}-01$	$.121_{10}+10$	$.386_{10}-13$	$.556_{10}-14$
257	$.501_{10}+01$	$.819_{10}+01$	$.132_{10}+02$	$.514_{10}-01$	$.122_{10}-10$	$.386_{10}-13$	$.560_{10}-14$
258	$.505_{10}+01$	$.848_{10}+01$	$.135_{10}+02$	$.524_{10}-01$	$.153_{10}+01$	$.677_{10}+00$	$.155_{10}-01$
510	$.855_{10}+01$	$.190_{10}+02$	$.276_{10}+02$	$.540_{10}-01$	$.107_{10}+20$	$.386_{10}+02$	$.155_{10}-01$
511	$.270_{10}+01$	$.158_{10}+02$	$.185_{10}+02$	$.361_{10}-01$	$.290_{10}-10$	$.388_{10}-13$	$.566_{10}-14$

Table 1 results for some q which illustrate the typical behaviour of *IBUD* with respect to the computation time and to inclusion properties. As predicted, we obtain optimal inclusions for the values of q given by Th. 4.2 (marked in bold in Table 1). For all other (nonoptimal) values of q, the inclusions are useless in examples 1 and 2. Depending on the additional width introduced in (3.19) by negative differences $\lambda_i^{(r)} - \beta_i^{(r)}$ the inclusions may be better in particular examples like for some q in example 3, but principally, they are drastically worse than optimal inclusions by several orders of magnitude in the measure (5.1).

Theoretically, we would expect $q_{rel}(\mathbf{x}, \mathbf{x}_{opt}) = 0$ for $q = 2^n(2^m + 1) - 1, n, m \in \mathbb{N}$. The computed values differ from the optimum because of the effect of rounding errors and its overestimation. We use a simulation of an interval arithmetic whose principles are described in [1]. Due to the lack of an interval arithmetic, we simulate it by computing interval bounds by the existing floating-point arithmetic (near-IEEE on IBM RS 6000). Principally, the floating point error in the computation of the bounds is compensated by the multiplication by suitable constants $1 \pm \varepsilon$, ε close to the relative machine precision, such that the resulting interval includes all rounding errors. In order to satisfy this requirement, one has also to take into account the effect of underflow. This can be implemented by particular case distinctions and the addition or subtraction of suitable constants close to the smallest positive machine number x. This simulation yields correct and satisfactory inclusions, but usually also a slight overestimation of the error.

Improved inclusions can be obtained by applying an implementation of an interval arithmetic which is independent of the floating point arithmetic which is available on the respective machine [6], [7]. The higher precision has to be paid for by significantly higher computation time. The simulation we used in the above tests requires several floating point operations and branches for every interval operation. The effect of the higher arithmetic complexity is less reflected by the computation times as the superscalar architecture of the IBM RS 6000 can execute a mix of different instructions (add, multiply, branchings, addressing and memory access) in roughly one cycle.

The (interval) arithmetic complexity of $IBUD$ is of order $O(\log_2(q)N)$, i.e as indicated by column 4 in Table 1 not proportional to N. Not surprisingly, we always observe a sharp reduction of the total (absolute and relative) CPU time for $q = 2^{n+1} - 1$ when compared to its neighbours. For these values of q, IBU and $IBUD$ coincide and the time consuming modifications which are the subject of $IBUD$ are not needed. The total computation time does not increase regularly with q due to different effects. In column 1 we note the CPU time which caused by the matrix dependent parts of $IBUD$, i.e. those parts of the algorithm which are independent of the righthand sides: the matrix transformations for the $\mathbf{A} - \alpha\mathbf{U}$ according to the chosen method LES (here IGA). The strong increase of the CPU time from $q = 2^{n+1} - 2$ to $q = 2^{n+1}$ is due to the increase of r_q by one, compare (3.20) and (3.21).

A similar effect can be observed with respect to the time for the treatment of the righthand sides. A drastical reduction appears from $q = 2^{n+1} - 2$ to $q = 2^{n+1} - 1$ which is again due to the reduction to the unmodified algorithm. In contrast to the matrix dependent parts, the times do not increase significantly from $q = 2^{n+1} - 2$ to $q \geq 2^{n+1}$. This is due to the fact, that for $q = 2^{n+1} - 2$ q is even, but all subsequent q_r are odd, while for $q = 2^{n+1}$ all q_r are even. The arithmetic complexity of the branch "q_r odd" in (3.16) is significantly higher than that of the branch "q_r even", as k_r additional partial systems have to be treated.

The comparison of columns 1 and 2 also illustrates the complexity of the matrix dependent parts of $IBUD$, which is only reduced for matrices with simpler coefficients like $(-\mathrm{I},(-1, 4, -1), -\mathrm{I})$. The present examples have been computed with an algorithm for matrices $(-d\mathrm{I}, (a_i, b_i, c_i), -e\mathrm{I})$.

We do not present comparisons with other interval methods with respect to computation times as the latter only reflect, analoguously to corresponding noninterval methods, the different orders of arithmetic complexity. As an example we mention the block Gauss algorithm with $O(p^3q)$ versus roughly $O(\log_2(q)N) = O(\log_2(q)pq)$ for $IBU(D)$.

6. Conclusion

The modification $IBUD$ of IBU improves the applicability by extending the range of admissible numbers of block equations. In the treatment of interval systems resulting, for example, from the application of difference methods, the mesh size and, therefore, also the system size can be chosen more flexibly. While the applicability can be shown for arbitrary values of q under appropriate conditions, optimal inclusions of the set of solutions can be guaranteed only under additional conditions for a restricted set of values of q. The admissible range for q is, however, significantly extended when compared to that of the original algorithm. The arithmetic complexity increases by roughly 50% due to the modification for $q \neq 2^{n+1} - 1$.

Acknowledgment

I want to express my gratitude to the referees for their helpful suggestions.

References

[1] Alefeld, G., Herzberger, J.: Introduction to interval computations. New York: Academic Press 1983.

[2] Barth, W., Nuding, E.: Optimale Lösung von Intervallgleichungssystemen. Computing 12, 117–125 (1974).

[3] Buneman, O.: A compact noniterative poisson solver, Institute for Plasma Research Report 294, Stanford University, 1969.

[4] Buzbee, B., Golub, G., Nielson, C.: On direct methods for solving Poisson's equation. SIAM J. Num. Anal. 7, 627–656 (1970).

[5] Frommer, A., Mayer, G.: Parallel interval multisplittings. Numerische Mathematik 56, 255–267 (1989).

[6] Kulisch, U., Miranker, W.: Computer arithmetic in theory and practice. New York: Academic Press 1981.

[7] Kulisch, U., Miranker, W.: A new approach to scientific computation. New York: Academic Press 1983.

[8] Mayer, G.: Enclosing the solution of linear systems with inaccurate data by iterative methods based on incomplete LU-decompositions. Computing 35, 189–206 (1987).

[9] Neumaier, A.: New techniques for the analysis of linear interval equations. Lin. Alg. Appl. 87, 155–179 (1987).

[10] Neumaier, A.: Interval methods for systems of equations. Cambridge: Cambridge University Press 1990.

[11] Schwandt, H.: An interval arithmetic approach for the construction of an almost globally convergent method for the solution of the nonlinear Poisson equation on the unit square. SIAM J. Sci. Stat. Comp. 5, 427–452 (1984).

[12] Schwandt, H.: Interval arithmetic for systems of nonlinear equations arising from discretizations of quasilinear elliptic and parabolic partial differential equations. Appl. Num. Math. 3, 257–287 (1987).

[13] Schwandt, H.: Cyclic reduction for tridiagonal systems of equations with interval coefficients of vector computers, SIAM J. Num. Anal. 26, 661–680 (1989).

[14] Swarztrauber, P.: Vector and parallel methods for the direct solution of Poisson's equation. J. Comp. Appl. Math. 27, 241–263 (1989).

[15] Swarztrauber, P.: The methods of cyclic reduction, Fourier analysis and the FACR algorithm for the discrete solution of Poisson's equations on a rectangle. SIAM Review 19, 490–501 (1977).

[16] Sweet, R.: A generalized cyclic reduction algorithm. SIAM J. Num. Anal. 11, 506–520 (1974).

[17] Sweet, R.: A cyclic reduction algorithm for solving block tridiagonal systems of arbitrary dimension. SIAM J. Num. Anal. 14, 706–720 (1977).

[18] Varga, R.: Matrix iterative analysis. Englewood Cliffs, New Jersey: Prentice-Hall 1962.

H. Schwandt
Fachbereich Mathematik
MA 6-4
Technische Universität Berlin
Strasse des 17. Juni 136
D-W-1000 Berlin 12
Federal Republic of Germany

Computing, Suppl. 9, 233–246 (1993)

© Springer-Verlag 1993
Printed in Austria

On the Existence and the Verified Determination of Homoclinic and Heteroclinic Orbits of the Origin for the Lorenz Equations*

H. Spreuer and **E. Adams**, Karlsruhe

Dedicated to Professor U. Kulisch on the occasion of his 60th birthday

Abstract — Zusammenfassung

On the Existence and the Verified Determination of Homoclinic and Heteroclinic Orbits of the Origin for the Lorenz Equations. For suitable choices of the parameters, the Lorenz ODEs possess (i) stable and unstable manifolds of the stationary point 0 at the origin, (ii) a homoclinic orbit of 0, and (iii) a heteroclinic orbit connecting a periodic orbit with 0. With the exception of only partial results regarding (iii), all addressed orbits are enclosed and verified as follows: (a) enclosures of truncated series expansions and of their remainder terms yield guaranteed starting intervals at some distance from 0, whose width is not more than two units of the last mantissa digit, and (b) a step-size controlled version of Lohner's enclosure algorithm for IVPs yields the continuations.

AMS Subject Classification: 34C35, 34C37

Key words: Enclosure methods, stable and unstable manyfolds.

Zur Existenz und verifizierten Bestimmung homokliner und heterokliner Orbits des Ursprungs für die Lorenz-Gleichungen. Für passende Werte der Parameter besitzen die DGln des Lorenzproblems (i) stabile und instabile Mannigfaltigkeiten des stationären Punktes $0 = (0, 0, 0)$, (ii) einen homoklinen Orbit zu 0 und (iii) einen heteroklinen Orbit, der eine periodische Lösung mit 0 verbindet. Mit Ausnahme von nur Teilergebnissen bzgl. (iii) sind alle erwähnten Orbits folgendermaßen eingeschlossen und verifiziert: (a) Einschluß abgebrochener Reihenentwicklungen samt Reihenrest liefert garantierte Startintervalle, die den stationären Punkt 0 nicht enthalten, und deren Weite nicht mehr als 2 Einheiten der letzten Mantissenstelle beträgt, und (b) eine schrittweitenkontrollierte Version des Lohnerschen Einschließungs-algorithmus für AWA liefert die Fortsetzung.

1. Introduction

Generally, the Lorenz equations (E. N. Lorenz [9], 1963) are believed to be a classical paradigm for Dynamical Chaos in continuous processes. These ordinary differential equations (ODEs) are given by

$$
\begin{aligned}
&\text{(i)} \quad x' = -\sigma x + \sigma y, \\
&\text{(ii)} \quad y' = rx - y - xz, \\
&\text{(iii)} \quad z' = -bz + xy;
\end{aligned}
\tag{1.1}
$$

here b, r, and σ are positive parameters. For the properties listed in the remainder of this paragraph, see Sparrow's monograph [12]. The ODEs (1.1) possess the following stationary points: (i) for all $b, r, \sigma \in \mathbb{R}^+$ the origin $0 := (0, 0, 0)^T$ and (ii) for

---

* Received September 28, 1992; revised December 16, 1992.

all b, $r - 1$, $\sigma \in \mathbb{R}^+$, the points C_1 and C_2 given by $(\pm\xi, \pm\xi, r - 1)^T$ with $\xi :=$ $\sqrt{b(r - 1)}$. For $r > 1$, the stationary point 0 possesses a two-dimensional stable manifold, M^s, and a one-dimensional unstable manifold, M^u. For $b = 8/3$, $\sigma = 10$, and $r > 24.74\ldots$, C_1 and C_2 possess two-dimensional manifolds M^u and one-dimensional manifolds M^s.

Remarks: 1) The global existence of these manifolds follows from the Center Manifold Theorem [4].

2) For arbitrary choices of $b = 2\sigma \in \mathbb{R}^+$ and $r \in \mathbb{R}^+$, W. F. Ames [1] has derived a representation of the set of solutions whose topographical simplicity excludes the properties of Dynamical Chaos or the existence of the manifolds to be discussed subsequently.

For $b = 8/3$, $r = 28$, and $\sigma = 6$, W. Kühn [6] (see also [2]) has (i) verified and enclosed five different periodic solutions of (1.1) and (ii) shown that their instability is consistent with the existence of (two-dimensional) manifolds M^s and M^u attached to these orbits.

For any true solution, the existence of manifolds M^s and M^u signifies its instability. The application of traditional numerical methods then is correspondingly unreliable, see [1] and [2] for possible consequences. So far in literature, quantitative work regarding solutions of (1.1) has generally been confined to applications of methods of this kind.

In the presence of manifolds M^s and M^u, there is the possibility of the existence of heteroclinic or homoclinic orbits with respect to either a stationary point or a periodic solution; they are characterized as follows:

- a heteroclinic orbit connects two different solutions of these kinds;
- a homoclinic orbit is a loop attached to one of these solutions.

For $b = 8/3$, $\sigma = 10$, and a certain value of r, the existence of a homoclinic orbit of the stationary point 0 is subsequently verified. Everywhere this orbit represents the intersection of the manifolds M^s and M^u of the point 0. Additional results (with only a partial verification) pertain to a heteroclinic orbit from one of the periodic solutions to the stationary point 0. This orbit is represented by an intersection of the manifold M^s of the point 0 with the manifold M^u of the periodic solution.

The following holds for manifolds M^s and M^u and heteroclinic and homoclinic orbits: it takes an unbounded time

- to depart from a stationary point or a periodic solution and
- to arrive at such a true solution.

For ODEs such as (1.1), the set of true solutions is said to be chaotic if there exists a transversal homoclinic orbit, e.g. [4]. In fact, an orbit of this kind is related to the existence of a horseshoe map [4]. So far in literature, the existence of an orbit of this kind has been verified only for a few low-order systems of ODEs, see [4] and [5]. They do not include the Lorenz equations (1.1).

Concerning systems of ODEs such as (1.1), a strange attractor consists of individual unstable true solutions, particularly,

- stationary points or periodic solutions possessing manifolds M^s and M^u;
- unstable portions of these manifolds or of homoclinic or heteroclinic orbits.

In mathematical literature, 'geometric models' (the 'Lorenz attractor') of the Lorenz Eq. (1.1) have been investigated, e.g. [4, p. 273–279]. According to Sparrow [12, p. 43]: "There is now a considerable rigorous mathematical literature on 'Lorenz attractors' ... but this literature is, unfortunately, not necessarily of direct relevance to the Lorenz flow. The 'Lorenz attractors' that are well understood ... occur in model flows that are constructed to have certain properties."

On the basis of traditional numerical methods, there are numerous conjecture-like assertions in Sparrow's monograph [12] on the Lorenz Eq. (1.1), particularly on

- the manifolds M^s and M^u of the origin 0,
- homoclinic or heteroclinic orbits, and
- the strange attractor of (1.1).

In view of the unreliability of these numerical results, it is desirable to apply the (computer-implemented) Karlsruhe Enclosure Methods for the execution of the following tasks:

(α) the totally error-controlled determination of enclosures of individual true solutions,

(β) merged with the automatic verification of their existence;

(γ) the performance of (α) and (β) especially in the case of a heteroclinic and a homoclinic orbit, both attached to the stationary point 0.

The Karlsruhe Enclosure Methods rest on the Kulisch Computer Arithmetic, e.g. [7]. For the subsequent applications of these methods with respect to (1.1), the enclosure algorithms to be employed have been developed as follows:

(a) for solutions of initial value problems (IVPs) by R. Lohner [8],

(b) for periodic solutions of IVPs by W. Kühn [6] (see also [2]);

(c) a supplemental step size control regarding (a) by W. Rufeger [10] (see also [11]), and

(d) the conversion of the codes for (a) and (c) to double precision by S. Beermann [3].

The stationary point 0 of (1.1) is not suitable as a starting vector for the enclosure of the manifold M^u of the point 0 or any orbit on the corresponding manifold M^s. The first author of the present paper has developed the absolutely convergent series expansions (2.2)–(2.7) whose (suitable) interval evaluations yield verified and arbitrarily small intervals at some distance from the point 0. These expansions will be presented in Section 2; they will be employed in Section 4 for the purpose of a determination of intervals which are guaranteed to contain at least one point of M^s or M^u, respectively. These intervals then serve as starting intervals for the determination of verified enclosures of orbits on M^s or M^u, respectively, making use of Section 3.

2. Local Series Expansions of the Manifolds M^s and M^u of the Origin

Regarding the stationary point $0 = (0, 0, 0)^T$, the linear variational system of (1.1) possesses the eigenvalues

$$-b < 0 \text{ and } \lambda_\pm := \tfrac{1}{2}(-(\sigma + 1) \pm \sqrt{(\sigma + 1)^2 + 4\sigma(r - 1)}) \Rightarrow \lambda_+ > 0 \text{ and } \lambda_- < 0. \tag{2.1}$$

Consequently, 0 is a saddle point provided $r > 1$. The Center Manifold Theorem [4, p. 127] then assures the existence of manifolds M^s and M^u. Because of (2.1), M^s is two-dimensional and M^u is one-dimensional.

For $t \geq 0$ and $\lambda = \lambda_-$, the following series are chosen as candidates for a representation of the true solutions $(x^*, y^*, z^*)^T \in M^s$ of (1.1):

$$x(t) := \sum_{m=0}^{\infty} \sum_{n=0}^{\infty} \alpha_{mn} \exp(((2m + 1)\lambda - nb)t) \text{ with free } \alpha_{mn} \in \mathbb{R}. \tag{2.2}$$

$$y(t) := \sum_{m=0}^{\infty} \sum_{n=0}^{\infty} \beta_{mn} \exp(((2m + 1)\lambda - nb)t) \text{ with free } \beta_{mn} \in \mathbb{R}, \tag{2.3}$$

and

$$z(t) := \gamma_{01} \exp(-bt) + \sum_{m=1}^{\infty} \sum_{n=0}^{\infty} \gamma_{mn} \exp((2m\lambda - nb)t) \text{ with free } \gamma_{01}, \gamma_{mn} \in \mathbb{R}. \tag{2.4}$$

Since $\lambda := \lambda_- < 0$, the exponents in (2.2)–(2.4) are non-positive for $t \geq 0$. The following properties (i)–(v) will be shown subsequently:

(i) all α_{mn}, β_{mn}, and γ_{mn} can be expressed in terms of α_{00} and γ_{01};
(ii) provided $|\alpha_{00}|$, $|\gamma_{01}| > 0$ are sufficiently small, the series (2.2)–(2.4) converge absolutely for $t = 0$ and, therefore, for all $t \geq 0$;
(iii) for $t \to \infty$, they approach the stationary point $0 = (0, 0, 0)^T$;
(iv) for t sufficiently large or $|\alpha_{00}|$, $|\gamma_{01}|$ sufficiently small, the series consist of terms which are linear in α_{00} or γ_{01} with nonlinear remainder terms of the order $0(\alpha_{00}^2)$, etc;
(v) consequently, variations of $(\alpha_{00}, \gamma_{01})^T$ with sufficiently small $|\alpha_{00}|$, $|\gamma_{01}| > 0$ yield points $(x(t), z(t), y(t))^T \in M^s$ covering a two-dimensional finite small neighborhood of $(0, 0, 0)^T$ with this point at its center.

Remark: Substitutions $\hat{\alpha}_{00} := \alpha_{00} \exp(\lambda_- t_0)$ and $\hat{\gamma}_{01} := \gamma_{01} \exp(-bt_0)$ assign an arbitrary time t_0 to any selected point of an orbit on the manifold M^s. The individual orbits on M^s are determined by the vector $(\alpha_{00}, \gamma_{01})^T$ provided the series converge.

For a corresponding representation of the true solution coinciding with M^u, the following series expansions are chosen for $t \leq 0$ and $\lambda = \lambda_+$:

$$x(t) := \sum_{m=0}^{\infty} \alpha_{m0} \exp((2m + 1)\lambda t), \tag{2.5}$$

$$y(t) := \sum_{m=0}^{\infty} \beta_{m0} \exp((2m + 1)\lambda t), \tag{2.6}$$

and

$$z(t) := \sum_{m=1}^{\infty} \gamma_{m0} \exp(2m\lambda t). \tag{2.7}$$

The coefficients in (2.5)–(2.7) are different from the ones in (2.2)–(2.4). In a forthcoming paper [13], it will be shown

(vi) that all α_{m0}, β_{m0} and γ_{m0} can be expressed in terms of α_{00};
(vii) provided $|\alpha_{00}| > 0$ is sufficiently small, the series (2.5)–(2.7) converge absolutely for $t = 0$ and, therefore, for all $t \leq 0$;
(viii) for $t \to -\infty$, they approach the stationary point $0 = (0,0,0)^T$;
(ix) for $t < 0$ with $|t|$ sufficiently large or for $|\alpha_{00}|$ sufficiently small, the series consist of terms which are linear in α_{00} and nonlinear remainder terms.

For (1.1) and (2.2)–(2.4), comparisons of coefficients are to be carried out with respect to

($\hat{\alpha}$) $\exp((2m + 1)\lambda - nb)t)$ in the ODEs (1.1i, ii) and $\left.\right]$ for $\lambda = \lambda_-$ and

($\hat{\beta}$) $\exp((2m\lambda - nb)t)$ in the ODEs (1.1iii). $\left.\right\}$ all $m + 1, n + 1 \in \mathbb{N}$.

In the *case of* ($\hat{\alpha}$), this yields the following linear systems:

$$\left\{ \begin{array}{l} (\lambda + \sigma)\alpha_{00} - \sigma\beta_{00} = 0 \\[2mm] -r\,\alpha_{00} + (\lambda + 1)\beta_{00} = 0 \end{array} \right\} \quad \text{for } m = n = 0 \tag{2.8}$$

and, by means of the Kronecker symbol δ_{0n} (in order to avoid a distinction of cases),

$$\left\{ \begin{array}{l} ((2m + 1)\lambda - nb)\alpha_{mn} - \sigma(\beta_{mn} - \alpha_{mn}) = 0 \\[3mm] ((2m + 1)\lambda - nb)\beta_{mn} - r\alpha_{mn} + \beta_{mn} = -\alpha_{m,n-1}\gamma_{01}(1 - \delta_{0n}) - \sum_{\mu=0}^{m-1}\sum_{\nu=0}^{n}\alpha_{\mu\nu}\gamma_{m-\mu,n-\nu} \\[3mm] \text{for } (m + 1), (n + 1) \in \mathbb{N} \text{ such that } (m, n) \neq (0, 0). \end{array} \right.$$

$$\tag{2.9}$$

Since $\lambda = \lambda_-$, the determinant of (2.8) vanishes and

$$\beta_{00} = \frac{\lambda + \sigma}{\sigma}\alpha_{00} \text{ with } \alpha_{00} \in \mathbb{R} \text{ free}. \tag{2.10}$$

Since $\lambda = \lambda_-$, the determinant of (2.9) is positive for all $m + 1, n + 1 \in \mathbb{N}$ such that $(m, n) \neq (0, 0)$. The first equation in (2.9) yields

$$\beta_{mn} = -\left(\frac{(2m + 1)|\lambda| + nb}{\sigma} - 1\right)\alpha_{mn} \qquad \text{for } m + 1, n + 1 \in \mathbb{N}. \tag{2.11}$$

The second equation in (2.9) then yields

$$\left\{ \begin{array}{l} \alpha_{mn} = \dfrac{-\alpha_{m,n-1}\gamma_{01}(1 - \delta_{0n}) - \sum_{\mu=0}^{m-1}\sum_{\nu=0}^{n}\alpha_{\mu\nu}\gamma_{m-\mu,n-\nu}}{((2m + 1)|\lambda| + nb - 1)[((2m + 1)|\lambda| + nb)/\sigma - 1] - r} \\[4mm] \text{for } m + 1, n + 1 \in \mathbb{N} \text{ with } (m, n) \neq (0, 0). \end{array} \right. \tag{2.12}$$

In the case of $(\hat{\beta})$ and $\lambda = \lambda_-$, a comparison of coefficients yields

$$\begin{cases} (2m\lambda - nb + b)\gamma_{mn} = -\sum_{\mu=0}^{m-1}\sum_{v=0}^{n}\alpha_{m-1-\mu,n-v}\left(\dfrac{(2\mu+1)|\lambda|+vb}{\sigma}-1\right)\alpha_{\mu v} \\ \text{for } m, n+1 \in \mathbb{N} \text{ with } \gamma_{01} \in \mathbb{R} \text{ free.} \end{cases} \qquad (2.13)$$

Lemma 2.14: *Provided* $\sigma > 1$ *and* $|\lambda|/\sigma - 1 \le 2|\lambda| - b$ *with* $\lambda := \lambda_-$,

$$2m\lambda - (n-1)b < 0 \qquad (2.14a)$$

and

$$|(|\lambda|/\sigma - 1)/(2|\lambda| - b)| \le 1. \qquad (2.14b)$$

Proof: Because of (2.1) and the assumptions, there follow

$$0 < 1/\sigma < |\lambda|/\sigma - 1 \le 2|\lambda| - b \qquad (2.14c)$$

and therefore (2.14b). Whereas (2.14a) is obvious for $n \ge 1$, the verification of (2.14a) for $n = 0$ makes use of (2.14c). $\qquad\square$

Theorem 2.15: *If* $|\alpha_{00}| \le K$ *and* $|\gamma_{01}| \le L$, *then*

$$\begin{cases} |\alpha_{mn}| \le A_{mn} := K^{2m+1}L^n\dfrac{(2m+1)^n}{n!}\dfrac{1}{A^n} \quad \text{for } m+1, n+1 \in \mathbb{N}, \\ \text{where } A := b\,\mathrm{Min}\{A^* + b/\sigma,\, 3A^* + 12|\lambda|/\sigma\} \overset{!}{>} 0 \text{ and} \\ A^* := 2|\lambda|/\sigma - (1 + 1/\sigma) \end{cases} \qquad (2.16)$$

and

$$|\gamma_{mn}| \le B_{mn} := K^{2m}L^n\dfrac{(2m)^n}{n!}\dfrac{m}{A^n} \qquad \text{for } m, n+1 \in \mathbb{N}. \qquad (2.17)$$

Proof by induction: There hold $|\alpha_{00}| \le K$ and, because of (2.13) and (2.14b),

$$|\gamma_{10}| = \alpha_{00}^2\left|\frac{|\lambda|/\sigma - 1}{2|\lambda| - b}\right| \le \alpha_{00}^2 \le K^2. \qquad (2.18)$$

These inequalities are consistent with (2.16), (2.17), respectively. The inequalities for $|\alpha_{mn}|$ and $|\gamma_{mn}|$ are assumed to be valid as follows:

$$|\alpha_{\mu v}| \le A_{\mu v} \qquad \text{for } 0 \le \mu \le m, \quad 0 \le v \le n \qquad \text{with } (\mu, v) \ne (m, n) \qquad (2.19)$$

and

$$|\gamma_{\mu v}| \le B_{\mu v} \qquad \text{for } 1 \le \mu \le m, \quad 0 \le v \le n \qquad \text{with } (\mu, v) \ne (m, n). \qquad (2.20)$$

In order to verify (2.17), an auxiliary parameter $C_{\mu v}$ satisfies

$$C_{\mu v} := \left|\left(\frac{(2\mu+1)|\lambda|+vb}{\sigma}-1\right)\Big/(2m|\lambda|+(n-1)b)\right| \le 1 \qquad (2.21)$$

because of (2.14). As a consequence of (2.13), and (2.19)–(2.20),

$$\begin{cases} |\gamma_{mn}| \le K^{2m}(L/A)^n \sum_{\mu=0}^{m-1} \sum_{\nu=0}^{n} \frac{(2(m-\mu)-1)^{n-\nu}}{(n-\nu)!} \frac{(2\mu+1)^{\nu}}{\nu!} C_{\mu\nu} \\ \qquad \le K^{2m}(L/A)^n m(2m)^n/n!. \end{cases} \tag{2.22}$$

Therefore, (2.17) has been verified. In order to verify (2.16) in the case of $m = 0$, there holds for an auxiliary parameter B

$$B := \left| \frac{nA}{(|\lambda| + nb - 1)((|\lambda| + nb)/\sigma - 1) - r} \right| \le 1 \tag{2.23}$$

because of (2.14c). As a consequence of (2.12), (2.19), and $|\gamma_{01}| \le L$,

$$|\alpha_{0n}| \le K(L/A)^n B/n! \le K(L/A)^n/n!. \tag{2.24}$$

Therefore, (2.16) has been verified in the case of $m = 0$. In order to verify (2.16) for $m, n + 1 \in \mathbb{N}$, an auxiliary parameter D is introduced as follows:

$$\begin{cases} D := \left(\frac{m(m+1)}{2} + \frac{An}{2m+1} \right) \Big/ \left(\left((2m+1)|\lambda| + nb - 1 \right) E - r \right) \le 1 \text{ with} \\ E := \frac{(2m+1)|\lambda| + nb}{\sigma} - 1. \end{cases} \tag{2.25}$$

The last inequality can be verified analogously to the one in (2.23). As a consequence of (2.12), and (2.18)–(2.20),

$$|\alpha_{mn}| \le K^{2m+1}(L/A)^n(2m+1)^n D/n! \le K^{2m+1}(L/A)^n(2m+1)^n/n!. \tag{2.26}$$

Therefore, (2.16) has been verified generally. $\square$

Provided the convergence of (2.2)–(2.4) has been shown for $t = 0$, then this property is valid for all $t > 0$.

Lemma 2.27: *If $K \exp(L/A) < 1$, then $\sum_{m=0}^{\infty} \sum_{n=0}^{\infty} \alpha_{mn}$ is absolutely convergent.*

Proof: Because of (2.26),

$$\sum_{m=0}^{\infty} \sum_{n=0}^{\infty} |\alpha_{mn}| \le \sum_{m=0}^{\infty} K^{2m-1} \sum_{n=0}^{\infty} ((L/A)^n(2m+1))^n/n! = \sum_{m=0}^{\infty} (K \exp(L/A))^{2m+1}.$$
$$\square \tag{2.28}$$

Remarks: 1) The condition of Lemma 2.27 is satisfied e.g. if

$$|\alpha_{00}| \le K \le 1/3 \text{ and } |\gamma_{01}| \le L \le A. \tag{2.29}$$

2) Provided the conditions on b and σ in Lemma 2.14 are satisfied, then this is also true for the condition $A \overset{!}{>} 0$; i.e., there is always a positive $A = A(b, r, \sigma)$. As an example, $A(8/3, 28, 6) > 12.9$ and $A(8/3, 28, 10) > 9.95$.

3) There are constants $a_{mn}, b_{mn}, c_{mn} \in \mathbb{R}$ not depending on α_{00} and γ_{01} such that

$$\begin{cases} \alpha_{mn} = a_{mn}\alpha_{00}^{2m+1}\gamma_{01}^n, \\ \beta_{mn} = b_{mn}\alpha_{00}^{2m+1}\gamma_{01}^n, \\ \gamma_{mn} = c_{mn}\alpha_{00}^{2m}\gamma_{01}^n. \end{cases} \tag{2.30}$$

4) The substitution $u := \alpha_{00} \exp(\lambda t)$ and $v := \gamma_{01} \exp(-bt)$ with $\lambda = \lambda_- < 0$ transforms (2.2)–(2.4) into series proceeding in powers of u and v. These series converge absolutely for $|u| \le 1/3$ and $|v| \le A$.

5) See [13] for more details of the presented analysis.

For practical purposes, the series (2.2)–(2.4) are truncated such that $m = 0(1)M$ and $n = 0(1)N$ with fixed $M, N \in \mathbb{N}$. The corresponding remainder terms can be estimated provided $|K| \le 1/3$ and $L \le A$. This estimate will now be presented for the case of (2.2), making use of the bound A_{mn} as introduced in (2.16). The following table displays the values to be determined:

$$
\begin{array}{cccccc}
\alpha_{00} & \cdots & \alpha_{0N} & A_{0,N+1} & A_{0,N+2} & \cdots \\
\vdots & & \vdots & \vdots & \vdots & \\
\alpha_{M0} & \cdots & \alpha_{MN} & A_{M,N+1} & A_{M,N+2} & \cdots
\end{array}
\tag{2.31}
$$

$$
\begin{array}{ccccc}
A_{M+1,0} & \cdots & A_{M+1,N} & A_{M+1,N+1} & \cdots \\
\vdots & & \vdots & \vdots &
\end{array}
$$

Since the exponents in (2.2) are non-positive for $t \ge 0$, there is the following estimate concerning (2.2) for $\mu \ge M + 1$, making use of (2.16):

$$
\left\{
\begin{aligned}
\sum_{\mu=M+1}^{\infty} \sum_{v=0}^{\infty} A_{\mu v} &= \sum_{\mu=M+1}^{\infty} K^{2\mu+1} \sum_{v=0}^{\infty} (L/A)^v (2\mu + 1)^v / v! \\
&= [K \exp(L/A)]^{2m+3} / [1 - (K \exp(L/A))^2].
\end{aligned}
\right.
\tag{2.32}
$$

The convergence of the series (2.32) will now be considered for any fixed $\mu \in \{0,\ldots,M\}$. Truncation of this auxiliary expansion at $v = N$ yields the following estimate of the remainder term:

$$
R_{N\mu} := \sum_{v=N+1}^{\infty} A_{\mu v} = K^{2\mu+1} \sum_{v=N+1}^{\infty} (L/A)^v (2\mu + 1)^v / v!.
\tag{2.33}
$$

Obviously, $R_{N\mu}$ is the remainder term of the power series representing $K^{2\mu+1} \exp(2\mu + 1)L/A$. The usual estimate for the corresponding remainder term yields

$$
R_{N\mu} \le K^{2\mu+1} [(2\mu + 1)L/A]^{N+1} [\exp((2\mu + 1)L/A)]/(N + 1)!.
\tag{2.34}
$$

Additionally to (2.32), there is consequently the following upper bound concerning the truncation of (2.2):

$$
\sum_{\mu=0}^{M} K^{2\mu+1} [(L(2\mu + 1)/A)^{N+1}]/(N + 1)! \exp(L(2\mu + 1)/A).
\tag{2.35}
$$

Remarks: 1) Regarding the truncation of (2.2), an upper bound of the corresponding error is represented by the sum of the expressions in (2.32) and (2.35). Provided *either* M and N are sufficiently large and $|\alpha_{00}| \le 1/3$, $|\gamma_{01}| \le A$ in (2.16), *or* $|\alpha_{00}|$ and $|\gamma_{01}|$ are sufficiently small, the sum of the estimates in (2.32) and (2.35) can be made arbitrarily small.

2) Concerning a determination of α_{mn} by means of (2.12) and γ_{mn} by means of (2.13), the cost is governed by two-dimensional convolutions with $m^2(n+1)^2$ multiplications and additions. Consequently, it is advantageous to determine the coefficients $\alpha_{\mu\nu}$ and $\gamma_{\mu\nu}$ within triangular tables instead of the rectangular one in (2.31).
3) The expansions (2.3) and (2.4) are treated analogously.

The preceding analysis can be summarized as follows:

Theorem 2.36: (A) *The series expansions* (2.2)–(2.4) *of the solutions of* (1.1) *converge for all* $t \geq 0$ *provided* (i) $\sigma > 1$ *and* $|\lambda|/\sigma - 1 \leq 2|\lambda| - b$ *where* $\lambda = \lambda_-$ *in* (2.1) *and* (ii) $|\alpha_{00}| \leq K \leq 1/3$ *and* $|\gamma_{01}| \leq L \leq A$.

(B) *As a consequence of* (i), *there is always a positive* $A = A(b, r, \sigma)$.
(C) *Upon truncation of the series expansions* (2.2)–(2.4), *the corresponding remainder terms can be made arbitrarily small through appropriate choices of* M *and* N *or* α_{00} *and* γ_{01}.

Remark: Condition (ii) of Theorem 2.36 is satisfied for all choices of

$$(\alpha_{00}, \gamma_{01})^T \text{ for instance in the domain } \left\{ (\alpha_{00}, \gamma_{01})^T \left| \left(\frac{\alpha_{00}}{K} \right)^2 + \left(\frac{\gamma_{01}}{L} \right)^2 \leq 1 \right. \right\}. \quad (2.37)$$

3. Enclosure Method for Orbits on the Manifolds M^s and M^u of the Origin

On the basis of Theorem 2.36, now

(a) a vector $(\alpha_{00}, \gamma_{01})^T$ consistent with (2.37) is chosen;
(b) intervals $[x(0)]$, $[y(0)]$, and $[z(0)]$ not containing zero are determined;
(c) at $t = 0$, these intervals contain that true solution on M^s which is determined by the choice of $(\alpha_{00}, \gamma_{01})^T$.

Under the conditions of Theorem 2.36, the error due to truncating the series (2.2)–(2.4) can be made arbitrarily small provided

 (i) M and N in (2.31) are sufficiently large and,
(ii) regarding (b), the rounding errors are sufficiently small by means of a corresponding floating point number format. For this purpose an arithmetic by W. Krämer has been used, which will appear in ZAMM, volume 73.

Starting intervals for applications of Lohner's enclosure algorithm can be determined as follows:

- the series expansions (2.2)–(2.4) are suitably truncated and enclosed, together with the corresponding remainder terms;
- this can always be executed such that the widths of the resulting intervals do not exceed one or at most two units of the last mantissa place, irrespective of the selected numerical precision.

For arbitrary but fixed and admissible choices of α_{00} and γ_{01}, therefore, portions of true orbits on M^s can be enclosed. Because of (v) subsequent to (2.4), the initial intervals $[x(t_0)]$ etc. can be determined everywhere on a closed curve on M^s surrounding $(0, 0, 0)^T$. The presented method for a totally error-controlled determination of (portions of) orbits on M^s will be applied in a few selected examples in Section 4. These examples demonstrate complicated topographical properties of M^s.

4. Computed Enclosures of Selected Orbits for the Lorenz Equations (1.1)

The purposes of the present section are

- a demonstration of the efficiency of the merged enclosure and verification method addressed before and
- its application to the (totally error-controlled) determination of selected orbits some of which have been approximated in literature, e.g., in Sparrow's monograph [12].

These orbits are on the manifolds M^s and M^u of the stationary point $0 = (0, 0, 0)^T$ and they pertain to suitable choices of the free parameters b, r and σ in (1.1). The enclosures to be presented rest on

- the analyzed series expansions (2.2)–(2.4) for orbits on M^s or
- the corresponding expansions (2.5)–(2.7) for M^u which have been analogously analyzed, see the forthcoming paper [13].

Figures 4.1, 4.3, and 4.4 display projections into coordinate planes of enclosures of orbits which have been determined by means of the enclosure algorithm with contributions by Lohner, Rufeger, and Beermann, see Section 1. The units of the scale are demarcated on the coordinate axes. Enclosures of the following kinds of orbits are presented in Figs. 4.1 or 4.3:

(P)   Kühn's two T_1-periodic orbits "starting" at intervals of width 10^{-10} as given by Kühn [6], see also [2];
(M) orbits on the manifold M^s or M^u of the stationary point 0, starting from intervals of widths less than 10^{-32} which have been determined by use of the series expansions (2.2)–(2.4) or (2.5)–(2.7), respectively.

For each one of the computed enclosures, the projections into the three coordinate planes are qualitatively similar; consequently, not all projections are presented here. In Figs. 4.1, 4.3, and 4.4, the computed enclosures are displayed by means of sequences of printed symbols, corresponding to the sequence of all executed time steps of the enclosure algorithm, with its control of the step size. For almost the total extensions of the displayed curves, the widths of the enclosures are smaller than 10^{-8}. Regarding the final displayed portions of several of the enclosures for orbits of type (M), the widths of the computed enclosures can be seen within graphical accuracy. Depending on more suitable choices of the artificial parameters of the employed enclosure algorithm, this growth can be delayed.

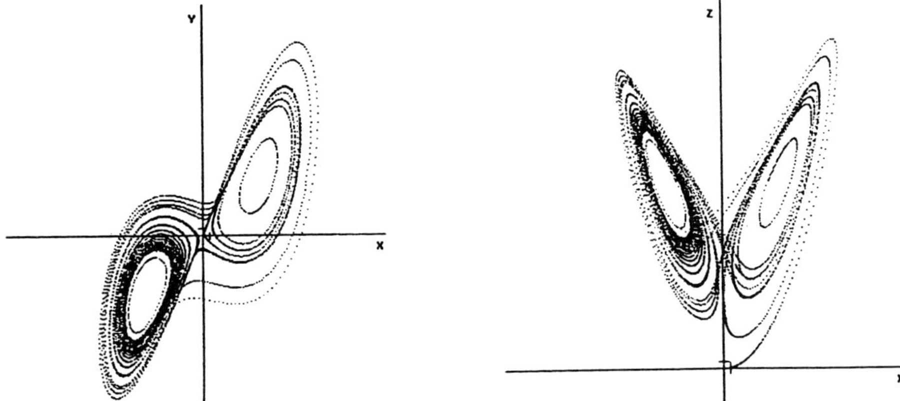

Figure 4.1a   **Figure 4.1b**

Projections of a leading portion of the (one-dimensional) unstable manifold M^u of the stationary point $(0,0,0)^T$

For the choices of $b = 8/3$, $r = 28$, $\sigma = 6$, and time t increasing, Figs. 4.1a and 4.1b display projections of a leading portion of one of the two branches of M^u, with, additionally, projections of the T_1-periodic orbits. Attached to the point 0, M^u in Fig. 4.1a is initially represented by the loop with the maximum value of y in the exhibited sequence of loops. Subsequently, M^u executes a "pseudo-random" winding pattern (see Sparrow [12, p. v] for this characterization) about the stationary points C_1 and C_2 defined in Section 1. Table 4.2 lists the numbers of consecutive penetrations of either the half-plane $y = y_{C1} > 0$ with $x \geq x_{C1}$ or the half-plane $y = y_{C2} < 0$ with $x \leq x_{C2}$:

Table 4.2. Numbers of penetrations regarding Fig. 4.1

Number of consecutive	$y = y_{C1}$	1	2	2	3	1	2	1	2	3	2	1
penetrations of half-planes	$y = y_{C2}$	20	1	7	1	1	6	1	2	1	2	

Remarks: *1*) Figures 4.1a and 4.1b do not exhibit all loops covered in Table 4.2.

2) Qualitatively similar projections of M^u were obtained for the choices of $b = 8/3$, $r = 28$, and $\sigma = 10$.

3) For the enclosures depicted in Fig. 4.1, a starting interval with a width less than 10^{-32} was determined by use of the series expansions (2.5)–(2.7); the distance of this interval from $(0,0,0)^T$ is characterized by a vector of the following magnitude: $(10^{-2}, 10^{-2}, 10^{-5})^T$.

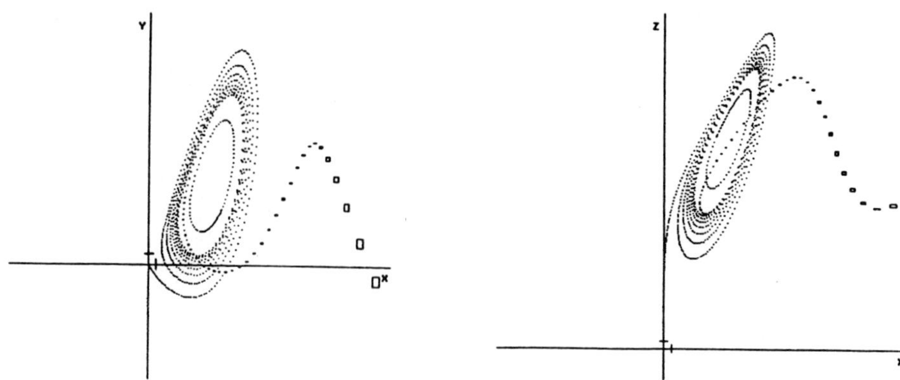

Figure 4.3a   **Figure 4.3b**
Projections of a final portion of an orbit on the (two-dimensional) stable manifold M^s of the stationary point $(0,0,0)^T$

For time t decreasing, Figs. 4.3a and 4.3b display projections of a final portion of an orbit on M^s that is determined by the following choices of the parameters: $b = 8/3, r = 28, \sigma = 6, \alpha_{00} = 3.2\ldots \times 10^{-6}$, and $\gamma_{01} = 1.87$. Additionally one of the T_1-periodic orbits is shown.

Remarks: 1) For the enclosures in the Figs. 4.1 and 4.3, a starting interval of width less than 10^{-32} was determined by means of the series expansions (2.2)–(2.4); the distance of this interval from $(0,0,0)^T$ is characterized by a vector of the following magnitude: $(10^{-6}, 10^{-6}, 10^{0})^T$.

2) Additional and still incomplete work suggests for $b = 8/3, r = 28$, and $\sigma = 6$ that there exists a heteroclinic orbit approaching

- the T_1-periodic orbit as $t \to -\infty$ and
- the stationary point 0 as $t \to \infty$.

The verification of this heteroclinic orbit involves series expansions corresponding to the ones in (2.2)–(2.7) in order to represent the (two-dimensional) manifold M^u of the T_1-periodic orbit, respectively.

For $b = 8/3, r \approx 13.926$, and $\sigma = 10$, Sparrow [12, p. 16] offers the conjecture that there is a homoclinic orbit attached to $0 = (0,0,0)^T$. If it exists and for its entire extension, this orbit is on the intersection of the manifolds M^s and M^u of the point 0. The existence of this homoclinic orbit has been verified as follows by the first author of the present paper:

(A) Any fixed choice of the free parameter α_{00} of the expansions (2.5)–(2.7) assigns a time t_0 to any selected point on M^u. For any choice of α_{00}, the coefficients in (2.5)–(2.7) are continuous functions of r. This is also true for the starting interval in $\mathbb{R}^3$ as determined by truncations of (2.5)–(2.7). The remainder terms can be made arbitrarily small; they can be estimated quantitatively.

(B) Any selected choices of the free parameters α_{00} and γ_{01} in the series expansions (2.2)–(2.4) determine a particular orbit on M^s and a time t_0 assigned to any selected point on this orbit. Corresponding to (A) and for any fixed choices of α_{00} and γ_{01}, truncations of the series expansion (2.2)–(2.4) are continuous functions of r.

(C) On the basis of (A), an interval can be determined which contains a portion of M^u. With this starting interval and by means of the enclosure methods addressed before, an enclosure can be determined which almost returns to the point 0, provided r is suitably chosen.

(D) On the basis of (B), the manifold M^s of the point 0 is a continuous function of r.

(E) Values $\underline{r}_0$ and $\bar{r}_0$ have been determined by trial and error such that, close to the point 0, the enclosures of the orbits, corresponding to the one addressed in (C), are on different sides of the manifold M^s. The continuous dependencies referred to in (A) or (B) imply the existence of an r^* such that the enclosed true solution coincides with a solution on M^s.

(F) Starting from $[\underline{r}_0, \bar{r}_0]$, a bisection method yields a sequence of intervals $[\underline{r}_v, \bar{r}_v]$ for $v \in \mathbb{N}$. Provided enclosure methods are used for the evaluations, $r^* \in [\underline{r}_v, \bar{r}_v]$ is true for all $v \in \mathbb{N}$. Consequently, this r^* can be enclosed by an interval of arbitrarily small width.

The execution of this bisection method has been terminated with the determination of the interval $[\underline{r}, \bar{r}] = 13.926557407520 + [1, 2]10^{-13} \ni r^*$. The (verified) enclosure of the homoclinic orbit is represented by a simple loop in $\mathbb{R}^3$, whose projection into the x-z-plane is displayed in Fig. 4.4. *This* homoclinic orbit does not possess the property of being transversal.

Remark: This homoclinic orbit is presumably unique, even though this property is not implied by the executed verification.

5. Concluding Remarks

Regarding heteroclinic or homoclinic orbits of the Lorenz Eq. (1.1), in literature there are only uncontrolled approximations without a verification of the existence. This paper presents

- a verification of the existence of a homoclinic orbit (and partial corresponding results for a heteroclinic orbit);
- the verification procedure is executed merged with a totally error-controlled quantitative determination of these orbits.

The employed verifying enclosure method consists of a combination of

- interval evaluations of the first author's series expansions (2.2)–(2.7) with
- applications of enclosure algorithms for ODEs which have been developed on the basis of the Kulisch Computer Arithmetic and as part of the Karlsruhe Enclosure Methods.

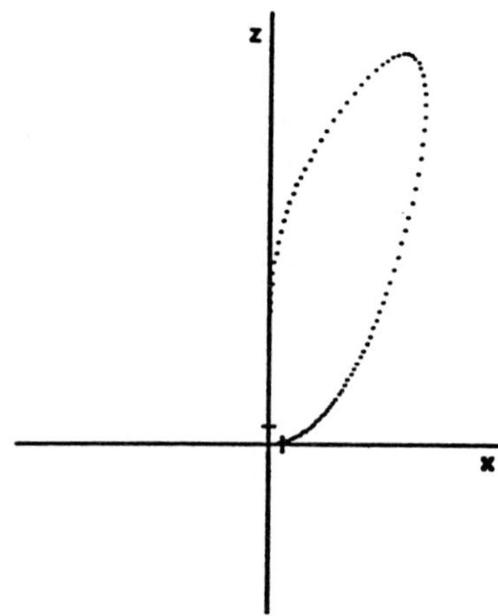

Figure 4.4. Projection into the x-z-plane of homoclinic orbit of the stationary point $(0, 0, 0)^T$. The units of the scale are demarcated on the coordinate axes

References

[1] Adams, E., Ames, W. F., Kühn, W., Rufeger, W., Spreuer, H.: Computational chaos may be due to a single local error. J. Comp. Physics *104*, 241–250 (1993).

[2] Adams, E.: The reliability question for discretizations of evolution problems. In: Adams, E., Kulisch, U. (eds.) Scientific computing with automatic result verification, pp. 423–526. Boston: Academic Press 1993.

[3] Beermann, S.: private communication.

[4] Guckenheimer, J., Holmes, P.: Nonlinear oscillations, dynamical systems, and bifurcations of vector fields, 2nd edn. New York: Springer 1983.

[5] Hale, J. K., Sternberg, N.: Onset of chaos in differential delay equations. J. Comp. Physics *77*, 221–239 (1988).

[6] Kühn, W.: Einschließung von periodischen Lösungen gewöhnlicher Differentialgleichungen und Anwendung auf das Lorenzsystem. Diploma Thesis. Karlsruhe, 1990.

[7] Kulisch, U. W., Miranker, W. L.: The arithmetic of the digital computer: A new approach. SIAM Review *28*, 1–40 (1986).

[8] Lohner, R.: Einschließung der Lösung gewöhnlicher Anfangs- und Randwertaufgaben und Anwendungen, Doctoral Dissertation, Karlsruhe, 1988.

[9] Lorenz, E. N.: Deterministic nonperiodic flow. J. Atmosph. Sc *20*, 130–141 (1963).

[10] Rufeger, W.: Numerische Ergebnisse der Himmelsmechanik und Entwicklung einer Schrittweitensteuerung des Lohnerschen Einschließungs-Algorithmus. Diploma Thesis, Karlsruhe, 1990.

[11] Rufeger, W., Adams, E.: A step-size control for Lohner's enclosure algorithm for ordinary differential equations with initial conditions. In: Adams, E., Kulisch, U. (eds.) Scientific computing with automatic result verifications, pp. 283–299. Boston: Academic Press 1993.

[12] Sparrow, C.: The Lorenz equations: Bifurcations, chaos, and strange attractors. New York: Springer 1982.

[13] Spreuer, H., Adams, E.: Existence and verified enclosures of heteroclinic and homoclinic orbits for the Lorenz equations (in press).

H. Spreuer
E. Adams
Institute for Applied Mathematics
The University of Karlsruhe
D-W-7500 Karlsruhe
Federal Republic of Germany

Computing, Suppl. 9, 247–263 (1993)

Verification in Computer Algebra Systems*

H. J. Stetter, Vienna

Dedicated to Professor U. Kulisch on the occasion of his 60th birthday

Abstract — Zusammenfassung

Verification in Computer Algebra Systems. In this paper, we have attempted to demonstrate that the question of *condition*, i.e. of the sensitivity of results w.r.t. perturbations of data, may play a role in algebraic algorithms, even if they are carried out in rational arithmetic. Poor condition is traced to a *near-degeneracy* of the situation specified by the data. Thus in a process called *verification* in this context, the presence of a genuinely degenerate problem *near* the specified problem should be discovered before or during the execution of the algorithm; the algorithm should switch to a stable modification in this case. Such modified versions are obtained by regarding the specified problem as a perturbation of the nearby degenerate one. Finally, it is indicated how these ideas may also lead to safe implementations of algebraic algorithms in floating-point arithmetic.

These ideas are developed considering the integration of rational functions, the choice of basis in multivariate polynomial interpolation, and the computation of zeros of multivariate polynomial systems.

AMS Subject Classification: 68Q40, 12Y05, 65H10, 65Y99

Key words: Computer algebra, sensitivity analysis, result verification.

Verifikation in Computeralgebra-Systemen. In dieser Arbeit versuchen wir zu zeigen, dass Fragen der Kondition, d.h. der Empfindlichkeit von Ergebnissen auf Datenstörungen, bei algebraischen Algorithmen eine Rolle spielen können, selbst wenn diese im Bereich der rationalen Zahlen ablaufen. Schlechte Kondition wird auf eine *Fast-Ausartung* der durch die Daten spezifizierten Situation zurückgeführt.

In einer, in diesem Zusammenhang als *Verifikation* bezeichneten Vorgehensweise sollte deshalb das Vorhandensein eines tatsächlich ausgearteten Problems *in der Nähe* des spezifizierten Problems vor oder während der Durchführung des Algorithmus entdeckt werden; in diesem Fall sollte der Algorithmus auf eine stabile Variante umschalten. Solche Varianten kann man erhalten, indem man das spezifizierte Problem als Störung des nahegelegenen ausgearteten Problems betrachtet. Schliesslich wird angedeutet, wie diese Überlegungen auch zu sicheren Gleitpunkt-Implementierungen von algebraischen Algorithmen führen können.

Diese Ideen werden entwickelt an Hand der Integration von rationalen Funktionen, der Basiswahl bei der multivariaten polynomialen Interpolation, und der Berechnung der Nullstellen von multivariaten polynomialen Systemen.

1. Introduction

At first sight, result verification and computer algebra appear to belong to two different worlds: result verification is an algorithmic tool to cope with the in-

---

* Received November 2, 1992; revised December 16, 1992.

securities of calculations in floating-point arithmetic, computer algebra systems employ symbolic manipulations or they work with rational numbers with potentially arbitrary large numerators and denominators so that rational operations are exact.

Yet, simple examples show that the results of computer algebra systems may have to be considered with care. Take, e.g., the integral

$$F(a) = \int_0^1 \frac{dx}{x^2 + 2ax + 2}; \tag{1.1}$$

any computer algebra system will deliver an arithmetic expression for $F(a)$ in no time, e.g. the expression (1.2) below. Naturally, this expression is fully exact in the mathematical sense. Let us now ask the system to evaluate (1.2) at a value of a near $\sqrt{2}$ and give us a decimal approximation of $F(a)$. Table 1 gives a list of results obtained on two current computer algebra systems at their basic numerical accuracy.

Table 1. Computed values of (1.2) at $a = \sqrt{2 + 10^{-2m}}$

m	DERIVE 2.01	Mathematica 2.0
4	.292894	.292893
6	.709546	.292893
8	.826640	.292893
10	0	.292893
12	0	.293043
14	0	.294209
16	0	0
18	0	0

Note that the values of the denominator of the integrand lie in $[2, 3 + 2a]$ and that F is arbitrarily smooth in the neighborhood of $\sqrt{2}$. Yet something goes wrong when a approaches $\sqrt{2}$, and a naive user who retrieves only one value of the integral may well be deceived into accepting a result which has no correct digit. In which sense could a concept of *verification* have been used in this context?

In our simple example, the cause of the inaccurate results is obvious: The *arithmetic expression*

$$\frac{1}{2\sqrt{a^2 - 2}} \left[\log \frac{1 + a - \sqrt{a^2 - 2}}{a - \sqrt{a^2 - 2}} - \log \frac{1 + a + \sqrt{a^2 - 2}}{a + \sqrt{a^2 - 2}} \right] \tag{1.2}$$

(or a similar one) which is generated as a *representation* of F is *not defined* at $a = \sqrt{2}$. Since it has the structure of a difference quotient, and since even a computer algebra system can only *approximate* logarithms of rational numbers by rational numbers, the large factor $1/\sqrt{a^2 - 2}$ amplifies these errors the more the closer a is to $\sqrt{2}$.

On the other hand, the character of computer algebra systems lets a naive user assume that accurate rational operations with rational numbers which approximate some exact values will contain the approximation error at the level of its generation.

A straightforward validation remedy would consist in the generation of lower and upper approximations of the logarithmic terms and the use of (rational) interval arithmetic in the subsequent operations. This would produce enormous intervals for the final result in the case $a \approx \sqrt{2}$ and thus reveal the difficulty.

However, such an *a-posteriori control mechanism* is not in the spirit of computer algebra systems which attempt to employ algebraic rather than arithmetic tools. This paradigm requests that the sources of potential difficulties should be *spotted a-priori* during the algebraic part of the computation. If no parameters are present, the computation should then be directed along an *alternate, safe route.* If the situation is parameter dependent, appropriate *caveats* should be generated (and displayed); they must then be checked by the system when values are substituted for the parameters so that the difficulty may be circumvented if necessary.

In computer algebra systems, such *a-priori control mechanisms* could play a role analogous to the role of the well-known verification tools in floating-point compution. We will therefore—tentatively—use the term *verification tools* for such mechanisms, which explains the title of this paper.

In the subsequent section 2, we will consider three typical contexts to show that the source of a potential ill-conditioning of an algebraic algorithm (into which an approximation of some sort is introduced in a later phase) is the *proximity of a degenerate situation,* and that a general remedy consists in regarding the problem at hand as a *perturbation* of the nearby degenerate problem. This approach is well-established in other areas of mathematics (singular perturbations, unfolding of singularities etc.).

In section 3, we will indicate how algebraic algorithms may be appended so that they produce the nearby degenerate situation if it exists. This is a first step towards the *verification* introduced above.

In the concluding section 4, we will consider how the introduction of such verification mechanisms may permit the safe use of floating-point arithmetic in algebraic algorithms.

2. Degeneracies in Algebraic Algorithms

2.1 Multiple Roots in Univariate Polynomials

The degeneracy responsible for the ill-behavior of the integration procedure exhibited in Table 1 is evident: For $a = \sqrt{2}$, the denominator of the integrand has a double root, i.e.

$$F(\sqrt{2}) = \int_0^1 \frac{dx}{(x + \sqrt{2})^2} = \frac{1}{2 + \sqrt{2}}, \tag{2.1}$$

which is integrated and evaluated by a computer algebra system without any difficulties.

For a value of a close to $\sqrt{2}$, $F(a)$ could have been represented as

$$F(a) = \int_0^1 \frac{dx}{(x+a)^2} + \int_0^1 \frac{dx}{(x+a)^4}(a^2 - 2) + \int_0^1 \frac{dx}{(x+a)^4(x^2 + 2ax + 2)}(a^2 - 2)^2.$$

$$(2.2)$$

The first two integrals in (2.2) are easily integrated symbolically and well-behaved. The remainder integral $R(a)$ can be bounded by bounding its integrand. If a is sufficiently distinct from $\sqrt{2}$ so that $R(a)$ is not negligible, the standard expression (1.2) for $F(a)$ will be sufficiently well-conditioned. Of course, the expression (2.2) is easily generated by algebraic computation, also with further terms if necessary.

Table 2. Values at $a = \sqrt{2 + 10^{-2m}}$

m	first two terms of (2.2)	$F(a)$
2	.292882	.292891
4	.292893	.292893
6	.292893	.292893
8	.292893	.292893
$\vdots$	$\vdots$	$\vdots$

The use of (2.2), *without* the remainder term, yields the values of $F(a)$ in Table 2, cf. Table 1.

Difficulties as experienced with (1.1) appear whenever the denominator p of a rational integrand has nearly multiple roots. For the integration of rational functions, verification in the sense of this paper will therefore consist in checking whether there exist polynomials $\tilde{p}$ *close* to p which have multiple roots, or roots of a multiplicity higher than that which is found for p.

Multiple roots of a univariate polynomial are algebraically exhibited through the computation of the greatest common divisor (g.c.d.) of the polynomial and its derivative. If we can establish a relation

$$p(x) = s(x) \cdot q(x) + r(x),$$
$$p'(x) = \bar{s}(x) \cdot q(x) + \bar{r}(x),$$

$$(2.3)$$

with 'small' polynomials r and $\bar{r}$ of a degree $< \deg(q)$, then the k-fold zeros of q are *nearly* $(k+1)$-fold zeros of p, $k \geq 1$, in the following sense:

- For each k-fold zero of q there exists a small neighborhood which contains $(k+1)$ zeros of p;
- In a small neighborhood of p there exists a polynomial $\tilde{p}$ which has $(k+1)$-fold zeros at each of the k-fold zeros of q.

In his forthcoming Ph.D. Thesis, V. Hribernig will prove a more formal and more precise statement of this fact. He will also consider ε-g.c.d.s of two unrelated univariate polynomials p_1 and p_2, viz. polynomials q_k, $\deg(q_k) = k \geq 0$, such that k

is the largest integer for which a relation

$$p_1(x) = s_1(x)q_k(x) + r_1(x),$$
$$p_2(x) = s_2(x)q_k(x) + r_2(x),$$

(2.4)

with r_1 and r_2 smaller than $\varepsilon > 0$ in a well-defined sense, holds. Such ε-$g.c.d.$s have previously been considered, in a rather vague way, by Sasaki-Noda [1] and Rahman [2].

For the purpose of symbolic integration, and probably also in other contexts, the polynomial p of (2.3) should be regarded as a perturbation of some neighboring polynomial $\tilde{p}$ which has $(k + 1)$-fold zeros close to each of the k-fold zeros of q. Essential for the success of this 'unfolding' is the use of a polynomial $\tilde{p}$ which represents the most degenerate situation to be found near p, not the precise choice of the coefficients (or zeros) of $\tilde{p}$.

E.g., in the case of (1.1), we could just as well have used

$$(x^2 + 2ax + 2)^{-1} = (x + \sqrt{2})^{-2} - 2x(a - \sqrt{2})(x + \sqrt{2})^{-4} + r(a)$$

(2.5)

or a perturbation of $(x + \alpha)^{-2}$, with $\alpha \in (\sqrt{2}, a)$.

2.2 Special Node Locations in Multivariate Interpolation

A linear space P of polynomials from $\mathbb{P}^s$ (real polynomials in s variables) is called *correct* for a set T of n points $t_i \in \mathbb{R}^s$ if the *multivariate interpolation problem*

$$\text{Find } p \in P \quad \text{such that } p(t_i) = w_i, \quad i = 1(1)n,$$

(2.6)

has a unique solution for arbitrary data $w \in \mathbb{R}^n$.

Multivariate polynomial interpolation has remained a strangely neglected subject in the numerical and analytical literature. A plausible reason may be the fact that the choice of a correct polynomial interpolation space P is far from unique and that it is difficult to specify natural conditions which restrict its choice. Furthermore, there are always special locations of the interpolation nodes for which an interpolation space which is correct for the same number of nodes in general position is no longer correct, and these locations may be far from obvious.

Quite recently, de Boor and Ron ([3], [4]) have presented an approach to multivariate polynomial interpolation which not only specifies a correct interpolation space P for any number and location of the nodes and for any dimension s, but also leads to interpolants which retain all the nice transformation and continuity properties which make univariate interpolation such a powerful tool.

While the approach of de Boor and Ron is strongly analytic in character, let us consider shortly the *algebraic structure* of the polynomial interpolation problem (2.6):

Let

$$\mathscr{F} := \{p \in \mathbb{P}^s : p(t_i) = 0, t_i \in T\}$$

(2.7)

be the *ideal* of all polynomials in s variables vanishing on T; then it is clear that, without the restriction $p \in P$, an interpolant is only determined modulo $\mathscr{F}$. Thus, uniqueness is guaranteed within the *residue class ring* R mod $\mathscr{F}$, and the choice of a particular correct interpolation space P is equivalent to the choice of a particular basis for the n-dimensional vector space R which implies a particular *representation* of R.

From this algebraic point of view, the apparent complications which arise for $s > 1$ dimensions are easily explained:

In $\mathbb{P}^1$, $\mathscr{F}$ is spanned by $\omega(t) = \prod_{i=1}^{n} (t - t_i)$ and thus contains only polynomials of a degree $\geq n$. R, on the other hand, may be represented as $\text{span}\{t^j, j = 0(1)n - 1\}$ so that P consists of all polynomials of degree $< n$. This clear separation of degrees makes any other choice of an interpolation space, e.g. $\{p(t) + \omega(t), p(t) \in \mathbb{P}_{n-1}\}$, unnatural.

In $\mathbb{P}^s$, $s > 1$, however, the structure of the ideal $\mathscr{F}$ may be rather complicated; in particular, it may contain polynomials of a total degree which is not higher or even lower than the largest total degree which must occur in R. Thus, in almost all cases, the selection of a basis for the representation of R takes on an arbitrary character because a 'shift' by some polynomial from $\mathscr{F}$ may not increase the highest total degree in R.

From the point of view of a constructive solution of (2.6), one should choose a basis $\{p_j, j = 1(1)n\}$ for R such the condition of the generalized *Vandermonde matrix*

$$V := (p_j(t_i), \, i, j = 1(1)n) \tag{2.8}$$

is minimal, i.e. one should orthogonalize *some* basis w.r.t. evaluation on T.

However, this would generally lead to curious basis polynomials. For $s > 1$, in particular for s not small, one will usually prefer a basis of *power products* (PPs)

$$x^j := x_1^{j_1} x_2^{j_2} \ldots x_s^{j_s}, \tag{2.9}$$

and one will simultaneously try to keep the highest total degree $|j| := \sum_\sigma j_\sigma$ small:

$$\max_{p_j \in P} (\text{total degree of } p_j) \quad \text{minimal}. \tag{2.10}$$

Under the assumption of rational nodes t_i, the following fully rational selection procedure will arrive at a correct basis satisfying (2.10):

In the set of *all* PPs (2.9), assume a linear order which is consistent with nondecreasing total degree. Consider the semi-infinite Vandermonde matrix V_∞ where the p_j are the PPs in the assumed order. By Gaussian elimination triangularize V_∞ from left to right; use row interchanges only where a pivot element is equal to zero. If no pivot element is available in the current column proceed to the next column which provides a non-zero pivot element (after row interchange if necessary). Use the PPs of the pivot columns to span the interpolation space P.

Example: Consider $s = 2$ and $n = 6$; the appropriate interpolation space for points in general position is

$$P = \mathrm{span}\{1, x_1, x_2, x_1^2, x_1 x_2, x_2^2\} = \mathbb{P}_2^2. \tag{2.11}$$

However, if all 6 nodes lie on a *conic* the columns of V_∞ pertaining to the above PPs must be linearly dependent and we will have to proceed further so that there will be at least one basis PP of degree 3 in the correct interpolation space.

In terms of our algebraic structure, we now have the *2nd* degree polynomial p_c defining the conic in the ideal (2.7) while any representation of R contains some *3rd* degree polynomials. Obviously, we may add multiples of p_c to the basis functions and/or to any specific interpolant without violating (2.6).

Although the fact that 6 points in 2-space lie on a common conic is only a slight degeneracy, it gives rise to the same *dilemma* that we have met previously:

If the 6 interpolation nodes lie *almost* on a common conic they are algebraically in general position and (2.11) may be used. However, since the associated Vandermonde matrix (2.8) would be singular if the t_i would move onto the common conic, V must be *extremely ill-conditioned* for the specified *near-conic* location of the nodes.

On the other hand, the *special* basis P_c which is selected for points *on* the conic will also serve very well for nearby points since the condition of V is a continuous function of the t_i except when V becomes singular.

Obviously, if multivariate interpolation is to be implemented in a computer algebra system and if it is ever to be used on data w with limited accuracy (cf. (2.6)) so that the condition of V is significant, an *a-priori control mechanism* or *verification step* must accompany the automatic selection of the basis PPs. This verification has to ascertain that the specified nodes are *not in a near-degenerate position*, where again degeneracy has nothing to do with a confluence or a similar singular behaviour but is simply characterized by the existence of some algebraic relation between the nodes. Such a degeneracy may not be recognizable geometrically, even in 2-space.

Let us substitute some concrete values into the above example ($s = 2$, $n = 6$):

Take the points of the unit hexagon $t_i = \begin{pmatrix} \cos i \frac{\pi}{3} \\ \sin i \frac{\pi}{3} \end{pmatrix}$, $i = 1(1)6$, and perturb the x_2-components of t_1, t_2, t_4, t_5 by approximately $\frac{1}{2} 10^{-4}$ into

$$\tilde{t}_{1,5} = \begin{pmatrix} \frac{1}{2} \\ \pm \frac{97}{112} \end{pmatrix}, \quad \tilde{t}_{2,4} = \begin{pmatrix} -\frac{1}{2} \\ \pm \frac{84}{97} \end{pmatrix}, \quad \tilde{t}_{3,6} = t_{3,6} = \begin{pmatrix} \pm 1 \\ 0 \end{pmatrix}. \tag{2.12}$$

With P from (2.11), which would be an algebraically correct interpolation space, we would obtain $\mathrm{cond}_\infty V = 56451$; for smaller perturbations, the condition could get arbitrarily bad.

If we perturb the data w_i alternately by $\pm\varepsilon$, the perturbation in the interpolant for the nodes (2.12) and the basis (2.11) becomes

$$\Delta p(x_1, x_2) \approx \varepsilon[-x + 18817(x_1^2 + x_2^2 - 1)]. \tag{2.13}$$

Note the occurrence of the critical conic polynomial p_c in Δp and the fact that its huge coefficient would completely destroy the validity of the interpolant away from the unit circle.

2.3 Multivariate Polynomial Systems of Equations

Consider a set of m (real or complex) polynomials in s variables

$$F := \{f_\mu \in \mathbb{P}^s, \mu = 1(1)m\}. \tag{2.14}$$

The constructive analysis of sets of multivariate polynomials is one of the fundamental tasks of computer algebra systems, e.g. the characterization and possible computation of the set of all *joint zeros* of the polynomials f_μ.

Most of the algebraic properties of a set F are really properties of the ideal $\mathscr{F}$ spanned by the polynomials in F. It has been found that related questions about $\mathscr{F}$, e.g. about the set of its joint zeros, may more easily be answered in a constructive fashion after a *special basis* for $\mathscr{F}$ has been formed, viz. a (reduced) *Groebner Basis*

$$G := \{g_\kappa \in \mathbb{P}^s, \kappa = 1(1)k\}, \tag{2.15}$$

see, e.g., [5]. From

$$\text{Ideal}(G) = \mathscr{F} = \text{Ideal}(F), \tag{2.16}$$

it follows that the 'solution' of the polynomial system

$$g_\kappa(x) = 0, \qquad \kappa = 1(1)k, \tag{2.17}$$

will produce the complete set of the joint zeros of the f_μ, i.e. the solution set of the original polynomial system

$$f_\mu(x) = 0, \qquad \mu = 1(1)m. \tag{2.18}$$

Given F, a Groebner Basis for $\mathscr{F}$ is uniquely defined for each linear order of the PPs x^j which is consistent with multiplication. For a specified order, the associated Groebner Basis G of $\mathscr{F}$ may be computed by the Buchberger Algorithm (cf. e.g. [5]) which uses rational operations only.

However, the determination of the zeros of (2.17) will generally require numerical procedures which use floating-point or other approximate computations. Thus, once more, the condition of (2.17) w.r.t. the computation of its zeros will play an important role. We distinguish two cases:

Case 1: $\mathscr{F}$ is 0-dimensional

One of the general approaches to the numerical solution of (2.17) which is independent of the PP-order for the generation of G, is the following (cf. e.g. [6]):

(i) Let $x^{j^{(\kappa)}}$ be the leading PPs in the g_κ of G, $\kappa = 1(1)k$. The PPs *not* contained in $\bigcup_\kappa \{x^j : j \geq j^{(\kappa)}\}$ form a PP-basis P for the *residue class ring R* mod $\mathscr{F}$ in $\mathbb{P}^s$, let $\dim(R) = n$. By $Z(x)$ we will denote the n-vector of the PPs in P, in the appropriate order.

(ii) Determine the *multiplication tables* for R w.r.t. the basis P. The multiplication tables are specified by those $n \times n$ matrices A_σ for which

$$x_\sigma \cdot Z(x) \equiv A_\sigma \cdot Z(x) \bmod \mathscr{F}, \qquad \sigma = 1(1)s. \tag{2.19}$$

The computation of the nontrivial elements of the A_σ from G requires only rational operations.

(iii) Compute the *joint eigenvectors* $z_v \in \mathbb{C}^n$, $v = 1(1)n$, of the A_σ. Except in the case of multiple eigenvalues (which can also be easily handled), all A_σ have the *same* set of n linearly independent eigenvectors z_v which may be normalized by 1 in the last component; cf. [6]. This task will generally require *numerical* (i.e. approximate) computation.

(iv) The interpretation

$$z_v = Z(t_v), \qquad v = 1(1)n, \tag{2.20}$$

yields the components of the n zeros $t_v^* \in \mathbb{C}^s$, $v = 1(1)n$, of $\mathscr{F}$.

Example: $s = 2$; let the PPs of x_1, x_2 be ordered by x_1 before $x_2(x_1 > x_2)$ and either

- total degree ordering (e.g. $x_2^3 > x_1 x_2$) or
- lexicographic ordering (e.g. $x_1 x_2 > x_2^3$).

For both orderings, the Groebner Basis of the ideal $\mathscr{F}$ with the corners of the unit hexagon as zeros t_v^*, $v = 1(1)6$, is

$$G = \{x_1^2 + x_2^2 - 1, \, x_2^3 - \tfrac{3}{4}x_2\}. \tag{2.21}$$

The leading terms x_1^2 and x_2^3 of G determine $P = \{x_1 x_2^2, x_1 x_2, x_2^2, x_1, x_2, 1\}$ as PP-basis for the residue class ring R associated with $\mathscr{F}$ so that[1]

$$Z(x) := \begin{pmatrix} x_1 x_2^2 \\ x_1 x_2 \\ x_2^2 \\ x_1 \\ x_2 \\ 1 \end{pmatrix}. \tag{2.22}$$

The related multiplication tables are immediately seen to be

$$x_1 \begin{pmatrix} x_1 x_2^2 \\ x_1 x_2 \\ x_2^2 \\ x_1 \\ x_2 \\ 1 \end{pmatrix} \equiv \begin{pmatrix} 0 & 0 & \tfrac{1}{4} & 0 & 0 & 0 \\ 0 & 0 & 0 & 0 & \tfrac{1}{4} & 0 \\ 1 & 0 & 0 & 0 & 0 & 0 \\ 0 & 0 & -1 & 0 & 0 & 1 \\ 0 & 1 & 0 & 0 & 0 & 0 \\ 0 & 0 & 0 & 1 & 0 & 0 \end{pmatrix} \begin{pmatrix} x_1 x_2^2 \\ x_1 x_2 \\ x_2^2 \\ x_1 \\ x_2 \\ 1 \end{pmatrix} \bmod \mathscr{F},$$

---

[1] For lexicographic order, the sequence of x_2^2 and x_1 would be inverted.

$$
x_2 \begin{bmatrix} x_1 x_2^2 \\ x_1 x_2 \\ x_2^2 \\ x_1 \\ x_2 \\ 1 \end{bmatrix} \equiv \begin{bmatrix} 0 & \frac{3}{4} & 0 & 0 & 0 & 0 \\ 1 & 0 & 0 & 0 & 0 & 0 \\ 0 & 0 & 0 & 0 & \frac{3}{4} & 0 \\ 0 & 1 & 0 & 0 & 0 & 0 \\ 0 & 0 & 1 & 0 & 0 & 0 \\ 0 & 0 & 0 & 0 & 1 & 0 \end{bmatrix} \begin{bmatrix} x_1 x_2^2 \\ x_1 x_2 \\ x_2^2 \\ x_1 \\ x_2 \\ 1 \end{bmatrix} \mod \mathscr{F}.
$$

A_1 has simple eigenvalues ± 1, and double eigenvalues $\pm\frac{1}{2}$. A_2 has 3 double eigenvalues at 0, $\pm\frac{1}{2}\sqrt{3}$. The *joint* eigenvectors of A_1 *and* A_2 form the matrix

$$
\begin{bmatrix} | & & | \\ z_1 & \cdots & z_6 \\ | & & | \end{bmatrix} = \begin{bmatrix} \frac{3}{8} & -\frac{3}{8} & 0 & -\frac{3}{8} & \frac{3}{8} & 0 \\ \frac{1}{4}\sqrt{3} & -\frac{1}{4}\sqrt{3} & 0 & \frac{1}{4}\sqrt{3} & -\frac{1}{4}\sqrt{3} & 0 \\ \frac{3}{4} & \frac{3}{4} & 0 & \frac{3}{4} & \frac{3}{4} & 0 \\ \frac{1}{2} & -\frac{1}{2} & -1 & -\frac{1}{2} & \frac{1}{2} & 1 \\ \frac{1}{2}\sqrt{3} & \frac{1}{2}\sqrt{3} & 0 & -\frac{1}{2}\sqrt{3} & -\frac{1}{2}\sqrt{3} & 0 \\ 1 & 1 & 1 & 1 & 1 & 1 \end{bmatrix}.
$$

The x_1-and x_2-components of the z_ν, $\nu = 1(1)6$, yield the *joint zeros* of $\mathscr{F}$. Note that the matrix of the z_ν is precisely the Vandermonde matrix V for the interpolation problem on the hexagon with the basis P of (2.22)!

Since the only genuine computation generally occurs in step (iii) of the above procedure, it is the *condition of the eigenproblems* for the A_σ which determines the sensitivity of the zeros of (2.17). For the *verification* of a safe computation, we must recognize situations in which these eigenproblems would be very ill-conditioned so that we can reroute the computation in such a case.

As is well-known (cf. e.g. [7]), the condition of an eigenproblem for some $n \times n$ matrix A is characterized by the quantity $\|X\|\,\|X^{-1}\|$ where the $n \times n$ matrix X is composed of n linearly independent eigenvectors x_ν of A:

$$
X = \begin{bmatrix} | & & | \\ x_1 & \cdots & x_n \\ | & & | \end{bmatrix}
$$

As exhibited in the above example, this matrix X becomes identical with the Vandermonde matrix for the zeros t_ν^* w.r.t. the basis P of the residue class ring R in the case of the procedure (i)–(iv).

It follows that the same situations which we have found to be critical in the interpolation case, cf. section 2.2, are also critical in determining the zeros of polynomial equations: It is the *near-degeneracy of the zero set of $\mathscr{F}$* in the sense of the *interpolation problem* of section 2.2 which leads to an ill-conditioning of the eigenproblems for the A_σ.

However, while the nodes $t_\nu \in T$ are the *data* in the interpolation problem and thus available for scrutiny, the zeros t_ν^* of (2.17) are the *results* of the computation. But the same paradigm can be used as previously:

Assume that the zeros $\tilde{t}_\nu^*$ of some given polynomial system (2.18) are in a *near-degenerate location*, i.e. they are very close to values t_ν^* which are in a genuinely degenerate location. Then the Groebner Basis $\tilde{G}$ which determines the $\tilde{t}_\nu^*$ must be a small perturbation of a Groebner Basis G which determines the t_ν^*, and G should be of a 'simpler' or 'more degenerate' form than $\tilde{G}$. From this G, the t_ν^* may be safely computed.

If the perturbed basis $\tilde{G}$ would have led to an ill-conditioned computation, it must call for a PP-basis $\tilde{P}$ of the residue class ring $\tilde{R}$ of $\tilde{\mathscr{F}}$ which is *different* from the basis P for R, the residue class ring of the ideal $\mathscr{F}$ generated by G; cf. section 2.2. There we simply used the basis P for the genuinely degenerate location of the nodes to interpolate on the nodes $\tilde{T}$.

Conversely, in our present situation, we have to determine the multiplication tables for $\tilde{R}$ w.r.t. the basis P. These multiplication tables must exist because their elements are given by the interpolation, in terms of P, of the components of $x_\sigma Z(X)$ on the perturbed zeros $\tilde{t}_\nu^*$. Their computation is straightforward if the 'perturbed' Groebner Basis $\tilde{G}$ is computed along with the (more) degenerate basis G.

An alternative approach is a local linearization of the problem about the degenerate situation, e.g. a Newton-like procedure. The efficient implementation of this approach is under development.

Example: (cf. section 2.2): The ideal $\tilde{\mathscr{F}}$ whose zeros $\tilde{t}_\nu^*$ are the perturbed nodes $\tilde{t}_\nu$ of (2.12) is spanned by the following Groebner bases:

$$- \text{total degree:} \quad \tilde{G}_t = \left\{ x_1^3 + \frac{177039745}{37634} x_1^2 + \frac{88519872}{18817} x_2^2 - x_1 - \frac{177039745}{37634}, \right.$$

$$x_1^2 x_2 - \frac{1}{4} x_2,$$

$$x_1 x_2^2 - \frac{88519872}{18817} x_1^2 - \frac{17039745}{37634} x_2^2 + \frac{88519872}{18817},$$

$$\left. x_2^3 - \frac{18817}{118026496} x_1 x_2 - \frac{177039745}{236052992} x_2 \right\}$$

$$- \text{lexicographic:} \quad \tilde{G}_l = \left\{ x_1^2 - \frac{4}{3} x_2^4 + \frac{177039745}{88519872} x_2^2 - 1, \right.$$

$$x_1 x_2 - \frac{118026496}{18817} x_2^3 + \frac{177039745}{37634} x_2,$$

$$\left. x_2^5 - \frac{177039745}{118026496} x_2^3 + \frac{9}{16} x_2 \right\}$$

The leading terms of these polynomial sets call for the following PP-bases for the residue class $\tilde{R}$ of $\tilde{\mathscr{F}}$:

$$\tilde{P}_t = \{x_1^2, x_1 x_2, x_2^2, x_1, x_2, 1\},$$
$$\tilde{P}_l = \{x_2^4, x_2^3, x_2^2, x_1, x_2, 1\}. \tag{2.23}$$

The associated multiplication table matrices $\tilde{A}_\sigma$ have more non-vanishing elements than the A_σ; but due to our construction of $\tilde{\mathscr{F}}$ we know what their joint eigenvectors are: They are the columns of the Vandermonde matrices $\tilde{V}_t$ and $\tilde{V}_l$ for the interpolation problem on $\tilde{T}$, w.r.t. the bases $\tilde{P}_t$ and $\tilde{P}_l$ resp.

In section 2.2, we have seen that $\tilde{V}_t$ is seriously ill-conditioned; $\tilde{V}_l$ is even more ill-conditioned $(cond_\infty(\tilde{V}_l) \approx 82000)$. Furthermore, a smaller deviation of the $\tilde{t}_v^*$ from the unit circle could have made the condition of the eigenproblems for the $\tilde{A}_\sigma$ arbitrarily bad.

After rescaling the polynomials with very large coefficients in $\tilde{G}_t$ and $\tilde{G}_l$, one discovers many possibilities to write them as perturbations of *two* polynomials of the type

$$\alpha_1 x_1^2 + \alpha_2 x_2^2 - \alpha_0, \qquad \text{with } \alpha_1 \approx \alpha_2 \approx \alpha_0 \approx 1,$$
$$\beta_1 x_2^3 - \tfrac{3}{4}\beta_2 x_2, \qquad \text{with } \beta_1 \approx \beta_2 \approx 1. \tag{2.24}$$

In particular, one can write them as small perturbations of the two g_κ in G. Consider the polynomials $\tilde{g}_\kappa$ of $\tilde{G}_t$ after rescaling:

$$\tilde{g}_1(x_1, x_2) = \frac{37634}{177039745}\left[x_1^3 + \frac{177039745}{37634} x_1^2 + \cdots\right]$$

$$= (x_1^2 + x_2^2 - 1) + \frac{1}{177039745}(37634(x_1^3 - x_1) - x_2^2),$$

$$\tilde{g}_2(x_1, x_2) = x_1^2 x_2 - \frac{1}{4}x_2,$$

$$\tilde{g}_3(x_1, x_2) = \frac{37634}{177039745}\left[x_1 x_2^2 - \frac{88519872}{18817} x_1^2 - \cdots\right] \tag{2.25}$$

$$= -(x_1^2 + x_2^2 - 1) + \frac{1}{177039745}(37634 x_1 x_2^2 + x_1^2 - 1),$$

$$\tilde{g}_4(x_1, x_2) = x_2^3 - \frac{3}{4}x_2 - \frac{1}{236052992}(37634 x_1 x_2 + x_2).$$

After dropping the 'perturbations' in $\tilde{g}_1, \tilde{g}_3, \tilde{g}_4$, we may reduce the remaining system to g_1 and g_2 of G. In the case of $\tilde{G}_l$, we may proceed analogously.

The Groebner Basis (2.21) leads to a PP-basis (2.22) and thus to a well-conditioned computation of the zeros of (2.25). In this example, we may easily obtain multiplication tables for $\tilde{R}$ w.r.t. (2.22) by simply considering x_1^2 as the leading term of the third polynomial in place of $\tilde{G}_t$ of $x_1 x_2^2$. The eigenproblems for these $\tilde{A}_\sigma$ yield the $\tilde{t}_v^*$ in a well-conditioned fashion.

Alternatively, we may look for minima of

$$Q(x_1, x_2) = \sum_{\kappa=1}^{4} \tilde{g}_\kappa(x_1, x_2)^2$$

near $t_1^*, t_2^*, t_4^*, t_5^*$ (t_3^* and t_6^* satisfy all the $\tilde{g}_\kappa$); because of the special structure of $\tilde{g}_2$ it suffices to vary x_2 for $x_1 = \pm\frac{1}{2}$. One Gauss-Newton step yields approximations to the corresponding zeros of $\tilde{\mathscr{F}}$ to 6 correct decimal digits.

Case 2: Ideal $\mathscr{F}$ d-dimensional, $d > 0$

For this even more interesting case, we will only indicate the application of our paradigm:

A multivariate polynomial set F may define an ideal $\mathscr{F}$ of positive dimension d

- because there are fewer polynomials than variables,
- because some coefficients have special values so that a degeneracy occurs.

In the first situation, a perturbation of F will generally leave d invariant, while d may decrease under perturbation in the second situation.

Hence, if we have a system $\tilde{F}$ of polynomials spanning an ideal $\tilde{\mathscr{F}}$ of dimension $\tilde{d} < d$ such that a system F spanning a d-dimensional ideal $\mathscr{F}$ is close to $\tilde{F}$, the determination of the $\tilde{d}$-dimensional *zero manifold(s)* of $\tilde{\mathscr{F}}$ will be *ill-conditioned*. It will then be worthwhile to approach the problem via the determination of the *more degenerate nearby system* F and its d-dimensional zero manifold(s). In any case, an *a-priori verification* step should be employed to monitor the potential occurrence of nearby configurations of a higher degree of degeneracy.

Besides, if $\tilde{F}$ has arisen from the modelling of a real-life scenario, it may well be that the more degenerate system F is the more appropriate model.

A more rigorous analysis of Case 2 is under development; results will be reported in due time.

3. Algorithmic Implementation of Algebraic Verification

In all problems discussed we may assume that a degeneracy is characterized by the vanishing of certain terms in specified phases of an algorithm. Thus it is natural to check for the occurrence of *small but non-vanishing* terms in relevant positions. Such terms are then candidates for an *artificial extinction* subject to properly chosen conditions and thresholds. If the extinction is found appropriate the local perturbation (residual) caused by this extinction may be computed and saved for later use in the algorithm or for later analysis.

Technical details of such a general approach will differ between application areas; they will not be discussed in this introductory paper. Here we will only indicate how this general recipe could have been applied in the situations considered in section 2.

3.1 Euclidean Algorithm with Degeneracy Verification

In symbolic integration, the occurrence of multiple zeros in the denominator polynomial p of a rational integrand may be checked by an application of the Euclidean algorithm to p and p'. In our trivial example (1.1), this leads to

$$x^2 - 2ax + 2 = (\tfrac{1}{2}x - \tfrac{1}{2}a)(2x - 2a) + (2 - a^2). \tag{3.1}$$

Thus, together with the generation of (1.2), the expression $2 - a^2$ should have been set aside for verification upon substitution of a value for a:

$$F(a) = \quad \ldots(1.2)\ldots \quad \text{if } |2 - a^2| \geq \varepsilon, \tag{3.2}$$

with a value of ε suitably adjusted to the relative accuracy used in the evaluation of the log-terms. For a value of $|2 - a^2|$ smaller than the threshold ε, the algorithm should have been branched to expression (2.2) for F, e.g. This is confirmed by Tables 1 and 2.

As mentioned previously, quantitative criteria for the treatment and interpretation of small terms in the Euclidean algorithm are presently developed by V. Hribernig in collaboration with the author of this paper.

3.2 Verification of a Special Location of Interpolation Nodes

In multivariate interpolation, an a-priori computation and *verification* must determine (cf. section 2.2)

- a potential special or *near-special location* of the interpolation nodes,
- a basis for an interpolation space P in which the interpolation problem (2.6) is well-conditioned.

The algorithm of de Boor and Ron ([4]) appears well-suited for that purpose if an appropriate form of pivoting is performed. Actually, [4] contains a MATLAB program with a parameter *tol* which controls the pivoting (in the sense of the algorithm). From our point of view, it controls which terms are neglected (or extinguished) during the course of the computation.

While the authors discuss some effects of the choice of *tol*, particularly in connections with finite-precision arithmetic, they do not explicitly touch the question of the condition of (2.6) in [4]; in fact, they say 'in exact arithmetic, we could choose $tol = 0$'. (In their earlier publication [3], however, they have included a short discussion along the lines of this paper.)

The algorithm of de Boor and Ron amounts essentially to a block-triangularization of V_∞ (cf. section 2.2), with column blocks determined by total degree and with the use of a special scalar product for the orthogonalization of row sections within a block column. For the node set T of the exact corners of the unit hexagon, V_∞ is transformed into

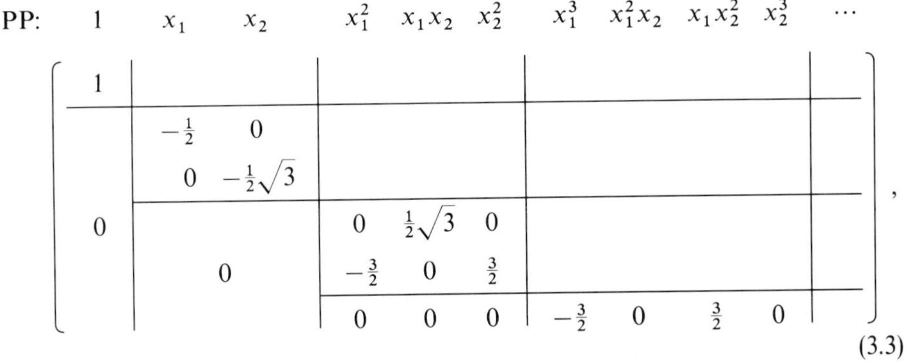

$$(3.3)$$

the elements in the empty blocks above are irrelevant. The form of (3.3) implies the appropriateness of the basis

$$\{1, x_1, x_2, x_1 x_2, x_1^2 - x_2^2, x_1^3 - 3x_1 x_2^2\} \tag{3.4}$$

according to the approach of de Boor and Ron. The zeros in the last row of the third block column exhibit the dependency of any further quadratic expressions on the nodes in T.

With the perturbed $\tilde{T}$ of (2.12), there appear small perturbations of some of the nonzero elements in (3.3) and two elements of equal sign and size $\approx 10^{-4}$ in the last component of the x_1^2 and x_2^2 columns. The acceptance or non-acceptance of these elements as *sufficiently different from zero* leads to $P = \mathbb{P}_2$ or to P spanned by (3.4) resp.

In section 2.2, we have seen how the first choice will considerably *destabilize* the subsequent solution of the interpolation problem (2.6) on $\tilde{T}$. Thus it appears worthwhile to study the effects of *tol* or of a more sophisticated control of the smallness of terms in the triangularization in more detail.

3.3 Multivariate Polynomial Systems

In section 2.3, the polynomial set $\tilde{F}$ with its joint zeros at the $\tilde{t}_\nu^*$ has been constructed artificially from its zeros; therefore, small terms have immediately been present in this system. Naturally, we could have concealed these small terms by forming linear combinations of the equations. The small terms in the Groebner Bases $\tilde{G}_t$ and $\tilde{G}_l$ would then have appeared in the course of the eliminations of the Buchberger algorithm.

In order to check for a potential near-degeneracy of a specified polynomial system, one will generally attempt to eliminate small terms as early as possible in the computation of the Groebner Basis—after appropriate analysis and with a proper notice of the effect. The technical details of such a *verification procedure* will be elaborated by V. Hribernig in his Ph.D. Thesis.

4. The Use of Floating-Point Arithmetic

The use of floating-point arithmetic in computer algebra systems has generally been considered inappropriate: In situations, where a distinction

$$D(y) < 0 \quad \text{or} \quad D(y) = 0 \quad \text{or} \quad D(y) > 0, \tag{4.1}$$

with y denoting some intermediate quantities in the algorithm, may be crucial for the qualitative structure of the result, small errors in y could completely upset the computation. (Of course, there may be a number of critical expressions in the algorithm, we will use D as a generic notation.)

Now, if $|D|$ is *never small* during the execution of some algebraic algorithm for some particular data, then the execution of some carefully implemented floating-point version of that algorithm will lead to the same qualitative result structure and possibly to good approximations of the exact result values. Most probably, the floating-point version will run significantly faster on any general-purpose computer.

In this paper, we have attempted to show that the *actual occurrence of a very small value* for some D during the execution of an algebraic algorithm should be monitored whenever some approximate computation is involved (input, use of output, etc.); we have emphasized that—under these circumstances—the algorithm should branch to a special version which is numerically stable at $D = 0$. With appropriate consideration in the implementation, it should be generally feasible to *discover* the occurrence of $|D| \approx 0$ in a floating-point version of the algorithm.

At this point, the decision about the further course of the computation will often depend on the sign of some *residuals* and, subsequently, accurate values of these residuals may be needed. In many cases quantities of residual type may be computed very accurately in floating-point arithmetic, with a tolerable amount of extra computation, particularly if an 'exact scalar product' operation is available; see, e.g., [8] and a large number of publications by U. Kulisch.

Thus, there is justified hope that carefully implemented floating-point versions of algebraic algorithms may be shown to be reliable and accurate even in rather sophisticated degenerate situations; the *achieved* accuracy could naturally be verified a-posteriori.

After all, *the* most widely used algebraic algorithm is *Gaussian elimination* for the solution of systems of linear algebraic equations. It has been established by proper analysis and proven by myriads of successful runs that good floating-point versions of this algorithm may be safely used, e.g. those in LAPACK [9]. Also, a-posteriori verification of the achieved accuracy may easily be supplemented.

Before undertaking the next stage in the Gaussian elimination process, each serious floating-point algorithm performs a verification to make sure that no degenerate ($=$ singular) matrix exists near the one which has been specified. If a near-degeneracy is found (no further sufficiently large pivot elements available) most implementations will return; but they could equally well switch to a QR-algorithm for the least squares solution of the remaining part of the system, or the like.

The fact that a computer algebra system will deliver a solution vector for an *arbitrarily ill-conditioned* nonsingular linear system with rational data *without warning* is a final demonstration of how exact algebraic computation without verification of the well-conditioning of the problem may often be a Danaers' gift[2].

References

[1] Sasaki, T., Noda, M. T.: Approximate square-free decomposition and rootfinding for ill-conditioned algebraic equations. J. Inf. Proc. *12*, 159–168 (1989).

[2] Rahman, A. A.: On the numerical solution of polynomial equations, Ph.D. Thesis, Univ. of Bradford, 1989.

[3] de Boor, C., Ron, A.: On multivariate polynomial interpolation. Constr. Approx. *6*, 287–302 (1990).

[4] de Boor, C., Ron, A.: Computational aspects of polynomial interpolation in several variables, CS Tech. Rep. #924, Univ. Wisc. 1990.

[5] Davenport, J. H., Siret, Y., Tournier, E.: Computer algebra: Systems and algorithms for algebraic computation. London: Academic Press 1988.

[6] Auzinger, W., Stetter, H. J.: An elimination algorithm for the computation of all zeros of a system of multivariate polynomial equations. Conference in Numerical Analysis, ISNM *86*, 11–30 (1988).

[7] Golub, G. H., van Loan, C. F.: Matrix computations 2nd edn. Baltimore: John Hopkins University Press 1989.

[8] Kulisch, U., Stetter, H. J. (eds.): Scientific computation with automatic result verification. Wien New York: Springer 1988 (Computing Suppl. 6).

[9] Anderson, E., et al.: LAPACK User's Guide; SIAM, 1992.

Prof. Hans J. Stetter
Institute of Applied and
Numerical Mathematics
Technical University Vienna
Wiedner Hauptstrasse 6–10
A-1040 Wien
Austria

---

[2] In Virgil's Aeneid, Laocoon warns his compatriots of the 'Trojan horse' because even the gifts of the Danaers (= Greek) should be distrusted.

Computing, Suppl. 9, 265–285 (1993)

© Springer-Verlag 1993
Printed in Austria

FORTRAN-XSC
A Portable Fortran 90 Module Library for Accurate and Reliable Scientific Computing*

W. V. Walter, Karlsruhe

Dedicated to Professor U. Kulisch on the occasion of his 60th birthday

Abstract — Zusammenfassung

FORTRAN-XSC: A Portable Fortran 90 Module Library for Accurate and Reliable Scientific Computing.
Fortran 90, the new international Fortran standard released in 1991 [14], offers a multitude of new
features and enhancements compared with FORTRAN 77 [2]. However, numerical problems persist
since the new standard contains no accuracy requirements for the arithmetic operators and mathematical
functions, making reliable computation extremely difficult. Even with IEEE arithmetic [3], results may
vary widely depending on code optimization and vectorization [9, 28].

FORTRAN-XSC, a portable Fortran 90 module library, aims at providing a flexible and versatile
toolbox for analyzing and improving the accuracy and reliability of numerical application programs. It
is particularly suited for the design of algorithms delivering automatically verified results of high
accuracy. The library features accurate scalar, vector and matrix arithmetic for real and complex numbers
and intervals, conversion routines for numeric constants, fully dynamic multiple precision arithmetic,
and much more. The result of every operation is optimal (accurate to 1 ulp) with respect to the rounding
selected by the user.

AMS Subject Classification: 65G10, 68N15

Key words: Fortran 90, automatic result verification, interval arithmetic, accurate dot product.

**FORTRAN-XSC, eine portable Fortran 90 Modulbibliothek für genaues und zuverlässiges wissenschaft-
liches Rechnen.** Fortran 90, die seit 1991 gültige neue internationale Fortran-Norm [14], bietet im
Vergleich zu FORTRAN 77 [2] eine Vielzahl von Neuerungen und Verbesserungen. Viele der numerischen Probleme bleiben jedoch bestehen, da die neue Norm keinerlei Genauigkeitsforderungen an die
arithmetischen Operatoren und mathematischen Funktionen stellt. Dies erschwert das zuverlässige
Rechnen extrem. Selbst mit IEEE-Arithmetik [3] können die Ergebnisse je nach Codeoptimierung und
Vektorisierung stark variieren [9, 28].

Mit FORTRAN-XSC soll ein anpassungsfähiges und vielseitiges Werkzeug zur Analyse und Ver-
besserung der Genauigkeit und Zuverlässigkeit numerischer Anwendungsprogramme zur Verfügung
gestellt werden. Für die Entwicklung von Algorithmen, welche hochgenaue, automatisch verifizierte
Ergebnisse liefern, ist FORTRAN-XSC besonders geeignet. Die Modulbibliothek bietet genaue Skalar-,
Vektor- und Matrixarithmetik für reelle und komplexe Zahlen und Intervalle, Konversionsroutinen für
numerische Konstanten, volldynamische Langzahlarithmetik, und vieles mehr. Jede Operation liefert
gemäß der vom Benutzer gewählten Rundung ein optimales (1 ulp genaues) Ergebnis.

---

* Received November 16, 1992; revised December 23, 1992.

1. Introduction

The common programming languages attempt to satisfy the needs of many diverse fields. While trying to cater to a large user community, these languages rarely provide specialized tools for specific areas of application. Thus the user is often left with ill-suited means to accomplish a task. In recent years, this has become quite apparent in numerical programming and scientific computing. Even though programming has become more convenient through the use of more modern language concepts, numerical programs have not necessarily become more reliable. This is true even if "good" floating-point arithmetic (e.g. which conforms to the IEEE Standard 754 for Binary Floating-Point Arithmetic [3]) is employed.

At the Institute of Applied Mathematics at the University of Karlsruhe there has been a long-term commitment to the development of programming languages suited for the particular needs of numerical programming. With languages and tools such as PASCAL-XSC [19], C-XSC, ACRITH-XSC [12, 34], and ACRITH [10, 11], the emphasis is on accuracy and reliability in general and on automatic result verification in particular.

In the 1980's, the programming language FORTRAN-SC [5, 25] was designed as a FORTRAN 77 extension featuring specialized tools for reliable scientific computing. It was defined and implemented at the University of Karlsruhe in a joint project with IBM Germany. The equivalent IBM program product *High Accuracy Arithmetic—Extended Scientific Computation*, called ACRITH-XSC for short, was released for world-wide distribution in 1990 [12]. Numerically it is based on IBM's *High-Accuracy Arithmetic Subroutine Library* (ACRITH) [10, 11], a FORTRAN 77 library which was first released in 1984.

When Fortran 90 finally became a standard in 1991, it was tempting to define another language extension for scientific computing. However, it was decided not to write another compiler because the Fortran 90 language is, on the whole, powerful enough to allow the implementation of most of the desired features within the language—something completely impossible in FORTRAN 77.

Therefore, FORTRAN-XSC is made up of a set of portable Fortran 90 modules—a versatile toolbox for accurate and reliable numerical computation. In particular, FORTRAN-XSC is designed to facilitate the development of numerical algorithms with *automatic result verification*. Such algorithms deliver results of high accuracy which are verified to be correct by the computer. For example, self-validating numerical techniques have been successfully applied to a variety of engineering problems in soil mechanics, optics of liquid crystals, ground-water modelling and vibrational mechanics where conventional floating-point methods have failed.

The elementary arithmetic operations for real and complex numbers are available with a choice of five different rounding modes: *nearest* (to the nearest floating-point number), *upwards* (towards $+\infty$), *downwards* (towards $-\infty$), *towards zero* (truncation), and *away from zero* (augmentation). Additionally, the FORTRAN-XSC package offers the corresponding interval operations which are essential for automatic result verification. Besides the elementary arithmetic operations, routines for the conversion of numerical constants and input/output data are provided.

The new Fortran 90 intrinsic functions SUM, DOT_PRODUCT, and MATMUL are usually unreliable because they may suffer from cancellation of leading digits. Therefore, an accurate and fully reliable alternative implementation using a long fixed-point accumulator is offered by FORTRAN-XSC. Accurate accumulation is essential in many algorithms to attain high accuracy. The accurate dot product provides the foundation on which all vector/matrix products are built. Furthermore, it may serve as a basis for a highly accurate implementation of the Basic Linear Algebra Subprograms (BLAS), an "industry standard" defining a set of commonly used vector/matrix operations [27]. For all of the aforementioned operations and functions, the results are always optimal with respect to the selected rounding. This implies that their error is never more than 1 ulp (1 unit in the last place).

The Fortran 90 code is designed to allow a certain degree of optimization by the compiler, especially vectorization. On certain architectures, code optimization can be further improved by using special compiler directives or language extensions.

FORTRAN-XSC is closely related to the following standards:

Fortran 90 International Standard: Information technology—Programming languages—Fortran, ISO/IEC 1539: 1991 (E) [14]

IEEE 754 IEEE Standard 754 for Binary Floating-Point Arithmetic, ANSI/IEEE Std 754–1985 [3]

IEEE 854 IEEE Standard 854 for Radix-Independent Floating-Point Arithmetic, ANSI/IEEE Std 854–1987 [4]

LIA-1 Draft International Standard: Information technology–Language independent arithmetic—Part 1: Integer and floating point arithmetic, ISO/IEC CD 10967-1:1992 (formerly called Language Compatible Arithmetic Standard (LCAS)) [16] *(in preparation)*

The design of FORTRAN-XSC has also been influenced by other research in the area of computer arithmetic. The simulation of double-precision or higher precision arithmetic in software has been the subject of a number of investigations. Some of the early work was done by Møller [26], Kahan [17], Dekker [8] and Linnainmaa [23]. A recent summary which appeals to hardware manufacturers to make *exact floating-point operations* accessible to the users is given in [6]. A vectorizable FORTRAN 77 version of the arithmetic runtime operations needed to run ACRITH-XSC [12] was designed and implemented by Schmidt [29]. A rigorous mathematical definition of computer arithmetic is given by Kulisch and Miranker in [21].

2. Principles and Requirements

The module library FORTRAN-XSC consists of a number of Fortran 90 modules providing various arithmetic operations and other fundamental tools for a given floating-point system (which is usually provided by the hardware). At the Fortran 90 level, this floating-point system is identified by its kind type parameter value `fpkind`. The corresponding intrinsic types are `REAL(fpkind)` and `COMPLEX(fpkind)`.

Since FORTRAN-XSC automatically adapts to the specified underlying floating-point system, it should be portable to any platform with minimal effort. Of course, a Fortran 90 compiler must be available on the target machine. All one has to do is to explicitly specify the kind type parameter value `fpkind` of the floating-point system to be used and recompile all the modules.

2.1 Floating-Point System

A **floating-point system** $F = F(b, p, minexp, maxexp)$ is defined by the following characteristics: its base (or radix) b, its precision (or mantissa length) p, and its exponent range, bounded by the minimal exponent $minexp$ and the maximal exponent $maxexp$. These characteristics can be obtained in a portable fashion in Fortran 90 via the intrinsic inquiry functions `RADIX`, `DIGITS`, `MINEXPONENT`, and `MAXEXPONENT`, respectively. The letter F will be employed indiscriminately for a floating-point system and for the set of floating-point numbers it defines.

A **floating-point number** x in F is either 0, or it consists of a sign, a fixed-length mantissa with digits $d_1, d_2, \ldots, d_p$, and an exponent e with $minexp \leq e \leq maxexp$:

$$x = \pm b^e \sum_{i=1}^{p} d_i b^{-i}$$

According to this definition, the (radix) point is just to the left of the first digit of the mantissa. This convention is not important, however, since this can always be achieved by formally shifting the exponent range. The mantissa is given in base b notation, i.e. all $d_i \in \{0, \ldots, b - 1\}$. A floating-point number is **normalized** if its leading mantissa digit d_1 is nonzero. Normalized floating-point numbers have a unique representation. The Fortran 90 standard uses "model numbers", which are normalized (or zero) by definition, to speak of floating-point numbers. A **denormalized** floating-point number is characterized by $d_1 = 0$ and $e = minexp$. Denormalized numbers are uniformly distributed around 0 (between the largest negative and the smallest positive normalized floating-point number).

The exponent of a floating-point number x (in the sense above) will be denoted by $e(x)$. It can be obtained via the Fortran 90 intrinsic function `EXPONENT(x)`. Also, the notation

$$\mathrm{ulp}(x) := b^{e(x)-p}$$

is used to mean 1 unit in the last place of a floating-point number x. In Fortran 90, the intrinsic function `SPACING(x)` can be employed to determine 1 ulp relative to x. For any floating-point number x, $x + \mathrm{ulp}(x)$ and $x - \mathrm{ulp}(x)$ are the neighboring floating-point numbers above and below x, unless one of them overflows or underflows, or unless x is an integral power of the base of the floating-point system. In the latter case, if $x = b^n$ for some integer n, then there are $b - 1$ floating-point numbers between $x - \text{ulp}(x)$ and x. By symmetry to zero, an analogous statement holds for $x = -b^n$. The immediate neighbor above/below any floating-point number x can be obtained in a portable way by calling the Fortran 90 intrinsic function `NEAREST(x,d)` with a positive/negative d, respectively.

Many common floating-point systems allow *denormalized* numbers, for example both IEEE Standards [3, 4]. However, the Fortran 90 standard always uses "model numbers"—which are *normalized* by definition—when speaking of floating-point numbers. Thus it does not specify how *denormalized* numbers are to be treated. Furthermore, Fortran 90 does not provide an inquiry function for determining whether a floating-point system allows denormalized numbers. This makes a uniform treatment of floating-point systems with denormalized numbers more difficult, but not impossible.

By default, FORTRAN-XSC assumes all floating-point numbers to be normalized, but the use of denormalized numbers is allowed if the underlying floating-point system supports them. The flag *denorm* is used to indicate whether denormalized numbers are allowed or not (LIA-1 [16] makes a similar provision). FORTRAN-XSC automatically determines whether denormalized numbers are supported and sets this flag accordingly.

2.2 Rounding and Computer Arithmetic

A **rounding** is a special mapping from the real numbers $\mathbf{R}$ onto the floating-point numbers F. Any rounding $\bigcirc$ should satisfy the following two properties:

$$\bigcirc(s) = s \qquad \text{for all } s \in F \qquad (projection) \qquad (1)$$

$$x \le y \Rightarrow \bigcirc(x) \le \bigcirc(y) \qquad \text{for all } x, y \in \mathbf{R} \qquad (monotonicity) \qquad (2)$$

In other words, a rounding should be a *monotonic projection* from $\mathbf{R}$ onto F. This ensures that a rounding is always accurate to 1 ulp (one unit in the last place of the mantissa of a floating-point number). In particular, the floating-point numbers F are invariant under any rounding. If a real number x is not representable in F, any rounding returns one of the two neighboring floating-point numbers (either the one above or the one below x).

Today, most computers provide at least so-called "faithful" arithmetic, which at least guarantees 1 ulp accuracy for the elementary floating-point operations. However, traditionally, the monotonicity rule (2) is not generally observed by the hardware arithmetic of computers.

Whenever an arithmetic operation requires a rounding, the following five *roundings* (or *rounding modes*) are provided in FORTRAN-XSC:

N	Nearest (to the nearest floating-point number)
Z	towards Zero (truncation)
A	Away from zero (augmentation)
U	Upwards Δ (upwardly directed, towards $+\infty$)
D	Downwards ∇ (downwardly directed, towards $-\infty$)

The rounding to *nearest* is not uniquely defined. In FORTRAN-XSC, it is defined as in the IEEE Standards for Floating-Point Arithmetic [3, 4], that is, with the "even in case of tie" rule: if the exact result is halfway between two floating-point numbers, the rounding is performed in such a way that the last digit of the rounded result is even. However, for floating-point systems with an *odd* base (which are nonexistent for all practical purposes), the rounding to *nearest* has to be defined differently.

Except for the rounding *away from zero*, all of the above roundings are required by the IEEE Standards as well. Also, the first three rounding modes are **antisymmetric**:

$$\bigcirc(-x) = -\bigcirc(x) \qquad \text{for all } x \in \mathbf{R} \qquad (antisymmetry) \qquad (3)$$

whereas the last two are not. Rather, the rounding modes *upwards* and *downwards* satisfy the symmetry rule: $\nabla(-x) = -\Delta(x)$ (or, equivalently, $\Delta(-x) = -\nabla(x)$).

The notion of **rounding** can be directly extended to complex numbers and the usual product spaces, the real and complex vectors and matrices. In all of these spaces, the order relation $\leq$ is defined componentwise, inducing a partial ordering on these sets. If T is any one of these sets and S is its computer-representable subset, a rounding function from T onto S fulfills the same properties (1), (2), (3) with $\mathbf{R}$ replaced by T and F by S. It can be shown that these roundings are equivalent to applying the analogous roundings to the individual components [21].

For the corresponding interval spaces, the order relation is the subset relation $\subseteq$ (so $\leq$ has to be replaced by $\subseteq$ in (2)). A rounding from any interval set T onto its computer-representable subset S is again defined by properties (1), (2), (3) plus the additional property

$$x \subseteq \bigcirc(x) \qquad \text{for all } x \in T \qquad (inclusion) \qquad (4)$$

Whenever an arithmetic operation in one of the common mathematical spaces (the real and complex numbers, vectors and matrices) is modelled on the computer, it should be defined by **semimorphism** [21]. Again, if T is one of the mathematical spaces under consideration, let S be its computer-representable subset. If $\circ$ is an arithmetic operation in T, the corresponding *computer operation* $\circledcirc$ in S should be defined in such a way that

$$r \circledcirc s := \bigcirc(r \circ s) \qquad \text{for all } r, s \in S \qquad (semimorphism) \qquad (5)$$

where $\bigcirc$ is the rounding in use. In other words, the computer operation should be performed *as if* the exact result were first computed and then rounded with the selected rounding mode. Note that the rounding $\bigcirc$ uniquely determines the (rounded) computer operation $\circledcirc$.

Property (5) guarantees that *all* arithmetic operations in the common mathematical spaces are accurate to at least 1 ulp, whereas the IEEE Standards for Floating-Point Arithmetic [3, 4] as well as most existing hardware platforms only provide the four (scalar) floating-point operations $+, -, *, /$ with this quality. A careful analysis leads to the conclusion that all of the remaining operations can essentially be implemented with the aid of just *one* additional operation: the **dot product** of two vectors with floating-point components. For more details, refer to [21] where the concept of **semimorphism** is introduced.

2.3 *System Requirements*

In order for FORTRAN-XSC to function properly, the system on which it is compiled and run must satisfy some minimal requirements. The most important of these is that the intrinsic floating-point arithmetic provided by the system (typically in hardware) be *faithful*. This means that the four elementary floating-point operations $+, -, *, /$ must be accurate to 1 ulp. Note that, by definition, 1 ulp (least bit) accuracy implies that whenever the mathematically exact result of an operation is representable as a floating-point number, the computed result is *exact*. Thus the maximum error is always *less than* 1 ulp. This guarantees that certain operations will be performed without error (e.g. addition of two floating-point numbers with the same exponent and opposite signs).

Least bit accuracy of the elementary machine operations is crucial in many places in the module library. On the other hand, a particular rounding mode is not required. In fact, it may well be that the machine "rounding" is not monotonic. In practice, the above accuracy requirement excludes only very few of today's machine architectures (e.g. Cray vector processors, which do not provide faithful arithmetic).

Another prerequisite concerns the Fortran 90 compiler (and compiler options) with which the FORTRAN-XSC module package is translated. The Fortran 90 compiler must respect parentheses in expressions, that is, the intended order of evaluation must be preserved. There are a number of places in the module library where any change to the order of evaluation would have numerically disastrous effects. Unfortunately, this requirement will sometimes preclude the use of high optimization levels. Note, however, that a Fortran 90 processor that violates the integrity of parentheses does not conform to the Fortran 90 standard. Despite all this, large parts of the library are vectorizable.

For FORTRAN-XSC to function properly, it is also important that the Fortran 90 intrinsic functions for numeric inquiry and floating-point manipulation plus some others work without fail. FORTRAN-XSC has to rely on these functions (see section 6).

3. Arithmetic Modules

3.1 Exact Floating-Point Operations

In FORTRAN-XSC, the design principle for elementary arithmetic is to implement the exact operations first since they are the most general, and to base the operators with rounding control on the exact operations. For the four arithmetic operations $+, -, *, /$ it is possible to define and implement **exact floating-point operations** via subroutines with four floating-point arguments each: two input and two output arguments, all in the same floating-point format. In this context, **exact** means that the computed result is mathematically correct, that is, no information about the true mathematical result of the operation is lost. As long as no exception (such as overflow, underflow, or division by zero) occurs, the result of each of these floating-point operations can be represented without error by a pair of floating-point numbers in the same format as the operands.

The set of subroutines for exact floating-point arithmetic is summarized in the following table:

subroutine name	arguments		mathematical specification
	in	out	
ADD_EXACT	(x, y, h, l)		$x + y = h + l$ with $e(l) \leq e(h) - p$ unless $l = 0$
SUB_EXACT	(x, y, h, l)		$x - y = h + l$ with $e(l) \leq e(h) - p$ unless $l = 0$
MUL_EXACT	(x, y, h, l)		$x * y = h + l$ with $e(l) \leq e(h) - p$ unless $l = 0$
DIV_EXACT	(x, y, q, r)		$x = qy + r$ with $|r| < |y| \cdot \mathrm{ulp}(q)$ unless $q = 0$

In the case of addition, subtraction and multiplication, the high-order part h and the low-order part l are floating-point numbers whose mathematical sum is the exact result of the operation and which do not overlap, i.e. their exponents differ by at least p (the precision or mantissa length), unless $l = 0$. In the case of division, a partial quotient q and the corresponding exact remainder r are produced, again both floating-point numbers. The exponent difference $e(x) - e(r)$ must sometimes be allowed to be $p - 1$, while at other times it is required to be at least p. Essentially, this means that the relationship must be $|r| < |y| \cdot \mathrm{ulp}(q)$ and not $|r| < \mathrm{ulp}(x)$.

A proof of the fact that the result of these operations can always be represented by two floating-point numbers (unless an exception occurs) can be found in [6].

3.2 Elementary Arithmetic

Fortran 90, as opposed to ACRITH-XSC [12] and FORTRAN-XSC, does not provide any means for automatic error control or for deliberate rounding control. In particular, the arithmetic operators with directed roundings +<, +>, -<, ->, *<, *>, /<, />, which are predefined in ACRITH-XSC, are not available in Fortran 90. Thus, regrettably, Fortran 90 does not provide access to the rounded floating-point operations defined by the IEEE Standards 754 and 854 [3, 4]. In view of the steadily increasing number of processors conforming to these standards, this is most unfortunate for the whole numerical community.

The elementary arithmetic operations +, −, *, / for real (and complex) numbers are available with the five rounding modes provided by FORTRAN-XSC. Alternatively, the rounding mode can be set by calling a special subroutine named `set_rounding`. The elementary operations are then accessible via the generic operators `.ADD.`, `.SUB.`, `.MUL.`, and `.DIV.`, applying the rounding specified in the last call to `set_rounding`. As long as the rounding has never been set by the user, it defaults to *nearest*.

Experience shows that a default rounding mode is generally only useful to compute *approximations*. The upwardly and downwardly directed roundings, on the other hand, are typically used in pairs to do interval arithmetic. Unfortunately, most IEEE floating-point processors do not provide the arithmetic operations with the rounding mode integrated into the instruction code. Rather, they require a special instruction to change the rounding mode to be used in subsequent operations. Note that switching the rounding mode to *upwards* before evaluating an arithmetic expression does not, in general, deliver an upper bound to the value of the expression by any means, nor will *downwards* deliver a lower bound.

The following table lists the elementary operators with rounding control:

generic	Nearest	Zero	Augment	Up	Down
`.ADD.`	`.ADDN.`	`.ADDZ.`	`.ADDA.`	`.ADDU.`	`.ADDD.`
`.SUB.`	`.SUBN.`	`.SUBZ.`	`.SUBA.`	`.SUBU.`	`.SUBD.`
`.MUL.`	`.MULN.`	`.MULZ.`	`.MULA.`	`.MULU.`	`.MULD.`
`.DIV.`	`.DIVN.`	`.DIVZ.`	`.DIVA.`	`.DIVU.`	`.DIVD.`

Of course, on any particular machine featuring IEEE arithmetic in hardware, the elementary arithmetic operations with one of the rounding modes *nearest, towards zero, upwards,* and *downwards* may be implemented using the hardware operations for greater efficiency. This cannot be done exclusively in standard-conforming Fortran 90, however. Thus FORTRAN-XSC does not currently provide any special support for IEEE hardware. However, an adaptation is certainly possible.

FORTRAN-XSC also provides rounding control for the conversion of numeric constants and input/output data. Besides the default rounding, the monotonic downwardly and upwardly directed roundings, symbolized by $<$ and $>$, respectively, are available. This ensures that the user knows exactly what data enters the computational process and what data is produced as a result. For further details, refer to section 4.

3.3 Complex Arithmetic

Complex addition and subtraction are trivially reduced to the corresponding real operations. Complex multiplication requires the accurate evaluation of an expression of the form $ab \pm cd$. Such expressions are sometimes called *short dot products*. Because of the danger of cancellation of leading digits, exact double-length products need to be computed and added/subtracted with sufficient accuracy to allow correct rounding. For reasons of efficiency and because it is a relatively frequent operation occurring in many applications, the short dot product is included as a special operation. Whenever a dot product involves only two terms, this operation can be employed for greater efficiency.

Complex division is rather intricate and requires careful implementation. A special algorithm is needed to obtain sufficient accuracy for correct rounding. For details, refer to [24].

3.4 Interval Arithmetic

By controlling the rounding error at each step of a calculation, it is possible to compute guaranteed bounds on a solution and thus verify numerical results on the computer. Enclosures of a whole set or family of solutions can be computed using interval arithmetic, for example to treat problems involving imprecise data or other data with tolerances, or to study the influence of certain parameters. Interval analysis is particularly valuable for stability and sensitivity analysis. It provides one of the essential foundations for reliable numerical computation.

FORTRAN-XSC provides complete interval arithmetic for the derived types `INTERVAL` and `COMPLEX_INTERVAL`, consisting of all the necessary arithmetic and relational operators and type conversion functions. The result of every arithmetic operation is accurate to 1 ulp.

An interval is represented by a pair of (real or complex) numbers, its infimum (lower bound) and its supremum (upper bound). For the infimum, the direction of rounding is always downwards, for the supremum, upwards, so that the inclusion property is never violated. By adhering to this principle, the computed result interval will and must always contain the true solution set.

The arithmetic interval operators $+$, $-$, $*$, $/$ as well as the operators `.IS.` (intersection) and `.CH.` (convex hull) are provided. The relational operators for intervals are the standard comparison operators and the operators `.SB.` (subset), `.SP.` (superset), `.DJ.` (disjoint), `.IN.` (point contained in interval), and `.INT.` (point or interval contained in interior of interval).

In order to be able to access interval bounds, to compose intervals and to perform various other data type changes, type conversion functions such as `INF` (infimum), `SUP` (supremum), and `IVAL` (composition of an interval, conversion to interval) are available. Other useful functions include `MID` (midpoint of an interval), `RADIUS`, and `DIAM` (diameter (width) of an interval).

An example of how a very simple module for real interval arithmetic may be written in Fortran 90 is given below. Note, however, that this module will only work if the computer provides faithful arithmetic, i.e. if the elementary floating-point operations are accurate to 1 ulp. The idea is to compute a 1 ulp approximation of each interval bound using standard floating-point arithmetic and then to add 1 ulp to the supremum and subtract 1 ulp from the infimum. Under these conditions, this simple and fast implementation will produce guaranteed inclusions, but the results will only be of 2 ulp accuracy. Furthermore, this idea cannot be extended to compound operations such as complex interval multiplication and division or the interval dot product.

```
MODULE REAL_INTERVALS ! real interval arithmetic
                      ! of 2 ULP accuracy
!!! this module requires 1 ULP floating-point arithmetic !!!

PRIVATE                   ! by default, nothing is exported

PUBLIC : : INTERVAL, INF, SUP, IVAL, MID, RADIUS, DIAM
PUBLIC : : OPERATOR(+), OPERATOR(-), OPERATOR(*),
PUBLIC : : OPERATOR(/), OPERATOR(.IS.), OPERATOR(.CH.)
   ...                    ! explicit exportation list

TYPE INTERVAL
  PRIVATE               ! structure components are not
  REAL : : INF, SUP    ! accessible outside this module
END TYPE INTERVAL

INTERFACE OPERATOR (+)
  MODULE PROCEDURE  ADD, MONADIC_PLUS
END INTERFACE

INTERFACE OPERATOR (-)
  MODULE PROCEDURE  SUB, MONADIC_MINUS
END INTERFACE

INTERFACE OPERATOR (*)
  MODULE PROCEDURE  MUL
END INTERFACE

   ...

INTERFACE IVAL
  MODULE PROCEDURE  IVAL
END INTERFACE
```

```
  CONTAINS

  FUNCTION ADD (L, R)
    TYPE (INTERVAL) L, R, ADD
    ADD%INF=NEAREST (L%INF+R%INF, -1.0)          ! approx. sum -1 ulp
    ADD%SUP=NEAREST (L%SUP+R%SUP, +1.0)          ! approx. sum +1 ulp
  END FUNCTION ADD
  ...

  FUNCTION MUL (L, R)
    TYPE (INTERVAL) L, R, MUL
    REAL II, IS, SI, SS
    II=L%INF*R%INF ; IS=L%INF*R%SUP ;
    SI=L%SUP*R%INF ; SS=L%SUP*R%SUP ;
    MUL%INF=NEAREST (MIN(II,IS,SI,SS), -1.0) ! -1 ulp
    MUL%SUP=NEAREST (MAX(II,IS,SI,SS), +1.0) ! +1 ulp
  END FUNCTION MUL
  ...

  TYPE (INTERVAL) FUNCTION IVAL (L, R)
    REAL L, R
    IVAL=INTERVAL(L, R)
  END FUNCTION IVAL

  END MODULE REAL_INTERVALS
```

Of course, FORTRAN-XSC also offers modules for *optimal* (1 ulp) interval arithmetic as defined by semimorphism (see section 2.2). The arithmetic operations for real and complex intervals are implemented using the elementary arithmetic operations with the *downwardly* and *upwardly* directed roundings to compute the infimum and the supremum. However, complex interval multiplication and division require more sophisticated algorithms [24]. FORTRAN-XSC also provides a special notation for real and complex intervals and routines for the conversion of interval constants and input/output data. For details, refer to section 4.

3.5 *Vector/Matrix Arithmetic*

In traditional programming languages such as FORTRAN 77, Pascal, or Modula-2, each vector/matrix operation requires an explicit loop construct or a call to an appropriate subroutine. Unnecessary loops, long sequences of subroutine calls, and explicit management of loop variables, index bounds and intermediate result variables complicate programming enormously and render programs virtually incomprehensible.

Fortunately, the situation has improved a lot with Fortran 90. Fortran 90 offers extensive array handling facilities such as allocatable arrays, array pointers, subarrays (array sections), various intrinsic array functions, and predefined array operators. All array operators are defined as element-by-element operations in

Fortran 90. This definition has the advantage of being uniform, but the disadvantage that highly common operations such as the dot product (inner product) of two vectors or the matrix product are not easily accessible. The Fortran 90 standard does not provide an operator notation for these operations, and it prohibits the redefinition of an intrinsic operator (e.g. $*$) for an intrinsically defined usage. Instead, the dot product is only accessible through the intrinsic function call DOT_PRODUCT(V,V), the other vector/matrix products through the intrinsic function calls MATMUL(V,M), MATMUL(M,V), and MATMUL(M,M), where V stands for any vector and M for any matrix. Clearly, function references are far less readable and less intuitive than operator symbols, especially in complicated expressions.

If one wants to reference the intrinsic functions DOT_PRODUCT and MATMUL via an operator notation, there are only two choices: either one defines a new operator symbol, say .MUL., for all possible type combinations that can occur in vector/matrix multiplication, or one defines new data types, e.g. RVECTOR, DRVECTOR, CVECTOR, DCVECTOR, RMATRIX, ... and then overloads the operator symbol $*$ for all possible type combinations of these new types. Both of these methods are quite cumbersome and seem to contradict one of the major goals of the Fortran 90 standard, namely to cater to the needs of the numerical programmer, in particular by providing extensive and easy-to-use array facilities. Note that both of these methods require a minimum of 64 operator definitions to cover all of the intrinsic cases. If more than two REAL and two COMPLEX types (single and double precision) are available, this number becomes even larger.

Unfortunately, the notational inconveniences caused by the intrinsic functions DOT_PRODUCT and MATMUL are surpassed by the *numerical difficulties* they cause. These Fortran 90 intrinsics and the intrinsic function SUM (for the summation of the components of an array) are *generally unreliable* numerically because of the accumulation of intermediate rounding errors and the possibility of *severe cancellation* in the summation process. Since the Fortran 90 standard lacks any kind of accuracy requirements, it seems inevitable that different implementors will implement these functions differently.

Even worse is the fact that *any* traditional floating-point summation technique is highly sensitive to the order of summation and is sure to fail in ill-conditioned cases because *leading-digit cancellation* may completely destroy the result. On vector processors, the problem is compounded by automatic vectorization. Typically, on pipelined processors, several partial sums are first computed and then added to form the final result. In the process, the summands are completely scrambled. The user has virtually no influence on the order in which the accumulation is performed.

Now that these critical functions have been "formally standardized" (but not numerically), the potential danger to the user becomes quite evident. For the Fortran 90 programmer, these functions appear to be very welcome since they seem to provide a portable way of specifying these highly common operations, especially as they are inherently difficult to implement. However, the user has no knowledge or control of the order in which the accumulation is performed. This makes any kind of realistic error analysis virtually impossible.

The inevitable consequence of this situation is that these three new intrinsic functions are unusable for many practical purposes—at least if one wishes to write portable Fortran 90 programs which deliver reliable results. Tests on large vector computers show that simple rearrangement of the components of a vector or a matrix can result in vastly different results [9, 28]. Different compilers with different optimization and vectorization strategies and different computational modes (e.g. scalar mode or vector mode with a varying number of vector pipes) are often responsible for incompatible and unreliable results.

As an example, consider the computation of the trace of the $n \times n$ product matrix C of a $n \times k$ matrix A and a $k \times n$ matrix B, which is defined by

$$\text{trace}(C) = \text{trace}(A \cdot B) = \sum_{i=1}^{n} C_{ii} = \sum_{i=1}^{n} \sum_{j=1}^{k} A_{ij} * B_{ji}.$$

In ACRITH-XSC [12] this double sum can be calculated by the following so-called *dot product expression*:

```
TRACE=#*( SUM(A(i,:)*B(:,i), i=1, n) )
```

The notation is simple and efficient and the computed result is guaranteed to be accurate to 1/2 ulp in every case.

In contrast, the corresponding Fortran 90 program looks something like this:

```
TRACE=0.0
DO I=1, N
  TRACE=TRACE+DOT_PRODUCT(A(I,:), B(:,I))
END DO
```

Here the computational process involves approximately $2nk$ rounding operations if, as is typical in the computation of dot products, the products are rounded before they are added and the accumulation is performed in the same floating-point format in which the elements of A and B are given. Far more critical is the fact that cancellation can, and often will, occur during summation. This leads to results of unknown accuracy at best, or to completely wrong and meaningless results if many leading digits cancel. Since the Fortran 90 standard does not impose any accuracy requirements on intrinsic functions such as SUM, DOT_PRODUCT, and MATMUL, there are no simple remedies.

Thus, unless an implementation gives explicit error bounds for these intrinsic functions, every Fortran 90 programmer should think twice before using them, especially if the possibility of leading digit cancellation cannot be excluded.

For the above reasons, FORTRAN-XSC provides an alternative, highly accurate implementation of these functions which are so fundamental to most branches of mathematics. In order to be able to compute the dot product of arbitrary vectors with 1 ulp accuracy, the summation of the exact double-length products is performed without error in a *long fixed-point accumulator*. Such an accumulator covers twice the exponent range of the underlying floating-point system in order to accommodate all possible double-length products. In the case of FORTRAN-XSC, it is implemented as a sequence of floating-point segments (instead of integer segments as is the case in ACRITH [10, 11], ACRITH-XSC [12], and PASCAL-XSC [19]) because it is assumed that this will improve the performance of the elementary dot product operations. However, this assumption may be false for certain modern RISC processors. The exact (unrounded) result of an accumulation process can be stored to full accuracy in a variable of type `DOT_PRECISION` or rounded to a floating-point number using one of the five available roundings.

The *dot product* for real vectors and for real interval vectors are both elementary in the sense that the interval case cannot be reduced to the real case in any simple way. The complex and complex interval dot products, on the other hand, are easily reduced to the corresponding real case. Analogously, two different routines are necessary for the *summation* of real numbers and real intervals. Again, the complex cases are reducible to their real analogues.

All vector/matrix products are implemented via the accurate dot product and produce results which are accurate to 1 ulp in every component. Additionally, FORTRAN-XSC provides the arithmetic element-by-element operations for vectors and matrices with interval and complex interval components.

4. Data Conversion for Input/Output

FORTRAN-XSC provides routines for accurate conversion of numerical data from one base to another and from a character string to one of the elementary data types of FORTRAN-XSC (for input) and vice versa (for output). All of the non-interval conversions are available with a choice of five roundings. For interval data, the rounding is always to the smallest possible representable interval enclosing the given interval.

On input, the constant given in the input string may be specified with an arbitrary number of digits in any base in the range 2–36. The letters A–Z are used to represent the digits 10–35, respectively. A provision for bases greater than 36 has not been made. It is assumed that any other base of interest is a power of one of those provided. The conversion uses as many digits as are necessary to determine the correctly rounded internal floating-point number or interval. On output, an arbitrary number of digits may be requested by the user. The length, the base and the rounding of the output constant can be chosen by the user.

The following functions are available for input:

```
REAL (string, rounding)
CMPLX (string, rounding)
IVAL (string)
CIVAL (string)
```

The functions for output are of the form

```
STR (number, base, length, rounding)
```

where number may be of type REAL, COMPLEX, INTERVAL, or COMPLEX_INTERVAL. Of course, for intervals, the rounding is predetermined and must not be specified. For the conversion from string to string, the following function is provided:

```
STR (string, base, length, rounding)
```

In some sense, this function is the ultimate conversion routine. It is capable of correctly converting a constant specified with an arbitrary number of digits in a given base to any other base, generating the prescribed number of digits in the target number system while respecting the rounding mode.

In the above functions, the type of the argument string and the result type of all functions with generic name STR is VARYING_STRING. This derived type for varying-length character strings is defined in module ISO_VARYING_STRING, whose functional definition is to become a collateral standard (currently draft international standard ISO/IEC CD 1539-1 [15]) to the Fortran 90 standard [14]. This module provides elementary tools for working with fully dynamic character strings of arbitrary length.

Non-advancing (stream) I/O is (finally!) available in Fortran 90. This enables reading and writing of partial records, making it possible to define one's own I/O in a portable and flexible way. In combination with the above routines, input/output with rounding control is straightforward.

Reading from left to right, the syntax of a constant is as follows. The constant may be optionally preceded by a $+$ or $-$ sign. The mantissa is specified as a sequence of digits which may or may not include a (radix) point. The base (radix) is given in decimal notation and is appended to the mantissa with a % sign as separator. The default for the base is 10 (decimal). If the base is greater than 10 and the first significant digit of the mantissa is represented by a letter, i.e. at least 10, then the mantissa should be preceded by an extra zero. The exponent is introduced by the letter E or D (for compatibility with Fortran), which may be uppercase or lowercase. The value of the exponent is given by an optionally signed integer in decimal notation. The sign, the (radix) point, the base, and the exponent are optional. The base and the exponent are always in decimal notation.

The parameter rounding is optional. If it is omitted on output, the rounding to be applied is the default rounding. If it is omitted on input, the rounding may instead be specified within the constant notation in the argument string. If a rounding is specified in the input string, the rounding symbol ($<$ for *downwards* or $>$ for *upwards*) must precede the constant. The whole notation must be parenthesized in this case. If no rounding is specified at all, the rounding used is the default rounding currently in effect.

Examples of the accepted syntax are:

```
1001101%2              ! binary constant=77 (decimal)
-76.50%8               ! octal constant=62.625 (decimal)
+0FFA000%16E-4         ! hexadecimal const=255.625 (dec.)
(<-3.14159265E+000)    ! decimal constant rounded downwards
(>0Z.YX4%36E2)         ! base 36 const rd upwards=46617.111...
```

For intervals, a rounding must not be specified. The rounding is always to the smallest enclosing interval. There is a special notation for intervals:

```
(<-2.00001,-1.99999>)   ! real interval enclosing -2
(<-0.0001%3>)           ! optimal (1 ulp) enclosure of -1/81
((<2.9,3.1>),(<1E0,1>)) ! complex interval, Re around 3, and Im=1
(<(2.9,1E0),(3.1,1)>)   ! same complex interval as above
```

Note that the last two examples define the same constant, that is, a complex interval may either be specified as a pair of real intervals (defining the real and the imaginary part) or as an interval with a pair of complex bounds (defining the lower left and the upper right corner of a rectangle in the complex plane).

5. Exception Handling

It is an unfortunate fact that Fortran 90 does not provide any exception handling facilities. Thus it is impossible to treat exceptions which occur during runtime in any simple and portable way—if at all. This is particularly detrimental to numerical applications, where exceptional cases are quite common. FORTRAN-XSC has to live with this situation and tries to preclude the possibility of numerical exceptions such as overflow, underflow, and division by zero in the floating-point operations it uses by preliminary testing and scaling as appropriate.

Among others, the following arithmetic exceptions are recognized by FORTRAN-XSC:

exception	meaning
invalid_op	invalid operation
div_by_zero	division by zero
overflow	exponent overflow
underflow	exponent underflow
inexact	result of an "exact" operation is inexact

The IEEE Standards 754 and 854 [3, 4] define the same five types of exceptions, whereas LIA-1 [16] does not distinguish *division by zero* and *invalid operation* and does not recognize the exception *inexact*.

Trap handling can be enabled or disabled for each exception individually. All exceptions are enabled by default. During runtime, the user may inquire about the occurrence of disabled exceptions at any point in time. It is also possible to define the manner in which an exception is to be treated by a user-written trap handling routine.

6. Useful Fortran 90 Intrinsic Functions

Besides the numeric inquiry functions RADIX, DIGITS, MINEXPONENT, and MAXEXPONENT which are needed to determine the main characteristics of a floating-point system, the following inquiry functions are sometimes useful:

function	result
EPSILON	Wilkinson epsilon (largest rel. dist. betw. 2 adjacent fl-pt numbers)
HUGE	largest floating-point number (maxreal)
TINY	smallest positive normalized floating-point number (minreal)

For the low-level treatment of floating-point numbers, the following floating-point manipulation functions are particularly helpful in ensuring better portability of the Fortran 90 code:

function reference	result
EXPONENT(x)	exponent of floating-point number x (base b)
FRACTION(x)	fraction (mantissa) of floating-point number x
SET_EXPONENT(x,e)	floating-point number x with exponent replaced by e
SCALE(x,k)	floating-point number x multiplied by b^k
NEAREST(x,d)	floating-point neighbor of x in direction of sign of d
SPACING(x)	1 ulp relative to floating-point number x

None of these intrinsic functions were available in FORTRAN 77. This made a portable implementation of floating-point software impossible in FORTRAN 77. Large parts of FORTRAN-XSC depend on these intrinsics. If they are not absolutely reliable, the module library cannot function properly.

7. Conclusion and Outlook

In its current state, FORTRAN-XSC provides only the essential foundations for accurate and reliable numerical computing. With the fundamental modules tested and running, things are only beginning to become interesting. Modules providing arithmetic for multiple-precision numbers (and intervals) of arbitrary length, for polynomials, and for truncated Taylor series are already being implemented.

An alternative, highly accurate implementation of the Basic Linear Algebra Subprograms (BLAS) [27] where every operation delivers results of 1 ulp accuracy is planned for the near future. Since a large set of mathematical functions is available in ACRITH [11] and in ACRITH-XSC [12] (for IBM /370 systems), and since PASCAL-XSC [19] provides a similar, fully portable set of functions, it was felt that the implementation of mathematical standard functions is not of high priority. However, fundamental problem solving routines for linear and nonlinear systems, for ordinary differential equations and integral equations, and others are being written to test the module library and to broaden the basis of FORTRAN-XSC.

In the foreword of the Fortran 90 standard [14], under the heading "Numerical computation", one finds the following statement: "Scientific computation is one of the principal application domains of Fortran, and a guiding objective for all of the technical work is to strengthen Fortran as a vehicle for implementing scientific software." In a modern programming language whose declared goal is to support scientific applications, a clear specification of the exact mathematical behavior and accuracy of the arithmetic operators and intrinsic functions should be an integral part of the language definition. Contrary to its claim, the Fortran 90 standard does not really provide much better support in this domain than did FORTRAN 77. Fortran 90 does, however, provide the necessary language tools which enable a software library such as FORTRAN-XSC to provide such numerical support. Unfortunately, as long as there is no adequate language and hardware support, a large performance penalty seems inevitable.

In the presence of a growing number of floating-point processors conforming to standards such as the IEEE Standard 754 for Binary Floating-Point Arithmetic [3] or the IEEE Standard 854 for Radix-Independent Floating-Point Arithmetic [4], the reluctance of programming language standardization committees to provide easy access to the elementary arithmetic operations with directed roundings by incorporating special operator symbols such as $+<, +>, -<, ->, *<, *>, /<, />$ is incomprehensible. Since none of the major programming languages provide any simple means to access these fundamental operations, it is not astonishing that they are seldom used. Interval arithmetic is one way of making these operations accessible and more widely accepted.

The IMACS-GAMM Resolution on Computer Arithmetic [13] requires that all arithmetic operations—in particular the compound operations of vector computers such as "multiply and add", "accumulate", and "multiply and accumulate"—be implemented in such a way that guaranteed bounds are delivered for the deviation of the computed floating-point result from the exact result. A result that differs from the mathematically exact result by at most 1 ulp (i.e. by just one rounding) is highly desirable and always obtainable, as demonstrated by various existing implementations [10, 11, 12, 34, 19]. The "Proposal for Accurate Floating-Point Vector Arithmetic" [7] essentially requires the same mathematical properties for vector operations as are required for the elementary arithmetic operations by the IEEE Standards. Hopefully, such user requests will influence the hardware design of computing machinery, especially of supercomputers, in the near future

[18, 20]. Accuracy requirements are also highly desirable in programming language standards to make scientific software more portable in the numerical sense.

Acknowledgements

My special thanks go to Enrik Berkhan and Arno Schaumann who did (and are still doing) most of the programming. The implementation of FORTRAN-XSC would not have been possible without their relentless efforts. I would also like to thank my colleagues at the Institute of Applied Mathematics for their willingness to discuss any problems and answer any questions. I am particularly indebted to my colleague Dr.-Ing. Lutz Schmidt for his invaluable advice on vectorization and the software simulation of computer arithmetic.

Above all, I would like to thank Professor Kulisch for creating a fruitful research atmosphere at the Institute of Applied Mathematics and for giving me the opportunity to pursue this project and to attend many national and international Fortran 90 standardization meetings.

References

[1] Adams, E., Kulisch, U. (eds.): Scientific Computing with Automatic Result Verification. Orlando: Academic Press 1992.

[2] American National Standards Institute: American National Standard Programming Language FORTRAN. ANSI X3.9–1978, 1978.

[3] American National Standards Institute/Institute of Electrical and Electronics Engineers: IEEE Standard for Binary Floating-Point Arithmetic. ANSI/IEEE Std 754–1985, New York, 1985.

[4] American National Standards Institute/Institute of Electrical and Electronics Engineers: IEEE Standard for Radix-Independent Floating-Point Arithmetic. ANSI/IEEE Std 854–1987, New York, 1987.

[5] Bleher, J. H., Rump, S. M., Kulisch, U., Metzger, M., Ullrich, Ch., Walter, W. (V.): FORTRAN-SC: A Study of a FORTRAN Extension for Engineering/Scientific Computation with Access to ACRITH. Computing 39, 93–110 (1987).

[6] Bohlender, G., Kornerup, P., Matula, D. W., Walter, W. V.: Semantics for Exact Floating Point Operations. Proc. of 10th IEEE Symp. on Computer Arithmetic (ARITH 10) in Grenoble, 22–26, IEEE Comp. Soc., 1991.

[7] Bohlender, G., Cordes, D., Knöfel, A., Kulisch, U., Lohner, R., Walter, W. V.: Proposal for Accurate Floating-Point Vector Arithmetic. In [1], 87–102 (1992).

[8] Dekker, T. J.: A floating-point technique for extending the available precision. Numerical Mathematics 18, 224–242 (1971).

[9] Hammer, R.: How reliable is the arithmetic of vector computers? In [30], 467–482 (1990).

[10] IBM System/370 RPQ, High-Accuracy Arithmetic. SA22-7093-0, IBM Corp., 1984.

[11] IBM High-Accuracy Arithmetic Subroutine Library (ACRITH), General Information Manual. 3rd ed., GC33-6163-02, IBM Corp., 1986.
..., Program Description and User's Guide. 3rd ed., SC33-6164-02, IBM Corp., 1986.

[12] IBM High Accuracy Arithmetic—Extended Scientific Computation (ACRITH-XSC), General Information. GC33-6461-01, IBM Corp., 1990.
..., Reference. SC33-6462-00, IBM Corp., 1990.
..., Sample Programs. SC33-6463-00, IBM Corp., 1990.
..., How to Use. SC33-6464-00, IBM Corp., 1990.
..., Syntax Diagrams. SC33-6466-00, IBM Corp., 1990.

[13] IMACS, GAMM: Resolution on Computer Arithmetic. In: Mathematics and Computers in Simulation 31, 297–298 (1989); In: Zeitschrift für Angewandte Mathematik und Mechanik 70/4, p. T5 (1990); In: Ullrich, Ch. (ed.) Computer Arithmetic and Self-Validating Numerical Methods, 301–302. San Diego: Academic Press 1990; In: [30], 523–524 (1990); In: Kaucher, E., Markov, S. M., Mayer, G. (eds.) Computer Arithmetic, Scientific Computation and Mathematical Modelling, IMACS Annals on Computing and Appl. Math. 12, 477–478, Basel: J. C. Baltzer 1991.

[14] International Standards Organization: Information technology—Programming languages—Fortran. Int. standard ISO/IEC 1539: 1991 (E), 1991.

[15] International Standards Organization: Varying length character strings in Fortran. Draft int. std. ISO/IEC CD 1539-1 (collateral std. to ISO/IEC 1539: 1991 (E)), 1992.

[16] International Standards Organization: Information technology—Language independent arithmetic—Part 1: Integer and floating point arithmetic. Draft int. std. ISO/IEC CD 10967-1: 1992, 1992.

[17] Kahan, W.: Further remarks on reducing truncation errors. Commun. ACM 8, 40 (1965).

[18] Kirchner, R., Kulisch, U.: Arithmetic for Vector Processors. Proc. of 8th IEEE Symp. on Computer Arithmetic (ARITH 8) in Como, 256–269, IEEE Comp. Soc., 1987.

[19] Klatte, R., Kulisch, U., Neaga, M., Ratz, D., Ullrich, Ch.: PASCAL-XSC Sprachbeschreibung mit Beispielen. Berlin, Heidelberg: Springer 1991.
...: PASCAL-XSC Language Reference with Examples. Berlin, Heidelberg: Springer 1992.

[20] Knöfel, A.: Fast hardware units for the computation of accurate dot products. Proc. of 10th IEEE Symp. on Computer Arithmetic (ARITH 10) in Grenoble, 70–74, IEEE Comp. Soc., 1991.

[21] Kulisch, U., Miranker, W. L.: Computer Arithmetic in Theory and Practice. New York: Academic Press 1981.
Kulisch, U.: Grundlagen des numerischen Rechnens: Mathematische Begründung der Rechnerarithmetik. Reihe Informatik 19, Bibl. Inst., Mannheim, 1976.

[22] Kulisch, U., Miranker, W. L. (eds.): A New Approach to Scientific Computation. Notes and Reports in Comp. Sci. and Appl. Math., Orlando: Academic Press 1983.

[23] Linnainmaa, S.: Analysis of some known methods of improving the accuracy of floating-point sums. BIT 14, 167–202 (1974).

[24] Lohner, R., Wolff v. Gudenberg, J.: Complex interval division with maximum accuracy. Proc. of 7th IEEE Symp. on Computer Arithmetic (ARITH 7) in Urbana, Illinois, 332–336, IEEE Comp. Soc., 1985.

[25] Metzger, M., Walter, W. (V.): FORTRAN-SC: A Programming Language for Engineering/ Scientific Computation. In [30], 427–441 (1990).

[26] Møller, O.: Quasi double-precision in floating-point addition. BIT 5, 37–50 (1965).

[27] NAG: Basic Linear Algebra Subprograms (BLAS). Oxford: The Numerical Algorithms Group 1990.

[28] Ratz, D.: The effects of the arithmetic of vector computers on basic numerical methods. In [30], 499–514 (1990).

[29] Schmidt, L.: Semimorphe Arithmetik zur automatischen Ergebnisverifikation auf Vektorrechnern. Ph.D. thesis, Univ. Karlsruhe, 1992.

[30] Ullrich, C. (ed.): Contributions to Computer Arithmetic and Self-Validating Numerical Methods. IMACS Annals on Computing and Appl. Math. 7, Basel: J. C. Baltzer 1990.

[31] Walter, W. V.: Flexible Precision Control and Dynamic Data Structures for Programming Mathematical and Numerical Algorithms. Ph.D. thesis, Univ. Karlsruhe, 1990.

[32] Walter, W. V.: Fortran 90: Was bringt der neue Fortran-Standard für das numerische Programmieren? Jahrbuch Überblicke Mathematik 1991, 151–175, Vieweg, Braunschweig, 1991.

[33] Walter, W. V.: A Comparison of the Numerical Facilities of FORTRAN-SC and Fortran 90. Proc. of 13th IMACS World Congress on Computation and Appl. Math. (IMACS '91) in Dublin, Vol. 1, 30–31, IMACS, 1991.

[34] Walter, W. V.: ACRITH-XSC: A Fortran-like Language for Verified Scientific Computing. In [1], 45–70 (1992).

Wolfgang V. Walter
Institut für Angewandte Mathematik
Universität Karlsruhe
Kaiserstrasse 12
D-W-7500 Karlsruhe
Federal Republic of Germany

Computing, Suppl. 9, 287–291 (1993)

Computing
© Springer-Verlag 1993
Printed in Austria

Implementation of Accurate Matrix Multiplication on the CM-2*

J. Wolff v. Gudenberg, Würzburg

Dedicated to Professor U. Kulisch on the occasion of his 60th birthday

Abstract — Zusammenfassung

Implementation of Accurate Matrix Multiplication on the CM-2. In this paper we report about a parallel implementation of matrix multiplication in the sense of Kulisch's arithmetic definition. Using the architectural support of the CM-2 the prize for accuracy compared to rounded software arithmetic is not too high.

AMS Subject Classification: 65Y05, 15-04

Key words: Computer arithmetic, parallel computers, BLAS.

Implementierung genauer Matrixmultiplikation auf der CM-2. In dieser Arbeit wird über eine parallele Implementierung der Matrixmultiplikation im Sinne von Kulischs Arithmetikdefinition berichtet. Es stellt sich heraus, daß der Preis für die garantierte Genauigkeit im Vergleich zur üblichen gerundeten Rechnung nicht zu hoch ist, wenn die gleiche Unterstützung durch die Architektur der CM-2 verwendet wird.

1. Accurate Basic Linear Algebra Operations

One basic arithmetic operation in the theory of computer arithmetic, as introduced in [3, 4] is the optimal scalar product. Its importance for the accuracy of linear algebra computations is evident. Scalar products together with interval arithmetic build the foundation for reliable self-validating methods.

Maximally accurate versions of the following operations should therefore belong to the equipement of each numerical library. (A, A_v, B, B_μ matrices, x, x_μ, y vectors)

- $x^T \cdot y,$
- $A \cdot x,$
- $A \cdot x + y,$
- $\sum_{v=1}^{r} A_v \cdot \sum_{\mu=1}^{s} x_\mu,$
- $A \cdot B,$
- $\sum_{v=1}^{r} A_v \cdot \sum_{\mu=1}^{s} B_\mu$

All the results shall be computed with only one rounding in each component of the product matrix or vector. The most relevant rounding modes are those to the nearest floating-point number and to the smallest enclosing interval.

---

* Received September 17, 1992; revised December 18, 1992.

The operations are accessible in several language extensions like PASCAL-XSC or ACRITH-XSC via an extended expression concept and have efficiently been implemented on sequential computers.

But nowadays numerical calculations are more and more carried out on supercomputers of different architectures as there are pipelined vectorcomputers and SIMD or MIMD parallel computers. In several papers algorithms for accurate parallel linear algebra operations have been discussed [1, 6, 7]. The matrix multiplication is clearly a good candidate to measure the performance of a parallel computer, since it is computation and communication intensive. It also highlights the difference between rounded and exact arithmetic.

In this paper we therefore report about our implementation of several versions of the matrix multiplication on the Connection Machine CM-2.

2. Matrix Multiplication Algorithms

We consider two algorithms for the multiplication of $2\,n \times n$ matrices A and B. Both of them are variants of the fast systolic algorithm proposed in [5]. The first one uses $p^2 = n^2/k^2$ processors arranged in a two dimensional torus and needs nk^2 scalar product operations and $4nk^2 + 4k^2 \log n$ transportations [6]. In this paper a scalar product operation means an accumulation of a product of two floating-point numbers. Generalisation for rectangular matrices is straightforward.

Before the computation starts, the matrix elements are distributed to the processors, so that each processor obtains a pair of elements which contributes to the corresponding result (see Fig. 1). After each step A rotates rowwise and B rotates columnwise.

We collect each scalar product in a long accumulator, so that no error is made before the final rounding. We therefore need one long accummulator for each matrix element which is affordable as long as there are more processors than matrix elements.

Otherwise each physical processor has to simulate k^2 logical processors. If the space requirement for k^2 long accumulators exceeds the available memory, we switch to our second algorithm where each processor calculates its $k \times k$ block componentwise and the rows of A and the columns of B are shifted blockwise. Therefore only one accumulator per processor is needed.

This algorithm also performs $n \cdot k^2$ scalar product operations but $(4nk^3 + 4k^2 \log n)$ transportations [6].

3. Architectural Support for the Algorithms

Both algorithms are communication and computation intensive. During the computation communication is performed via a two dimensional torus, whereas for the

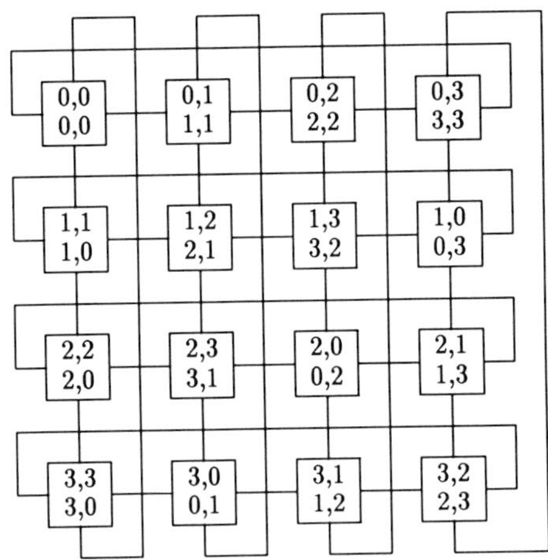

Figure 1. Initial distribution of 4 × 4 matrices subscripts of A in top line, B in bottom line

initial staggering of the matrices other processor topologies, e.g. hypercubes, may be favorable.

Let us now estimate the prize we have to pay for the accurate scalar product.

Double length floating-point products are needed in full precision. Their computation causes no overhead compared to software multiplication but can not efficiently utilize existing hardware for the most precise floating-point format.

The addition of that product to the long accumulator in the sequential case is usually only a double length fixed-point addition, since the long accumulator may be addressed by the exponent of the product. Again available floating-point hardware is not suitable for this case. So if both algorithms are performed in software, the accurate scalar product usually is not slower than the rounded one. This also holds for parallel computation of several scalar products if each processor has an instruction to access the appropriate part of its long accumulator individually. If this kind of indexed addressing is not available as on pure SIMD computers or very slow as on the Connection Machine, the addition has to take the full length of the accumulator into account. This may slow down the time for one scalar product operation by a considerable factor, since all products have to be adjusted in a second long accumulator according to their exponent before they are added to the scalar product accumulator [1].

Therefore an efficient index relative addressing mode for individual processors or at least fast fixed-point addition of many digits is crucial for long accumulator addition. The multiply and add instructions should be efficiently combined.

The Connection Machine CM-2 is a hypercube of 2^q processors supporting fast communication for a two dimensional grid (or torus) of $2^{\lceil q/2 \rceil} \times 2^{\lceil q/2 \rceil}$ elements (the NEWS grid). Many virtual processors may be allocated on one physical processor, so programs are portable from one configuration and problem size to another. The allocation and communication of virtual processors is mainly transparent to the user but limited by the individual memory.

Each processor has a one bit arithmetic unit. Floating-point accelerators for 32- or 64-bit IEEE format are provided but these do not supply the double length product.

Index relative addressing is available but extremely slow, so that it was discarded.

Fixed-point addition of large integers is supported and is faster than the floating-point hardware instruction for according mantissa lengths.

4. Implementation and Performance Measurements

The algorithm was programmed in $C*$ using mainly its shape data type for parallel arrays and the left-indexing facility for fast grid communication. All crucial arithmetic operations are directly called from the Paris machine-level instruction set [2].

The length of the long accumulator depends on the actual exponent range and is precalculated for each matrix product.

The hardware configuration used was a 7 MHz CM-2 with 16K processors with 32 KB memory each and a 32-bit floating-point acceleration unit for each 32 processors. Some of the computations were performed on a 8K block so that the fast grid communication could not be used for a virtual processor rate less than 1.

The virtual processor rate for the first algorithm was limited to 8 due to the available memory, so the second algorithm was implemented for larger matrices. Fixed-point addition of integers of length 4096 allowed long accumulator additions with one or two instructions.

Randomly generated 64-bit floating-point matrices were multiplied. Each multiplication was repeated 20 times in batch mode to obtain reliable results. For comparison we quote the times for hand-coded rounded matrix multiplication which were nearly identical to those obtained from the CMSSL library.

The following table comprises the results for the first algorithm. All entries are CM-2 execution times in seconds. rc denotes rounded computation, ac accumulator computation and red accumulator computation for a reduced exponent range (-32 through 31).

No. of proc.	matrix dimen.	rc	ac	red
8K	16	2.0	9.7	5.9
	32	3.9	16.1	9.6
	64	8.4	27.4	12.2
	128	21.9	90.0	30.8
	256	169	662	202
16K	128	10.7	46.1	15.8
	256	81.1	307	101

We see the dependence on n and the virtual processor rate k^2.

The second algorithm which saves storage space is indeed a factor of k slower.

For the very large exponent range of the IEEE 64-bit floating-point format the accurate computation is slower by a factor of 4, but in examples with restricted exponent range and thus restricted accumulator length, we observe smaller factors.

We will not conceal that this comparison is worse for the 32-bit format where rounded computation is supported by a floating-point accelerator. Here we get slow-down factors of about 10.

Acknowledgements

We thank Thinking Machines Corp. and the supercomputer department of the GMD Schloß Birlinghoven for the access to the CM-2 and additional support and cand. math. D. Parr who programmed the algorithms.

References

[1] Bohlender, G., Wolff v. Gudenberg, J.: Accurate matrix multiplication on the array computer AMT-DAP. In: Kaucher, E., Markov, S. M., Mayer, G. (eds.) Computer arithmetic, scientific computation, and mathematical modelling. IMACS Annals on Computing and Applied Mathematics *12*, 133–150 (1991).

[2] The Connection Machine System. Thinking Machine Corp, Cambridge, 1990/91.

[3] Kulisch U.: Grundlagen des Numerischen Rechnens—Mathematische Begründung der Rechnerarithmetik, Mannheim: Bibliographisches Institut 1976.

[4] Kulisch, U., Miranker, W. L.: Computer arithmetic in theory and practice. New York: Academic Press 1981.

[5] Tichy, W.: Parallel matrix multiplication on the connection machine. Int. Journal of High Speed Computing *1*(2), 247–262 (1989).

[6] Wolff v. Gudenberg, J.: Accurate matrix operations on hypercube computers. In: Herzberger, J., Atanassova, L. (eds) Computer arithmetic and closure methods, 167–176. Amsterdam: North-Holland 1992.

[7] Wolff v. Gudenberg, J.: Grundroutinen für sichere, numerische Algorithmen auf Parallelrechnern. Wiss. Berichte der TH Leipzig *1991*, 421–426.

J. Wolff v. Gudenberg
Lehrstuhl für Informatik II
Universität Würzburg
Am Hubland
D-W-8700 Würzburg
Federal Republic of Germany

A Min Tjoa, I. Ramos (eds.)

Database and Expert Systems Applications

Proceedings of the International Conference
in Valencia, Spain, 1992

1992. 324 figures. XIV, 551 pages.
Soft cover DM 148,–, öS 1036,–
ISBN 3-211-82400-6
Prices are subject to change without notice

The Database and Expert Systems Applications (DEXA) conferences are mainly oriented to establish a state-of-the-art forum on database and expert systems applications. But practice without theory has no sense, as Leonardo said five centuries ago. Therefore, as presented in this book, a compromise has been aimed at these two complementary aspects. Five sessions are application-oriented, ranging from classical applications to more unusual ones in software engineering. Actual research aspects in databases, such as activity, deductivity and/or object orientation are also presented in DEXA '92, as well as the implications of the new "data models" such as OO-model, deductive model, etc. are included in the modelling sessions. Other areas of interest, such as hypertext and multi-media applications, together with the classical field of information retrieval are also considered. Finally, implementation aspects are reflected in very concrete fields.

Springer-Verlag Wien New York

Applied Data and Knowledge Engineering

The main goal of this journal is to provide an international forum for the presentation of application-oriented research in the field of database and expert systems.
In order to achieve this goal, the journal publishes (exclusively in English):

- Articles describing application-oriented development of databases and expert systems
- Articles on the coupling of knowledge and database systems
- Articles on expert systems and databases for research and development in the area of scientific and engineering applications
- Articles on expert systems and database applications in the humanities
- Surveys on new knowledge-representation techniques
- Surveys on new data modelling techniques

Subscription Information
1993. Vol.1 (4 issues each):
DM 182,–, öS 1.274,–
plus carriage charges.
ISSN 0942-251X   Title No. 728

Prices are subject to change without notice

Springer-Verlag Wien New York